Advanced Condensed Matter Physics

This graduate textbook includes coverage of important topics that are not commonly featured in other textbooks on condensed matter physics, such as treatments of surfaces, the quantum Hall effect, and superfluidity. It avoids complex formalism, such as Green's functions, which can obscure the underlying physics, and instead emphasizes fundamental physical reasoning. Intended for classroom use, it features plenty of references and extensive problems for solution based on the author's many years of teaching in the Physics Department at the University of Michigan. This textbook is suitable for physics, chemistry and engineering graduate students, and as a reference for research students in condensed matter physics. Engineering students will find the treatment of the fundamentals of semiconductor devices and the optics of solids of particular interest.

Leonard M. Sander is Professor of Physics at the University of Michigan. His research interests are in theoretical condensed matter physics and non-equilibrium statistical physics, especially the study of growth patterns.

Advanced Condensed Matter Physics

Leonard M. Sander

Department of Physics, The University of Michigan

CAMBRIDGE
UNIVERSITY PRESS

CAMBRIDGE
UNIVERSITY PRESS

University Printing House, Cambridge CB2 8BS, United Kingdom

One Liberty Plaza, 20th Floor, New York, NY 10006, USA

477 Williamstown Road, Port Melbourne, VIC 3207, Australia

314-321, 3rd Floor, Plot 3, Splendor Forum, Jasola District Centre, New Delhi - 110025, India

103 Penang Road, #05-06/07, Visioncrest Commercial, Singapore 238467

Cambridge University Press is part of the University of Cambridge.

It furthers the University's mission by disseminating knowledge in the pursuit of
education, learning and research at the highest international levels of excellence.

www.cambridge.org
Information on this title: www.cambridge.org/9780521872904

First published 2009

A catalogue record for this publication is available from the British Library

ISBN 978-0-521-87290-4 Hardback

To Mae & Evelyn

Contents

Preface

This book is intended as a textbook for a graduate course in condensed matter physics. It is based on many years' experience in teaching in the Physics department at The University of Michigan. The material here is more than enough for a one-semester course. Usually I teach two semesters, and in the second, I add material such as the renormalization group.

In this book advanced techniques such as Green's functions are not used. I have tried to introduce as many of the concepts of modern condensed matter physics as I could without them. As a result, some topics that are of central importance in modern research do not appear.

The problems are an integral part of the book. Some concepts that are used in later chapters are introduced as problems.

Students are expected to have a good background in statistical physics, non-relativistic quantum theory, and, ideally, know undergraduate Solid State physics at the level of Kittel (2005).

I decided to write this book as a result of coming back to teaching Condensed Matter after a number of years covering other subjects. I had hoped to find a substitute for the grand old standards like Ziman (1972) or Ashcroft & Mermin (1976) which I used at the beginning of my teaching career. Though there are newer texts that are interesting in many ways, I found that none of them quite fit my needs as an instructor. It is for the reader to decide how well I have succeeded in giving a modern alternative to the classics – they are very hard acts to follow.

Many people have helped me in writing this book. Craig Davis and Cagilyan Kurdak have been remarkably generous with their time, and found many errors. Jim Allen and Michal Zochowski have given valuable advice. I would like to particularly thank Brad Orr, Andy Dougherty, Dave Weitz, Jim Allen, Roy Clarke, and Meigan Aronson for figures. And, of course, my students have given invaluable feedback over more than three decades.

1 The nature of condensed matter

Condensed matter physics is the study of large numbers of atoms and molecules that are "stuck together." Solids and liquids are examples. In the condensed state many molecules interact with each other. The physics of such a system is quite different from that of the individual molecules because of *collective effects*: qualitatively new things happen because there are many interacting particles. The behavior of most of the objects in our everyday experience is dominated by collective effects. Examples of materials where such effects are important are crystals and magnets.

This is a vast field: the subject matter could be taken to include traditional solid state physics (basically the study of the quantum mechanics of crystalline matter), magnetism, fluid dynamics, elasticity theory, the physics of materials, aspects of polymer science, and some biophysics. In fact, condensed matter is less a field than a collection of fields with some overlapping tools and techniques. Any course in this area must make choices. This is my personal choice.

In this chapter I will discuss orders of magnitude that are important, review ideas from quantum mechanics and chemistry that we will need, outline what holds condensed matter together, and discuss how order arises in condensed systems. The discussion here will be qualitative. Later chapters will fill in the details.

1.1 Some basic orders of magnitude

To fix our ideas, consider a typical bit of condensed matter, a macroscopic piece of solid copper metal. As we will see later it is best to view the system as a collection of cuprous (Cu^+) ions and conduction electrons, one per atom, that are free to move within the metal. We discuss some basic scales that will be important for understanding the physics of this piece of matter.

Lengths A characteristic length that will be important is the distance between the Cu atoms. In a solid this distance will be of order of a chemical bond length:

$$L \approx 3 \text{ Å} \approx 3 \times 10^{-8} \text{cm}. \tag{1.1}$$

Note that this is very tiny on the macroscopic scale. The whole art of condensed matter physics consists in bridging the gap between the atomic scale and the macroscopic properties of condensed matter.

Energies We can ask about the characteristic energy scales for the sample. One important energy scale is the binding energy of the material per atom. A closely related quantity is the melting temperature in energy units:

$$1357 \text{ K} = 0.11 \text{ eV}. \tag{1.2}$$

This is a typical scale to break up the material. If we probe at much larger energies (KeV, for example) we will be probing the inner shells of Cu, namely the domain of atomic physics, or at MeV, the Cu nucleus, i.e. nuclear physics.

Cu has an interesting color (it is copper colored, in fact), so we might expect something interesting at the scale of the energy of ordinary light, namely,

$$E \approx \hbar\omega_{\text{opt}} \sim 3 \text{ eV} \tag{1.3}$$

which is also the strength of a typical chemical bond. A somewhat larger, but comparable scale is that of the Coulomb interaction of two electrons a distance L apart:

$$E \approx e^2/L \approx 5 \text{ eV}. \tag{1.4}$$

These energies are low even for atomic physics. This means that in our study of condensed matter we will always be interested only in the outer (valence electrons) which are least bound.

Speeds When a piece of Cu carries an electrical current of density, $\mathbf{j}$, the conduction electrons move at a drift velocity $\mathbf{v}_d$:

$$\mathbf{j} = ne\,\mathbf{v}_d \tag{1.5}$$

where n is the number density of conduction electrons and e is the charge on the electron. For ordinary sized currents we find a very small speed, $v_d \approx 0.01$ cm/sec.

There is another characteristic speed, the mean thermal speed, v_T of the Cu ions when they vibrate at finite temperature. We estimate v_T as follows. From the Boltzmann equipartition theorem the mean kinetic energy of an ion is:

$$Mv_T^2/2 \sim k_{\text{B}}T. \tag{1.6}$$

Here T is the absolute temperature, k_{B} is Boltzmann's constant, M is the mass of a Cu ion, and v_T is the mean thermal velocity. At room temperature we get $v_T \sim 3 \times 10^4$ cm/sec.

There is a larger speed associated with the electrons, namely the quantum mechanical speed of the valence electrons. We estimate this speed as [frequency of an optical transition] x length:

$$v \sim (E/\hbar)(L) \approx 10^7 \text{ cm/sec}. \tag{1.7}$$

As we will see below, there is another relevant speed, the magnitude of the Fermi velocity, which is of the same order.

In any case, all of these speeds are small compared to the speed of light. Thus, we seldom need the theory of relativity in condensed matter physics. (An exception is the spin-orbit interaction of heavy elements.)

Large numbers and collective effects The essential point of the subject is that we deal with very large *numbers* of ions and electrons, $\approx 10^{27}$ in a macroscopic sample. In a famous essay P. W. Anderson (1972) pointed out the significance of this fact. When many things interact we often generate new phenomena, sometimes called emergent phenomena. Or, as Anderson put it, "more is different." Some examples of collective effects that we will emphasize in this book are the existence of *order* of various types, e.g. crystalline order, magnetic order, and superconducting order.

1.2 Quantum or classical

We have seen that we are interested in non-relativistic physics. We can go further: for the case of Cu there are conduction electrons and Cu^+ ions. What type of physics is applicable to each? In particular, do we need quantum mechanics? A useful criterion is to compare the de Broglie wavelength of the relevant particle, $\lambda = h/mv$, to the interparticle spacing.

For the ions, the relevant speed is v_T which we estimated above. Thus:

$$\lambda = h/(2Mk_B T)^{1/2} \approx 10^{-9} \text{ cm} << L. \tag{1.8}$$

This is smaller than the spacing by an order of magnitude. For all ions in solids (except for He and H at very low temperatures) we can use classical mechanics. (As we will see, for vibrations of ions at low T, we need quantum mechanics too.)

For the electrons the situation is different because the electron mass, m, is is 63×1800 times smaller than the mass of a Cu ion, so we get

$$\lambda = h/(2mk_B T)^{1/2} \approx 3 \times 10^{-7} \text{ cm} >> L. \tag{1.9}$$

Electrons are quantum mechanical for all temperatures.

1.3 Chemical bonds

Matter condenses because atoms and molecules attract one another. In the condensed state they are connected by chemical bonds. This is the "glue" that holds condensed matter together. We will summarize here some notions from chemistry which we will need in the sequel.

van der Waals' bonds At long ranges the dominant interaction between neutral atoms or molecules is the van der Waals interaction which arises from the interaction of fluctuating induced dipoles. For two neutral molecules (or atoms) a distance d apart this effect gives rise to a potential energy of interaction given by:

$$V(r) \sim -1/r^6. \tag{1.10}$$

This equation is universally true if the molecules are far apart compared to the size of of their electronic clouds. For closed shell atoms and molecules such as Ar and H_2 that do not chemically react, the van der Waals' interaction is the attractive force that causes condensation. Since this is a weak, short-range force, materials bound this way usually have low melting points.

A rough argument for the r^{-6} dependence is as follows: suppose there is a fluctuation (a quantum fluctuation, in fact) on one of two molecules so that an instantaneous dipole moment, p_1, arises. This gives rise to an electric field of order $E \sim p_1/d^3$ at the other molecule. This electric field polarizes the other atom. To understand this, we introduce a concept that we will use later, the *polarizability*, α, of the molecule. It is defined by:

$$\mathbf{p}_{\text{ind}} = \alpha \mathbf{E}, \tag{1.11}$$

where $\mathbf{p}_{\text{ind}}$ is the induced dipole moment. Note that in our system of units the polarizability, α, has units of volume. It is roughly the molecular volume. Thus $p_2 \sim \alpha p_1/d^3$. This finally gives for identical molecules the fluctuating dipole-dipole interaction:

$$V \sim p_1 p_2/d^3 \sim \alpha p_1^2/d^6. \tag{1.12}$$

Since this expression depends on p_1^2 there is a time-averaged value for the potential. It is easy to show that the dipoles will be antiparallel so that the interaction is attractive. An actual calculation of the coefficient of r^{-6}, that is, of the average of p_1^2, can be done (in simple cases) using quantum mechanical perturbation theory.

Ionic bonds The chemistry of the valence electrons in a compound can lead to charge transfer, e.g.:

$$\text{Na} + \text{Cl} \rightarrow \text{Na}^+\text{Cl}^-. \tag{1.13}$$

In this case there will be strong forces due to the charges, and the ions will be bound by the Coulomb interaction:

$$V(r) = Zq_1q_2/r.$$

This is called ionic binding. Solid NaCl, table salt, is bound in this way. Ionic solids often have very large binding, and very large melting points.

Covalent bonds In elements with s and p electrons in the outer shell, covalent sp^3 orbitals give rise to directed bonds where electrons between ions glue together the material. Semiconductors such as Si, Ge, are bonded this way, as well as polymers and many biological materials. There are intermediate cases between the covalent and ionic materials, such as III-V semiconductors like GaAs.

Hydrogen bonds These arise in materials that contain H such as ice. The proton participates in the bonding. This is very important in biological materials.

Metallic bonding For most light metals like Cu or Na, the outer valence electrons are delocalized for quantum mechanical reasons which we will discuss in great detail, later. The electrons act as glue by sitting between the positively charged ions. These essentially free electrons give rise to the electrical conduction of metals such as Cu.

1.4 The exchange interaction

We have talked about bonds between atoms in terms of spatial degrees of freedom of the electrons, but we have not mentioned electron spin. There is another effect, very important for magnetism, which arises from the interplay between the Pauli exclusion principle, the spin degrees of freedom, and the electrostatic repulsion of electrons. It occurs, for example, for atoms which have unpaired spins.

We recall from quantum mechanics that the Pauli principle says that electron wavefunctions must be antisymmetric in the exchange of any two electrons. This implies that when we bring two atoms together the many-electron wavefunction must vanish when two electrons with parallel spins are at the same position. Therefore electrons with parallel spins are likely to be *farther apart* in space than antiparallel ones, and therefore have a smaller electrostatic repulsion. As a result, if the two atoms have parallel spins the energy is lower. Thus spins and therefore magnetic moments tend to line up when electrons from adjacent atoms overlap. This is called the exchange interaction. This is discussed in considerable detail below, Section 9.2.1, or in standard texts on quantum mechanics, e.g. (Landau & Lifshitz 1977, Schiff 1968, Baym 1990).

There are a few comments we should make about this. One is that there needs to be overlap of wavefunctions to have the effect work. The difference in energy between states with parallel and antiparallel spins on adjacent atoms (the strength of the interaction) is dependent on the overlap; the exchange interaction is very short range. Also, the size of the energy difference is basically the electrostatic energy of two electrons an atomic distance apart, a few electron volts.

Spin and symmetry effects need not favor parallel spins; it depends on the nature of the wavefunctions and what energies are most important. A simple example of favoring antiparallel spins is the hydrogen molecule, two electrons and two protons. In one approach to the problem (the Heitler–London approximation) we build up the wavefunction for the molecule from atomic wavefunctions centered on each proton. We can then form symmetric and antisymmetric combinations of these functions, as above. However, since the total wavefunction must be antisymmetric, parallel electron spins (total spin 1) go with the antisymmetric spatial function, and antiparallel spins (total spin 0) go with the symmetric spatial function; for more details see (Baym 1990). The electrostatic interaction with the hydrogen nuclei favors the symmetric state since the electrons spend more time between the nuclei, and the kinetic energy of the symmetric state is lower. As a result the ground (bonding) state of H_2 has total spin 0, and is symmetric in space.

Suggested reading

There are many excellent references and textbooks for this subject that the student can explore. The classic undergraduate text is by

Kittel (2005).

Successive editions of this book (the current one is the eighth) have been used by generations of physicists and engineers.

At the graduate level the following old standards are highly recommended:

Ashcroft & Mermin (1976)

Ziman (1972)

Kittel (1963)

More modern treatments can be found in:

Anderson (1997)

Marder (2000)

Grosso & Pastori Parravicini (2000)

Phillips (2003)

Taylor & Heinonen (2002)

Chaikin & Lubensky (1995)

The last book is particularly good on soft condensed matter such as polymers and liquid crystals, which are not treated in detail in this book.

Problems

1. Calculate the van der Waals' interaction between two H atoms in their ground state. Use the Hamiltonian for two single atoms as a reference: $\hat{\mathcal{H}}_0 = p_1^2/2m - e^2/|\mathbf{r}_1 - \mathbf{R}_1| + p_2^2/2m - e^2/|\mathbf{r}_2 - \mathbf{R}_2|$. You can put one nucleus at the origin and the other at distance d along the x axis. Use the rest of the energy as a perturbation in second-order perturbation theory:
   $\hat{\mathcal{H}}_1 = -e^2/|\mathbf{r}_1 - \mathbf{R}_2| - e^2/|\mathbf{r}_2 - \mathbf{R}_1| + e^2/r_{12}$, where $r_{12} = |\mathbf{r}_1 - \mathbf{r}_2|$. See (Schiff 1968) Assume $|\mathbf{r}_i| << d$. You may use only the first excited state of H in your perturbation theory.

2. Work out the exchange splitting between the singlet and triplet ($1s2s$) states of He. (a) Use hydrogenic $1s$ and $2s$ states as a basis. Write down symmetric and antisymmetric 2-electron wavefunctions. (b) Show which belongs to the triplet spin state, and which to the singlet. (c) Figure out the energy difference between the two states in terms of the direct and exchange integrals (you need not work out the integrals):

$$I = \int d\mathbf{r}_1 d\mathbf{r}_2 \psi_{1s}^*(\mathbf{r}_1) \psi_{2s}^*(\mathbf{r}_2) \frac{e^2}{r_{12}} \psi_{1s}(\mathbf{r}_1) \psi_{2s}(\mathbf{r}_2)$$

$$J = \int d\mathbf{r}_1 d\mathbf{r}_2 \psi_{1s}^*(\mathbf{r}_1) \psi_{2s}^*(\mathbf{r}_2) \frac{e^2}{r_{12}} \psi_{1s}(\mathbf{r}_2) \psi_{2s}(\mathbf{r}_1).$$

We have seen in the previous chapter that chemical bonds are the glue for condensed matter. If the temperature is low enough so that thermal fluctuations do not break the bonds, it is no surprise that atoms and molecules condense, i.e. stick together, so that there are large pieces of matter.

However, the precise structure of condensed matter is often quite surprising. For example, we might guess that the typical result of attractive chemical bonds would be a disorderly mass of molecules. This does occur; such materials are called glasses. However, very commonly something else happens: at low enough temperatures the atoms or molecules form a remarkable ordered structure, a *crystal*. A crystal is an ordered, periodic array of atoms or molecules. In the next chapter we will give a precise definition of this concept. For our purposes, it is enough to understand that crystals are made up of identical building blocks that are repeated many times. See Figure 2.1 for an example, the face-centered cubic (fcc) crystal structure.

Chemistry tells us that atoms or ions can have a magnetic moment, either from orbital currents or unpaired spins. However, you might expect that when large numbers of such ions are stuck together that the orientation of the moments would be random. This is not always the case. For some elements, e.g. Fe, Ni, Co, and many compounds the moments line up in regular arrays of various kinds due to the exchange interaction, discussed above.

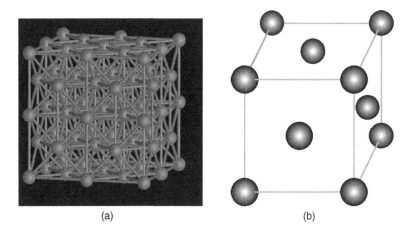

(a)   (b)

Fig. 2.1   **(a) A visualization of the face-centered cubic crystal structure. The nearest and next-nearest neighbor bonds are shown. (b) The structure may be thought of as a collection of cubes with atoms at the corners and the middle of all the faces.**

The large magnetic moment leads to the familiar phenomenon of magnets that pick up nails or stick to your refrigerator. This is called ferromagnetism because it was first noticed in iron and its compounds.

Complicated organic compounds sometimes form liquid crystals. These are liquids that are nevertheless ordered in some way. A nematic liquid crystal, for example, consists of long, rod shaped molecules. In certain temperature ranges the orientation of the molecules lines up, but the positions are random, as in a classical liquid. Nematics are the essential part of many liquid crystal displays such as those in laptop computers.

We will now discuss examples of ordered states in some detail.

2.1 Ferromagnets

Michael Faraday classified materials into three classes according to their magnetic state: there are diamagnets, paramagnets, and ferromagnets. The state can be characterized by the value of the magnetization, $\mathbf{M}$, which is defined to be the magnetic moment per unit volume.

In paramagnets there can be non-zero magnetization only if induced by an external field, e.g., by aligning magnetic moments on ions. Diamagnetism is usually a weak effect which gives an induced magnetization antiparallel to the magnetic field due to induced shielding currents described by the Lenz law. The ordered states we are interested in are in the third class where there is a non-zero spontaneous magnetization in the absence of an external magnetic field.

Ordered states of this sort are not uncommon. There are a handful of ferromagnetic elements, and many ferromagnetic compounds. To illustrate a simple case of macroscopic order in condensed matter, we concentrate on magnetic insulators, which may be thought of as a collection of atoms with spin and/or orbital magnetic moments arranged in a crystal. In the ordered state a finite fraction of *all* the moments line up, i.e. point in the same direction, because of strong interactions between the ions. Magnetic metals are a much more subtle phenomenon, and will be discussed later. The interaction that causes ferromagnetism is exchange. The strength of a bond between two magnetic atoms or ions depends on the relative orientation of the spins, as we have seen. If the energy is lower when the spins are parallel, we can have alignment.

A way to parameterize this energy difference was proposed by W. Heisenberg. He noted that the energy is a scalar, but the spins of the atoms are vectors. The simplest way to make a scalar from two vectors is to take their dot product. We expect to be able to write the energy as:

$$E = -2J\mathbf{s}_1 \cdot \mathbf{s}_2. \tag{2.1}$$

This is called the Heisenberg Hamiltonian, and J is called the exchange constant. The minus sign and the factor of 2 are conventional, and J can be positive or negative. It is easy to see for H_2, mentioned above, that J is related to the singlet-triplet energy splitting.

2.1.1 Magnetic order and energies

Suppose we have many magnetic ions in a crystal. We can extend Eq. (2.1) to this case by writing:

$$\mathcal{H} = -\sum_{i \neq j} J_{ij} \mathbf{s}_i \cdot \mathbf{s}_j. \tag{2.2}$$

J_{ij} is the exchange between ion i and ion j. Recall that the exchange interaction is very short-ranged, so we can suppose that $J_{ij} = 0$ unless i and j are nearest neighbors. In that case we can write:

$$\mathcal{H} = -\sum_{j,\delta} J \mathbf{s}_j \cdot \mathbf{s}_{j+\delta}, \tag{2.3}$$

where $j + \delta$ runs over the nearest neighbors of j.

Suppose that $J > 0$. Now it is clear that the lowest energy state of our model system has all the spins aligned in some direction, so that the magnetization is:

$$\mathbf{M} = n\gamma\hbar\langle \mathbf{s} \rangle \tag{2.4}$$

where $n = N/\Omega$ is the number of ions per unit volume, and $\gamma\hbar\langle \mathbf{s}\rangle$ the average moment of a single spin in the crystal. In this expression γ is the gyromagnetic ratio, the ratio between the moment of a single ion and its angular momentum, $\hbar\mathbf{s}$. Note that $\gamma\hbar = -g\mu_B = -ge\hbar/2mc$ where μ_B is the Bohr magneton, and g the g-factor of the ion. The state will be something like that pictured in Figure 2.2(a).

The difference in energy between the ground state of the system and a random one where the exchange interaction averages out is, from Eq. (2.2), $E_o = -NJs^2z/2$ where z is the *coordination number*, the number of nearest neighbors of a given ion. For example, in Figure 2.2, $z = 4$.

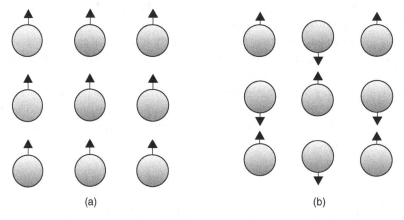

(a)   (b)

Fig. 2.2 **Magnetic order. (a) Ferromagnet (b) Antiferromagnet. The arrow denotes the direction of the magnetic moment.**

However, suppose that $J < 0$ so that antiparallel spins are preferred. Then, provided the arrangement of atoms permits, spins can rearrange into a Néel or antiferromagnetic state with every down spin surrounded by up spins, see Figure 2.2(b). This can happen if the crystal can be divided into two sublattices, A and B, such that the nearest neighbors of sites on lattice A are on B.

This structure was proposed by L. Néel. In this case there is no observable magnetic moment, but there is a macroscopic magnetization on every other site. This can be measured, as we will see below. The state pictured is not an eigenstate of the quantum mechanical Hamiltonian in Eq. (2.2). It is an approximation which is better in the classical limit, i.e., the limit of large spin.

In a real crystal there are effects due to spin-orbit interactions and the electric fields of ions. One result which is commonly observed is that magnetic moments align easily with certain crystal directions. These are called *easy axes*. Further, an effect of spin-orbit interactions can be to make the exchange constants, J, different for different directions (White 1970). For example, it can happen that the Heisenberg Hamiltonian should be replaced by:

$$\mathcal{H} = -\sum_{i \neq j} J_{ij} s_i^z s_j^z. \tag{2.5}$$

This is called the Ising model. It was originally introduced in order to make calculations easier. However, it turns out that there are real Ising magnets.

Ferro- and anti-ferromagnetism are not the only types of magnetic order. Very complicated magnetic structures are possible, including helical arrangements of moments in rare earth metals.

2.1.2 Ferromagnets and paramagnets

For $J > 0$ the ferromagnetic state has the lowest energy of any spin configuration. However, as temperature increases all ferromagnets become disordered, and at a temperature, T_c, the Curie temperature, they lose their macroscopic magnetism altogether. Instead of the ordered array of Figure 2.2 we have a random array of directions for the spins. This is called a paramagnet: any magnetic moment that arises is induced by an external field, rather than being permanent, as in a ferromagnet.

Why does this occur? In fact, the lowest energy state is not the thermal equilibrium state except at $T = 0$. Statistical physics tells us that the equilibrium state is that of minimum *free energy*: $F = E - TS$, where E is the energy and S is the entropy. The paramagnetic state has many different random orientations of spins, and thus has high entropy. (To see this recall Boltzmann's formula: $S = k_B \ln(W)$ where W is the number of equivalent configurations in a macrostate.) As T goes up, the entropy wins, and magnetic order is lost.

As an order-of-magnitude guess, we can turn a ferromagnet into a paramagnet when a typical thermal fluctuation (whose energy is $k_B T$) breaks an exchange bond, i.e., $k_B T_c \sim J$. We will refine this estimate in the next section.

2.1.3 Magnetic phase transition

As temperature goes up, disorder increases, and the magnetic moment decreases for the reasons given in the previous paragraph. Only a fraction of the time will spins point along the macroscopic moment. That is the thermally averaged moment is less than $N\gamma\hbar s$. Above T_c there is no order left. It is an observed fact that the magnetic order disappears *continuously*. In the language of statistical physics the magnetic phase transition is second order (or, more properly, continuous). In contrast, in crystals, which we will discuss next, the positional order disappears suddenly at the melting point, a first-order transition.

Molecular field theory

There is an approximate treatment of the magnetic phase transition, molecular field theory, proposed by P. Weiss. The idea is to compare Eq. (2.2) to the Zeeman interaction of a magnetic dipole with an external field:

$$-\gamma\hbar\mathbf{s_j} \cdot \mathbf{H}. \tag{2.6}$$

Weiss thought of each spin as seeing an effective (or molecular) field, $\mathbf{H}_M$, due to to all the other spins. From Eq. (2.3) we see that the relevant term is:

$$-J\mathbf{s}_j \cdot \sum_\delta \mathbf{s}_{j+\delta} \equiv -\gamma\hbar\mathbf{s}_j \cdot \mathbf{H_M}. \tag{2.7}$$

In order to make this expression tractable, we make a mean-field assumption: we replace the sum by its statistical average so we have:

$$\mathbf{H}_M = Jz\langle\mathbf{s}\rangle/\gamma\hbar = Jz\mathbf{M}/n(\gamma\hbar)^2. \tag{2.8}$$

We have used the fact that for ferromagnets the average of $\mathbf{s}_j$ is independent of j and we have used Eq. (2.4). Note that the molecular field is proportional to the magnetization: $\mathbf{H}_M = \lambda\mathbf{M}, \lambda = Jz/n(\gamma\hbar)^2$. However, we can also calculate $\mathbf{M}$ from the statistical mechanics of a spin in a magnetic field. Let the number of up spins be given by N_+ and the number of down spins by N_-. We consider the z-component of the magnetization. For spin $1/2$ we have, using Eq. (2.6):

$$
\begin{aligned}
M &= \frac{\gamma\hbar}{2\Omega}(N_+ - N_-) \\
&= \frac{1}{2}n\gamma\hbar\left(\frac{\exp(\beta\gamma\hbar H/2) - \exp(-\beta\gamma\hbar H/2)}{\exp(\beta\gamma\hbar H/2) + \exp(-\beta\gamma\hbar H/2)}\right) \\
&= \frac{n\mu}{2}\tanh(\beta\mu H/2),
\end{aligned}
\tag{2.9}
$$

where $\beta = 1/k_B T$.

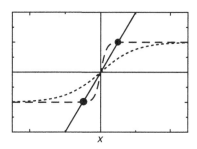

Fig. 2.3

Graphical solution of Eq. (2.10); $x = 2M/n\gamma\hbar$. **The functions plotted on the vertical axis are: solid** $y = x$ **(LHS of the equation); dotted, RHS for** $T < T_c$; **dashed, RHS for** $T > T_c$.

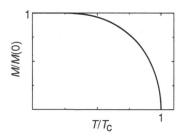

Fig. 2.4

Solution of Eq. (2.10) as a function of temperature.

The magnetic field in this case is the molecular field. This gives:

$$M = \frac{n\gamma\hbar}{2}\tanh(\beta\gamma\hbar\lambda M/2)$$

$$2M/n\gamma\hbar = \tanh[(\beta Jz/4)(2M/n\gamma\hbar)]. \tag{2.10}$$

An equation of this type is easy to solve numerically. In Figure 2.3 we show a graph of the two sides of the equation. Note from the figure that if the slope of the right-hand side near the origin is less than 1, there will only one solution, $M = 0$. Since the hyperbolic tangent has slope 1 near $M = 0$, this condition is equivalent to:

$$\beta Jz/4 = 1; \quad k_B T_c = Jz/4. \tag{2.11}$$

We have labeled the temperature in this expression T_c because it is the mean-field theory estimate of the Curie temperature. Figure 2.3 shows the following: for $T > T_c$ there is only one solution, $M = 0$, i.e. a paramagnet. Below T_c there are three solutions to Eq. (2.10), one corresponding to magnetic moment up, another to moment down, and the trivial solution, $M = 0$. One of the non-trivial solutions is shown in Figure 2.4. This should be interpreted as the temperature dependence of the spontaneous magnetization. It is quite easy to show that near T_c there is a square root singularity: $M(T) \propto (T_c - T)^{1/2}$. Eq. (2.11) can be generalized to other values of the spin:

$$k_B T_c = zJs(s+1)/3. \tag{2.12}$$

The estimate of the transition temperature and of $M(T)$ are in qualitative agreement with experiment. In detail the agreement is not very good since we have neglected thermal fluctuations.

It is interesting to return to Eq. (2.9) in the case where there is an external magnetic field in the z-direction so that $H = H_e + H_M$. Then we have:

$$M = \frac{n\gamma\hbar}{2}\tanh(\beta\gamma\hbar(H_e + \lambda M)/2). \tag{2.13}$$

Suppose that $T > T_c$ and that H_e is small. We are in the paramagnetic regime, and only M is that induced by the external field. The argument of the hyperbolic tangent is small, and we can expand it. After simple algebra we find:

$$M = \frac{\beta n(\gamma\hbar)^2}{4}[H_e + \lambda M],$$

$$M/H_e = \frac{n(\gamma\hbar)^2}{2k_B(T - T_c)}. \tag{2.14}$$

The quantity in the last equation is the magnetic susceptibility, χ.

In the case that there is negligible exchange we can put $T_c = 0$, and we recover the famous Curie law for paramagnets:

$$\chi = A/T; \qquad A = n(\gamma\hbar)^2/2k_B. \tag{2.15}$$

In the case where $T_c > 0$ we find that χ diverges at T_c. This is confirmed by experiment, but the form of the divergence is different close to T_c.

Landau theory

The fact that $M(T)$ goes to zero at T_c means that the magnetic phase transition is not a first-order transition like boiling or melting, but continuous. In 1937, L. Landau gave an empirical theory of such transitions which is widely used (Landau & ter Haar 1965). We will briefly review it here: it is very well discussed in many books on statistical physics.

Landau noted that since M is small near T_c, it is useful to do a Taylor expansion in M of relevant functions. He called M the order parameter; this means a quantity that marks the transition in the sense that the order parameter is zero above the transition temperature. Suppose we look at the probability to have various values of the order parameter in a region of the sample. This involves averaging with a Boltzmann distribution over many statistical states that give the same M. Thus the probability will depend on temperature. Landau wrote it as $\exp(-\beta F(M, H, T))$, where F is called the (Landau) free energy. The minimum of F with respect to the order parameter is the thermodynamic free energy.

We can write down a candidate form for F simply from symmetry. We work with an Ising model for which the magnetization is along the z-axis, and is a scalar. In the absence

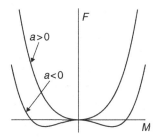

Fig. 2.5

The Landau free energy as a function of M.

of an external field F must be even in M. So we put, for the first terms in the power series expansion:

$$F = F_\circ + a(T)M^2 + bM^4 - \gamma\hbar MH. \qquad (2.16)$$

The last term is the Zeeman interaction. A plot of $F - F_\circ$ is given in Figure 2.5 for $H = 0$. The minimum of the free energy is at $M = 0$ unless a is negative. The simplest assumption for the temperature dependence of a is:

$$a(T) = a_\circ(T - T_c).$$

We will assume that b is not temperature dependent. Minimizing the free energy gives:

$$M(T) = \left(\frac{-a}{2b}\right)^{1/2} = \left(\frac{a_\circ[T_c - T]}{2b}\right)^{1/2} \quad T < T_c$$
$$= 0 \qquad\qquad T > T_c. \qquad (2.17)$$

This is the result we got already near T_c. Further, we see that the trivial solution, $M = 0$ is a maximum of the free energy in the ordered state. If we turn on the external magnetic field and assume that $T > T_c$ we can easily rederive the Curie law, Eq. (2.15).

The Landau theory is a very powerful tool in discussing phase transitions of different types. We have neglected fluctuations in this discussion. That is, we could have states with different values of M in different parts of the sample, and in thermal equilibrium there will be a statistical ensemble of such states. A generalized Landau theory taking spatial fluctuations into account is the starting point of modern theories of phase transitions. Even if fluctuations are not important, we can have spatial variations in M due to external fields and surfaces. In this case we need to deal with a free energy density, f, defined by:

$$F = \int d\mathbf{r}\, f(M(\mathbf{r})).$$

We will look in detail at a theory of this type in Chapter 10.

2.2 Crystals

Many pure substances condense into a crystalline structure in the solid state. All elements (except He) do this, and many simple compounds. The most commonly observed case is a polycrystal, that is, an arrangement that is locally crystalline with many rather small domains with grain boundaries between.

The physics of this is that each atom is in a minimum of the potential energy due to all the other atoms. If the potential produces a short-ranged central force then the lowest energy will be that of the best "packing," that is, when the atoms are closest together. The crystal structure shown in Figure 2.1 is close packed in the sense that it can be formed by spheres touching each other in an efficient way. It is a structure often observed at the grocery store: a pile of grapefruits will be best packed if the first layer is packed as triangles with each grapefruit surrounded by a hexagon of others touching it. The next layer is put in the triangular array of alternate holes in the first layer. Then the third layer is put above the first. This turns out to be the fcc structure. See Figure 2.6. There is another close-packed crystal structure that we will meet later.

Covalent bonds are directional, and do not lead to close packing, but rather to open structures which do not distort the natural bond angles too much. Ionic solids need to alternate positive and negative charges, so they are not close packed, either.

2.2.1 Crystalline order and cohesive energies

We now compute the ground-state energy of a crystal, as we did for a magnet. Consider a simple molecular-bonded solid such as Ar. For low enough temperature Ar condenses into a

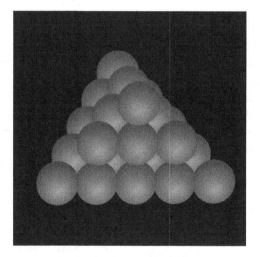

Fig. 2.6   **The same array of atoms as in Figure 2.1 sliced in half and displayed as a pile of close packed spheres.**

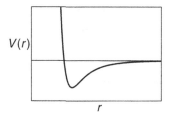

$V(r)$

r

Fig. 2.7 **A sketch of the Lennard-Jones potential. The minimum is given by** $r_{min} = 2^{1/6}\sigma$, $V_{min} = \epsilon$.

fcc crystal. In this case, the dominant attractive interaction is the van der Waals' interaction. The difference in energy between a solid made up of N atoms or molecules and that of the isolated molecules at the same T is called the *cohesive energy*. We will denote it by E. The solid is bound for $T \to 0$ if $\mathsf{E} < 0$.

The binding arises from the attractive forces. However, we should note that the attractive van der Waals force is not the whole story. At short range, when the closed shell electrons approach one another, there is repulsion (related to the exchange interaction discussed above) which arises from electrostatics and the Pauli principle. This effect is often modeled (for no good reason) by a potential $\propto -1/r^{12}$. Other functional forms would do as well, e.g., an exponential dependence on distance. However, the power law choice is both convenient and traditional. The total interaction can be written (the Lennard-Jones potential):

$$V(r) = 4\epsilon[-(\sigma/r)^6 + (\sigma/r)^{12}]. \tag{2.18}$$

Here ϵ is an energy scale, and σ a length scale. For Ar, $\sigma = 3.4$ Å; $\epsilon = 1.7 \times 10^{-14}$ ergs. These parameters can be measured directly in the gas phase by measuring the (non-ideal) equation of state, for example, the virial coefficients (Hirschfelder, Curtiss & Bird 1965).

We can now compute the cohesive energy for Ar in the fcc structure. We note that even if we know the crystal structure we do not know the lattice constant a, the distance between nearest neighbors. The crystal will adjust to have the lowest energy by adjusting a. That is, to find the equilibrium structure we compute $\mathsf{E}(a)$ and minimize it with respect to a. To this end we need to sum up the potential $V(r)$ over all pairs of atoms at distances r_{ij}:

$$\mathsf{E} = (1/2) \sum_{i \neq j} 4\epsilon[(\sigma/r_{ij})^{12} - (\sigma/r_{ij})^6]. \tag{2.19}$$

We need not consider the kinetic energy because, as is well known from statistical physics, it is the same for the solid and the gas if classical mechanics is valid. To see how to perform the sum, consider only the attractive terms. These terms can be scaled by a:

$$\frac{\mathsf{E}_{att}}{2N\epsilon} = -\frac{\sigma^6}{a^6}\left[12 + \frac{6}{\sqrt{2}^6} + \cdots\right] \equiv -S_6 \frac{\sigma^6}{a^6} \approx -12.75 \frac{\sigma^6}{a^6} \tag{2.20}$$

where N is the total number of atoms. Note that in this case we need to go beyond nearest neighbors since the potential is a power law. In the magnetic case the dependence is exponential and nearest neighbors often suffice.

We have summed over the 12 nearest neighbors and the 6 next nearest neighbors of an atom; see Figure 2.1. With a bit of work, more terms can be added. The sum converges quickly: the value after adding many neighbors is $S_6 = 14.45$ (instead of 12.75). The other term can be treated similarly; its value is $S_{12} \; \sigma^{12}/a^{12}$ where $S_{12} = 12.13$. Minimizing the energy amounts to putting $d\mathsf{E}/da = 0$, that is, putting each atom at the minimum of the potential due to the neighbors. It is easy to show that at equilibrium $a_o/\sigma = 1.09$ and $\mathsf{E}(a_o) = -2.15(4N\epsilon)$.

To decide if the fcc structure is the right one, calculations can be done for other candidate structures and the results compared. Calculations of this type work well for molecular crystals.

For ionic solids we need to do a more difficult sum:

$$\mathsf{E} = (1/2) \sum_{i \neq j} [\pm Ze^2/r_{ij} + v_{\text{repl}}]. \qquad (2.21)$$

The last term is a short-range repulsion between nearest neighbors which could be taken to be of the same form as the last term in Eq. (2.19). The sum in this expression is only conditionally convergent. We need to rearrange the terms in neutral bits to make sure that we have a meaningful value. For this, and other cases, a good reference is Kittel (2005).

For covalent and metallic binding the cohesive energy needs to be calculated by careful analysis of the quantum mechanics of the electrons. This will be treated in later chapters.

2.2.2 Solids, liquids, and gases

We have noted that crystalline solids are ubiquitous at low enough temperatures. A typical phase diagram for a simple substance looks like Figure 2.8.

Liquids and gases generally have higher energy than a solid: the atoms are farther apart, and do not sit at the minimum of the intermolecular potential as they do in a crystal. Why, then do liquids and gases occur? Again, the answer is entropy. In a gas and most liquids there are more rearrangements possible for the same number of atoms for the state of given V and T. Thus the entropy is large.

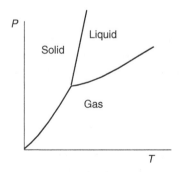

Fig. 2.8   A sketch of the phase diagram for a typical pure substance.

2.2.3 Melting

Estimating the melting temperature of a solid seems difficult. Since we are dealing with a first-order transition where the crystalline nature disappears abruptly, we need to compare two states, the solid and the liquid: a solid melts when its chemical potential is equal to that of its liquid. This criterion seems to indicate that we need to look at both phases to determine when melting occurs.

However, there is a remarkable observation due to F. Lindemann which involves only simple properties of the solid. He noted that an atom in a crystal at finite temperature will not remain at its "official" position. Instead it will vibrate in the potential well of the neighbors, by an amount **u**. A simple way to look at this, due to A. Einstein, is to assume that one atom vibrates as if the others remain fixed.

Later we will consider this in great detail. For the moment we can say the following: the chosen atom is in equilibrium due to the forces of its neighbors. If it vibrates away from equilibrium there will be forces. However, since the atom is at the bottom of a potential well, the leading terms in the potential will be quadratic in the components of the displacement. That is, the atom will act like a simple harmonic oscillator. In a sufficiently symmetric environment there will be one effective spring constant, or, equivalently, one natural frequency of vibration, ω_E, the Einstein frequency. We will see in Chapter 5 how to estimate ω_E.

The statistical mechanics of a harmonic oscillator is well known. From the equipartition theorem the mean-square displacement is given by:

$$u_{\mathrm{rms}} = \sqrt{\langle u^2 \rangle} = \sqrt{3 k_B T / m \omega_E^2}. \tag{2.22}$$

Note that this is an increasing function of temperature. Lindemann guessed that when this quantity increased so that it was an appreciable fraction of the lattice constant, a, the order of the crystal would be lost, and melting would occur. The fraction is rather small, about 10%. This works pretty well as an empirical rule. See, for example (Gilvarry 1956).

2.2.4 Colloidal Crystals

There is a strange and instructive example of crystallization, that of hard-sphere solids. These are realized in experiments as colloidal crystals, namely confined systems of plastic (usually polystyrene) spheres with a radius $b \approx 1 \mu m$ suspended in water. Note that the "atoms" in these cases are the colloidal particles that are made of many atoms. The size scales for this system are huge compared to that of an ordinary solid.

To proceed we need to know the nature of the interactions. To a good approximation the interactions are purely repulsive, and can, in most cases, be thought of as hard sphere-like:

$$V(r) = 0 \quad \text{for} \quad r > b; \quad V(r) = \infty \quad \text{for} \quad r \le b. \tag{2.23}$$

There is no energy for non-overlapping spheres, and T plays no role. Of course, the system must be confined by a container to prevent the spheres from wandering apart. It is observed

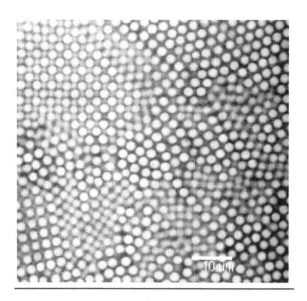

Fig. 2.9   **A micrograph of the surface of a colloidal crystal, courtesy of D. Weitz.**

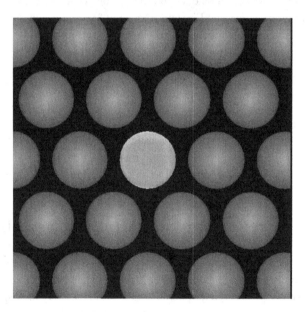

Fig. 2.10   **An illustration of one plane of a close-packed structure showing the volume available for vibration.**

(and confirmed in numerical experiments) that as the system is compressed, there is a transition from a disordered collection of spheres to a regular arrangement, i.e. from "liquid" to "solid"; see (Alder & Wainwright 1957). Figure 2.9 is an experimental picture.

What is going on? Even if there is no effect of temperature, entropy is still maximized for any state of fixed number and volume. For low densities there is lots of room for spheres to move around, and the disordered state has large entropy. However, as the system is compressed, the spheres tend to jam: they cannot move and sample different configurations. The key point is that jamming appears to occur at a density above the close-packing density for the solid. Thus, in the solid there is actually more room for the individual spheres to move around. The light-colored sphere in Figure 2.10 moves in the cage of its neighbors and has more freedom than if they were jammed in a disordered array.

2.3 Other ordered states

Ferromagnets and crystals are only two examples of ordered states in condensed matter. Others include liquid crystals where there is order in the orientation of large molecules, but their positions in space are random, as in a liquid. In a nematic liquid crystal the molecules are rod-like, and simply line up. There is a preferred direction, as in a magnet, called the *director*, $\mathbf{n}$. However, there is a subtle difference. In most nematics the molecules have a center of inversion so that $\mathbf{n}$ and $-\mathbf{n}$ are indistinguishable; see (Chaikin & Lubensky 1995). There are other kinds of order such as ferroelectricity where electric dipole moments line up. The most subtle kind of order that we will discuss occurs in superconductivity. In this case, the nature of the order is not in the least obvious, and took half a century to discern.

2.4 Order parameters

In statistical physics it is customary to describe ordered states by an *order parameter*, that is some quantity whose thermal average is non-zero in the ordered state and characterizes the order. We have seen an example in the case of magnetism in Section 2.1.3. In statistical physics the major use of order parameters is in formulating Landau theories of the ordered states. The conceptual advantage of the concept is that it allows us to say immediately what we mean by an ordered state.

2.4.1 Ferromagnets

In a ferromagnet there is a non-zero macroscopic magnetic moment which characterizes the order. We can generalize the definition used above for non-uniform situations: a reasonable definition would be the local density of magnetic moment for a region of the sample large compared to the atoms but small compared to the sample. We call it $\mathbf{M}(\mathbf{r})$. The conventional magnetization, $\mathbf{M}$, is the volume average of $\mathbf{M}(\mathbf{r})$:

$$\mathbf{M} = \frac{1}{\Omega} \int \langle \mathbf{M}(\mathbf{r}) \rangle d\mathbf{r}, \qquad (2.24)$$

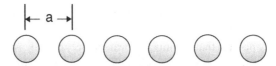

Fig. 2.11

A one dimensional periodic structure.

where the integral is over the sample volume. The brackets $\langle \cdot \rangle$ denote a thermal average. A non-zero order parameter of this type means that the system has lost a symmetry compared to the disordered state. Here, rotational invariance is lost. As the temperature rises above T_c the rotational order is restored and $\langle \mathbf{M} \rangle = 0$.

An order parameter for antiferromagnets is the *staggered magnetization* which is defined in terms of the magnetization on one of the sublattices minus that on the other: $\mathbf{M}_A - \mathbf{M}_B$.

2.4.2 Crystals

What should we use to characterize crystalline order? Clearly the array of atoms in Figure 2.1 is orderly in some sense. It has lost the translational invariance of a liquid. A clue to how to proceed is to consider a one-dimensional periodic structure, Figure 2.11. The density of matter along the line is not uniform; it is periodic with period a. In the simplest view it is the superposition of atomic densities:

$$n(x) = \sum_j n_a(x - ja), \tag{2.25}$$

where $n_a(x)$ is the density we associate with an atom at the origin.

The density, $n(x)$ is a periodic function, and we can expand it in a Fourier series:

$$n(x) = \sum_Q e^{iQx} n(Q)$$

$$n(Q) = \int_0^a e^{-iQx} n(x) dx / a; \qquad Q = 2\pi k / a. \tag{2.26}$$

Here k is an integer. Note that if a changes, so do the Q. And, in a liquid with uniform density $\tilde{n}(Q) = 0$ if $Q \neq 0$. For nearly uniform density the $\tilde{n}$'s will be small. Thus $\tilde{n}(Q)$ for non-zero Q is a reasonable choice for an order parameter. For the case of a three-dimensional crystal we need a little more mathematics to find which Q's we should use; these turn out to be vectors called reciprocal lattice vectors. This will be discussed in the next chapter.

2.5 Disordered condensed matter

For simple substances such as elemental solids, crystals are often the low-temperature state, as we have mentioned. For more complex molecules, there are a huge variety of states, even in equilibrium. For example, polymers are chain molecules with repeating units, which can be very long. Long-chain molecules of this sort can form a gel, that is, a linked, disordered mesh of molecules which form a solid without regular order; see de Gennes (1979).

Many materials form glasses, amorphous arrangements of molecules which can be thought of as a liquid which has such high viscosity that it stops flowing. Such materials look and act like solids for many macroscopic purposes.

There is a very interesting class of materials, usually metallic alloys, that form quasi-crystals. These are not amorphous but they are not periodic either. They are *quasi-periodic*. The interested student should consult DiVincenzo & Steinhardt (1999).

Alloys are mixtures of materials which are stabilized by the entropy of mixing. Often, in the case of substitutional alloys, the atoms occupy a crystal structure, but with a random occupancy of the sites. Si and Ge form such alloys, as do many metals.

The magnetic analog of a glass is a spin glass. This is a state observed in some magnetic alloys such as Fe in Au. In this case the iron ions have moments, but they are random in orientation down to the lowest temperatures. This is not the same as a paramagnet: there is evidence that below a transition temperature there is negligible thermal fluctuation of the moments, but they are randomly oriented. See Bolthausen & Bovier (2007).

Suggested reading

Magnetism and the magnetic phase transition is treated in books on statistical physics such as:
    Landau, Lifshitz & Pitaevskii (1980),
    Huang (1987),
or in monographs on magnetism:
    White (1970),
    Mattis (1988).
For an elementary treatment see
    Kittel (2005).
For Landau theory see:
    Chaikin & Lubensky (1995),
    Landau et al. (1980), or
    Huang (1987).
Crystal binding is in all the general references in Chapter 1. For order parameters in liquid crystals, see
    Chaikin & Lubensky (1995).

Problems

1. Argon has $\epsilon = 1.7 \times 10^{-14}$ ergs and $\sigma = 3.4$ Å. Find the distance between atoms and the binding energy per atom for (hypothetical) simple cubic Ar.

2. When spheres are as closely packed as possible they form either a face-centered cubic (fcc) lattice or another structure called hexagonal close-packed (hcp); see Figures 3.2, 3.7. Prove that when spheres are packed they fill the following percentages of the available volume (the packing fraction): simple cubic, 52% ; fcc 74%. The packing fraction for hcp is the same as for fcc.

3. Work out the Curie–Weiss theory for an antiferromagnet. Suppose that there are two sublattices, A and B. Write the effective fields in the following form:

$$H_A = H - \lambda_1 M_B - \lambda_2 M_A \quad H_B = H - \lambda_1 M_A - \lambda_2 M_B$$

(a) Identify $\lambda_{1,2}$ in terms of the nearest and next-nearest neigbor exchange constants. (b) Write down self-consistent field equations for $M_{A,B}$. (c) Solve under the assumption that $M_A = -M_B$. (d) Show that $\chi = C/(T + \Theta)$, where $\Theta \propto \lambda_1 + \lambda_2$. (e) Find the temperature, T_N, below which there is a finite value of M for zero H. You should find that T_N, which is called the Néel temperature, is proportional to $\lambda_1 - \lambda_2$.

4. Prove Eq. (2.12).

5. (a) Derive the Curie law by minimizing the Landau free energy for $H \neq 0$ and $T > T_c$. (b) Show that $F - F_o = -a_o^2(T_c - T)^2/4b$ for $T < T_c$.

3 Crystals, scattering, and correlations

We have seen in the previous chapter that crystals are common in nature. In this chapter we will investigate in more detail how to think about such three-dimensional periodic structures. Then we will turn to the interaction of waves with such structures. This will lead us to a discussion of correlation functions in condensed matter.

3.1 Crystals

In the previous chapter we defined a crystal as a structure which repeats periodically in space. There is a mathematical framework for dealing with physical quantities in perfect crystals; it is the science of crystallography. We will review some of the elementary concepts from this subject.

Of course, any real material is an imperfect realization of a perfect crystal; real materials always have impurities and defects. Even if a crystalline solid is very close to being strictly periodic in bulk, all materials have a surface where the periodicity fails. However, consider a large chunk of matter, say a cube of edge L where the distance between the atoms is a. The number of atoms in the bulk is of the order of $(L/a)^3$, but the number on the surface is of order $(L/a)^2$. If $L >> a$ the fraction on the surface is negligible.

3.1.1 Lattices

The first step to defining a crystal is to define a lattice. This is a set of points in d dimensions which are generated by taking linear combinations of d linearly independent vectors called generators: $\mathbf{a}_k, k = 1, \ldots d$ with integer coefficients. In three dimensions we generate points by:

$$\mathbf{R} = n_1 \mathbf{a}_1 + n_2 \mathbf{a}_2 + n_3 \mathbf{a}_3 \quad n_k = 0, \pm 1, \pm 2, \ldots \tag{3.1}$$

Note that the same set of $\mathbf{R}$'s can be made by different generators. For example, $\mathbf{a}_1 + \mathbf{a}_2, \mathbf{a}_2, \mathbf{a}_3$ will generate the same lattice as in Eq. (3.1).

The set of points so generated are periodic with periodicity defined by the generators:

$$\{\mathbf{R}\} = \{\mathbf{R} + \mathbf{a}_k\}. \tag{3.2}$$

Table 3.1 List of the 14 lattice types in three dimensions. The symbols P, C, I, F stand for primitive, base-centered, body-centered, and face centered, respectively. a, b, c are the lengths of the edges of the conventional unit cell, and α, β, γ are the angles between the edges.

System	Lattice Type	Restrictions	
Triclinic	P	$a \neq b \neq c;$	$\alpha \neq \beta \neq \gamma$
Monoclinic	P,C	$a \neq b \neq c;$	$\alpha = \gamma = \pi/2 \neq \beta$
Orthorhombic	P,C,I,F	$a \neq b \neq c;$	$\alpha = \beta = \gamma = \pi/2$
Trigonal	P	$a = b = c;$	$\alpha = \beta = \gamma \neq \pi/2$
Tetragonal	P,I	$a = b \neq c;$	$\alpha = \beta = \gamma = \pi/2$
Hexagonal	P	$a = b \neq c;$	$\alpha = \beta = \pi/2, \gamma = 2\pi/3$
Cubic	P,I,F	$a = b = c;$	$\alpha = \beta = \gamma = \pi/2$

In one dimension there is only one *lattice type*, a set of points equally spaced along a line; see Figure 2.11. Two lattices are said to belong to the same lattice type if they have the same symmetry, or, more precisely, if they can be mapped onto one another by a continuous deformation in which the symmetry does not change. The classification of lattice types in three dimensions is due to A. Bravais; another term for lattice type is Bravais lattice.

In two dimensions there are five Bravais lattices shown in Figure 3.1. It is a good exercise to work out the generators for the lattices shown. Also, the student should convince herself that though there is a separate lattice type for centered rectangular, there is no need for centered square. In three dimensions there are fourteen Bravais lattices. Figure 3.2 shows some of them.

3.1.2 Bases and crystal structures

A lattice is an abstract collection of points. A crystal structure is a lattice plus a *basis*, namely a collection of physical objects, e.g. atoms, of given relative position and orientation attached to each lattice point. Thus every atom in a crystal is labeled with two vectors, the lattice point, $\mathbf{R}_n, n = 1 \cdots N$ where N is the number of lattice points, and the displacement of the atom with respect to the lattice point, $\mathbf{s}_m, m = 1 \cdots B$, where B is the number of atoms in the basis. Figure 3.3 shows an example based on the square lattice with two atoms in the basis.

3.1.3 A few examples of crystal structures

Figure 3.4 shows an actual crystal structure, that of graphite. Graphite is made up of sheets of carbon atoms in a honeycomb pattern of covalent bonds. Note that a honeycomb pattern

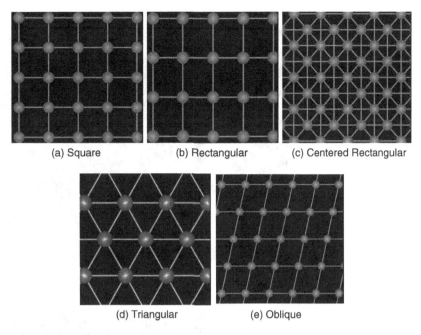

(a) Square   (b) Rectangular   (c) Centered Rectangular

(d) Triangular   (e) Oblique

Fig. 3.1   **The five Bravais lattices in two dimensions.**

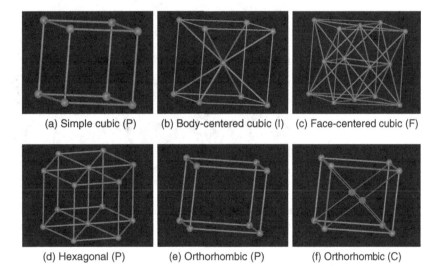

(a) Simple cubic (P)   (b) Body-centered cubic (I)   (c) Face-centered cubic (F)

(d) Hexagonal (P)   (e) Orthorhombic (P)   (f) Orthorhombic (C)

Fig. 3.2   **Some of the fourteen Bravais lattices in two dimensions. Parts of several unit cells are shown for the hexagonal lattice. The orthorhombic lattice has a unit cell with three different edge lengths.**

was not in the list of Bravais lattices. The reason is that we should regard it as a triangular lattice with a basis of two identical atoms. In the picture we show this by artificially shading the atoms at the lattice points and the displaced ones differently. Three dimensional graphite is made up of sheets of this type held together by the van der Waals' interaction. The sheets

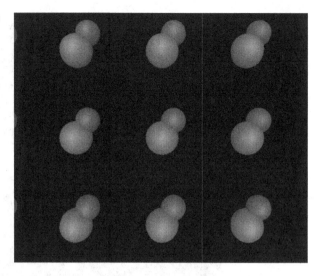

Fig. 3.3   **A square lattice with a basis of two atoms.**

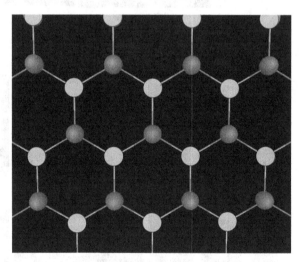

Fig. 3.4   **The crystal structure of a sheet of graphite showing the covalent bonds. The two atoms in the basis are shaded differently. Note that the atoms of the same color form a triangular lattice.**

slip easily; thus graphite is a lubricant. It is possible to introduce other atoms between the sheets to make intercalated graphite.

There is another form of carbon, crystalline diamond. In this case the bonds are tetrahedral sp^3 bonds, so that the crystal is a network of regular tetrahedrons and the angle between the bonds is $120°$. This may be viewed as a fcc lattice with a basis of two C atoms, one at the lattice point, and one shifted 1/4 of the body diagonal of the cube. There are compounds with a similar structure such as GaAs. In this case the two atoms in the basis are different. The diamond lattice is very important because it is the structure of the useful semiconductors Si and Ge.

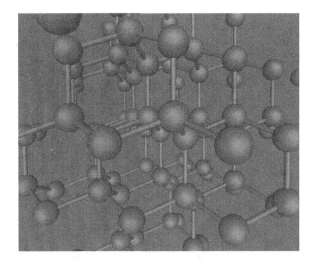

Fig. 3.5   The diamond structure. Note the tetrahedral coordination of the C atoms. It is a good exercise
to try to see how every other C atom is on the fcc lattice.

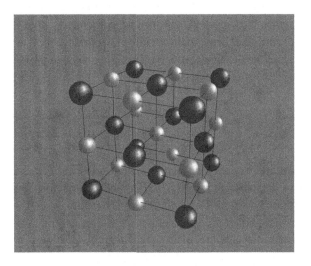

Fig. 3.6   The rocksalt structure. The larger atoms are Na$^+$, and the others Cl$^-$.

Another common crystal structure is the rocksalt crystal, NaCl. This may be visualized
as a simple cubic structure with every other atom Na, and the others Cl; see Figure 3.6. You
should be able to see that it is two interpenetrating face-centered cubic lattices. The basis
is one Na$^+$ and one Cl$^-$ shifted by 1/2 the cube edge.

Some of the most important structures are the close-packed crystals; see Figures 2.1, 2.6,
and 3.7. It is easy to show that if you want to have all atoms as close together as possible,
there are two ways to proceed. It is exactly the process of stacking grapefruit: the first layer
is triangular, and the second goes in half the holes. Then there are two choices for the third
layer: above the other holes in first layer, or above the fruit. The first gives a face-centered

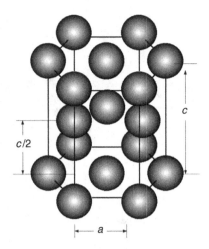

Fig. 3.7 **The hexagonal close packed crystal structure. If** $c/a = 1.613$ **the structure is has the same density as the fcc lattice.**

cubic lattice, fcc, which we have seen before and the second a lattice with a basis called hexagonal close packed, hcp. The hcp structure is a hexagonal lattice with a basis of two atoms, the other being in the middle of the prism cell. It is easy to show that the ratio of the length of the hexagonal rotation axis, the c axis, to the edge of the base of the prism, a, is $c/a = 1.613$; see Figure 3.7.

3.1.4 Unit cells

It is useful to divide up the lattice into identical cells each containing one lattice point. These are called unit cells. They can be put together to fill space. A familiar example is the tiling of a floor by identical tiles. There are several ways to construct such cells. The most straightforward is to define a *primitive* cell by using the generators, that is, we use the generators as edges of a a parallelepiped, see Figure 3.8. By construction the volume of this cell is:

$$v_c = |[\mathbf{a}_1 \times \mathbf{a}_2] \cdot \mathbf{a}_3|. \tag{3.3}$$

Any other unit cell will have the same volume. A more convenient choice is called the proximity or Wigner–Seitz (WS) cell. It is constructed by taking each lattice point and finding the part of space closer to it than to anywhere else. A constuction that does this is to draw lines to all other lattice points, construct the perpendicular bisector plane, and then take the smallest region bounded by the planes. This is most easily seen in two dimensions; see Figure 3.9.

In three dimensions, some unit cells are complicated. For example, the simplest primitive cell for the fcc lattice is the strange shape shown in Figure 3.8. The proximity cells for fcc

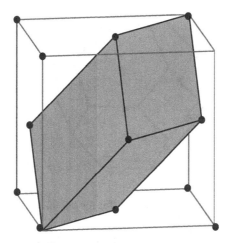

Fig. 3.8   **A primitive unit cell of the fcc crystal. Its edges are generators of the lattice.**

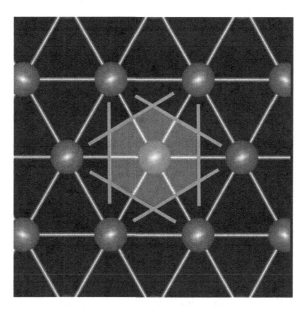

Fig. 3.9   **Construction of the proximity cell for the triangular lattice. The hexagonal region inside the lines is the unit cell.**

and bcc are shown in Figure 3.10. Of course, the unit cell for the simple cubic lattice is a simple cube, and the proximity cell the same cube shifted to be centered on a lattice point.

It is often useful to talk about larger, non-primitive unit cells to keep things simple. Thus, we often think of the body-centered cubic lattice as a simple cubic with a basis of two atoms, or, equivalently, two interpenetrating simple cubic lattices. The proximity cell of this lattice is a cube twice a big as the cell for the real bcc lattice. The fcc lattice can be considered to be four interpenetrating sc lattices.

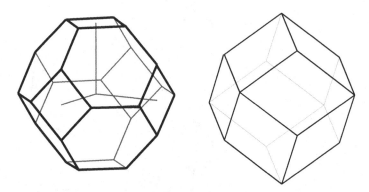

Fig. 3.10   The proximity cells for the bcc lattice, left, and the fcc lattice, right.

3.2 Fourier analysis and the reciprocal lattice

If atoms are arranged in a crystal, the electron density, for example, is periodic in three dimensions. For any periodic function the tool of choice is Fourier analysis. We will see that if we learn how to do Fourier analysis on a lattice we will be able to deal very elegantly with X-ray scattering from crystals.

3.2.1 Fourier series

Consider a periodic function $f(x)$ whose period is a. We can write, using Fourier's theorem (for functions that are smooth enough):

$$f(x) = \sum_G F(G)e^{iGx}; \quad G = \frac{2\pi j}{a}; \quad j = 0, \pm 1, \pm 2, \dots \tag{3.4}$$

The G's are discrete and have the units of inverse length. The coefficients are given by:

$$F(G) = \frac{1}{a} \int dx\, f(x) e^{-iGx}. \tag{3.5}$$

The integral is over a unit cell, e.g. $[0, a]$, or $[-a/2, a/2]$, or whatever is convenient.

For a function in three dimensions which is periodic with in x, y, z with periods a_1, a_2, a_3, it is easy to see that Eq. (3.4) and (3.5) imply:

$$f(\mathbf{r}) = \sum_\mathbf{G} F(\mathbf{G})e^{i\mathbf{G}\cdot\mathbf{r}}$$

$$F(\mathbf{G}) = \frac{1}{a_1 a_2 a_3} \int d\mathbf{r}\, f(\mathbf{r}) e^{-i\mathbf{G}\cdot\mathbf{r}},$$

$$\mathbf{G} = 2\pi \left(\frac{j_1}{a_1}, \frac{j_2}{a_2}, \frac{j_3}{a_3}\right); \quad j_i = 0, \pm 1, \pm 2, \dots \tag{3.6}$$

The integral is over the region $x \in [0, a_1], y \in [0, a_2], z \in [0, a_3]$.

3.2.2 The reciprocal lattice

We can look at Eq. (3.6) in the following useful way: The original lattice points on which $f(\mathbf{r})$ is periodic is the set $n_1\mathbf{a}_1 + n_2\mathbf{a}_2 + n_3\mathbf{a}_3$, where the vector $\mathbf{a}_1$ is given by $(a_1, 0, 0)$, and likewise for $\mathbf{a}_2, \mathbf{a}_3$. That is, f lives on an orthorhombic lattice whose generators are the $\mathbf{a}$'s. The vectors $\mathbf{G}$ also live on a lattice whose generators are:

$$\mathbf{g}_1 = (2\pi/a_1, 0, 0), \quad \mathbf{g}_2 = (0, 2\pi/a_2, 0), \quad \mathbf{g}_3 = (0, 0, 2\pi/a_3).$$

This is called the reciprocal lattice. Note that short edges of the direct lattice give rise to long edges of the reciprocal lattice.

What we have done so far will only work for a lattice whose generators are mutually perpendicular. For the general case we need to do something else. The proper generalization is to define the generators of the reciprocal lattice in the general case:

$$\mathbf{g}_i = \frac{2\pi}{v_c} \mathbf{a}_j \times \mathbf{a}_k, \tag{3.7}$$

where i, j, k are in cyclic order, i.e., 123, 231, 312, and v_c is the volume of the unit cell, see Eq. (3.3).

It is clear from the definition that:

$$\mathbf{a}_m \cdot \mathbf{g}_n = 2\pi\delta_{m,n}. \tag{3.8}$$

The generators of the direct and reciprocal lattice are mutually perpendicular. Now set

$$\mathbf{G} = j_1\mathbf{g}_1 + j_2\mathbf{g}_2 + j_3\mathbf{g}_3; \quad j_i = 0, \pm 1, \pm 2, \ldots \tag{3.9}$$

Since $\mathbf{G}$ is a linear combination of the generators, it is a reciprocal lattice vector. All this implies that we can write, for any lattice:

$$f(\mathbf{r}) = \sum_{\mathbf{G}} F(\mathbf{G}) e^{i\mathbf{G}\cdot\mathbf{r}}. \tag{3.10}$$

Where:

$$F(\mathbf{G}) = \frac{1}{v_c} \int d\mathbf{r}\, f(\mathbf{r}) e^{-i\mathbf{G}\cdot\mathbf{r}}. \tag{3.11}$$

The integral is over the unit cell in the direct lattice. Thus, any function periodic on the lattice can be written as a series in the reciprocal lattice vectors. The pair of equations, Eq. (3.10) and Eq. (3.11) will be used repeatedly in what follows.

Properties of the reciprocal lattice

Here are some useful properties of the reciprocal lattice.

- The volume of the unit cell in the reciprocal lattice is, Eq. (3.3):

$$v_{BZ} = |[\mathbf{g}_1 \times \mathbf{g}_2] \cdot \mathbf{g}_3| = (2\pi)^3/v_c. \tag{3.12}$$

The proximity cell in the reciprocal lattice is called the Brillouin zone. The expression above is the volume of the Brillouin zone.

- The reciprocal of the reciprocal lattice is the direct lattice. This is easy to show by looking at the generators. The fcc and bcc lattices are reciprocals of each other. The simple cubic lattice is its own reciprocal.

    The shape of the Brillouin zone for fcc is exactly that of the proximity cell for bcc in Figure 3.10.

- Each reciprocal lattice vector, $\mathbf{G}$, is perpendicular to a set of parallel planes in the direct lattice. The length of $\mathbf{G}$ is given by the spacing of the planes in the set. To be precise, if $\mathbf{G}_\circ$ is the shortest reciprocal lattice vector parallel to $\mathbf{G}$, then

$$|\mathbf{G}_\circ| = 2\pi/d, \tag{3.13}$$

where d is the spacing between the members of the set. The Miller indices of a plane, [mnl], are defined to be the inverses of the intercept of the plane with the coordinate axes, reduced by any possible common factor. So, for example, in a simple cubic lattice the cube faces have indices [100], the bisector of a cube face, [110], and the plane perpendicular to the body diagonal of the cube has indices [111]. The Miller indices are the coefficients of expansion of $\mathbf{G}_o$ in the generators :

$$\mathbf{G}_o = m\mathbf{g}_1 + n\mathbf{g}_2 + l\mathbf{g}_3. \tag{3.14}$$

- Since $\mathbf{G}$ is a linear combination of the generators, $\mathbf{g}_n$, with integer coefficients, and any lattice vector, $\mathbf{R}$, is a similar linear combination of the $\mathbf{a}_m$:

$$\exp(i\mathbf{G} \cdot \mathbf{R}) = \exp(2\pi i l) = 1 \tag{3.15}$$

where l is some integer. This follows immediately from Eq. (3.8).

A few useful identities Here are two identities that will be useful below. These are easy to prove. The first follows immediately from the fact that $\exp(i\mathbf{G} \cdot \mathbf{R}) = 1$, so that the phases add coherently if $\mathbf{k} - \mathbf{k}' = \mathbf{G}$. For $\mathbf{k} - \mathbf{k}' \neq \mathbf{G}$ it is not hard to show that the phases cancel, and give 0. The second is similar.

$$\sum_{\mathbf{R}} e^{i\mathbf{k} \cdot \mathbf{R}} e^{-i\mathbf{k}' \cdot \mathbf{R}} = N \sum_{\mathbf{G}} \delta_{\mathbf{k}, \mathbf{k}'+\mathbf{G}} \tag{3.16}$$

$$\sum_{\mathbf{k} \in BZ} e^{i\mathbf{k} \cdot \mathbf{R}_i} e^{-i\mathbf{k} \cdot \mathbf{R}_j} = N \delta_{\mathbf{R}_i, \mathbf{R}_j} \tag{3.17}$$

3.2.3 Non-periodic functions

In what follows we will need to deal with functions that are not necessarily periodic in the lattice, such as the wavefunctions of electrons or the displacements of atoms in sound waves. We will also use Fourier analysis for them. It is conventional in such a case to use a continuous Fourier transform. We go another route, which is more convenient for what follows.

To do this we impose boundary conditions on the whole macroscopic sample. For a large system if doesn't make much difference what we do at the outer boundary, so we use the standard trick (that saves us subsequent trouble) of assuming that such functions are periodic in a (huge) cell. We might imagine, for a one-dimensional system, that we are thinking of a ring, or in two dimensions, of a torus. In three dimensions there is no such visualization available, but we proceed anyway.

Now we can say that the whole sample is a huge replica of the unit cell, M spacings on an edge, where M is a large integer. Thus we assume, for all the functions that we consider:

$$h(\mathbf{r}) = h(\mathbf{r} + M\mathbf{a}_i),$$

for $i = 1, 2, 3$.

We can apply the construction of the reciprocal lattice for the big lattice. We have generators which are very short (since the generators of the "direct lattice," $M\mathbf{a}_i$ are very long):

$$\mathbf{b}_n = \mathbf{g}_n/M.$$

For example, for a simple cubic lattice of edge a, $|\mathbf{b}_n| = 2\pi/(Ma) = 2\pi/L$, where L is the size of the whole crystal. These reciprocal lattice vectors are very closely spaced, so that there are many of them in the Brillouin zone. Any linear combination of the $\mathbf{b}$'s is called an "allowed" vector, $\mathbf{k}$. (They are the wavevectors allowed by periodic boundary conditions).

From Eq. (3.10) we can write:

$$h(\mathbf{r}) = \sum_{\mathbf{k}} H(\mathbf{k}) e^{i\mathbf{k} \cdot \mathbf{r}} \tag{3.18}$$

$$H(\mathbf{k}) = \frac{1}{\Omega} \int d\mathbf{r}\, e^{-i\mathbf{k} \cdot \mathbf{r}} h(\mathbf{r}), \tag{3.19}$$

where the integral is over the whole crystal, and Ω is the volume of the sample.

There is an important relation which we will use several times. Suppose $g(\mathbf{k})$ changes slowly, i.e., $g(\mathbf{k} + \mathbf{b}_n)$ differs from $g(\mathbf{k})$ by a small enough amount that we can replace the sum in Eq. (3.19) by an integral. Recall that the $\mathbf{k}$ are very closely spaced. Then we can show that:

$$\sum_{\mathbf{k}} g(\mathbf{k}) \approx \frac{\Omega}{(2\pi)^3} \int d\mathbf{k}\, g(\mathbf{k}). \tag{3.20}$$

In order to see this, we need to know v_p, the volume per point in $\mathbf{k}$ space . Note first:

$$\Omega = |[M\mathbf{a}_1 \times M\mathbf{a}_2] \cdot M\mathbf{a}_3| = M^3 v_c = N v_c$$
$$\mathbf{k} = (1/M)(j_1\mathbf{g}_1 + j_2\mathbf{g}_2 + j_3\mathbf{g}_3). \tag{3.21}$$

Here, $N = M^3$ is the number of unit cells in the sample. We count the number of $\mathbf{k}$'s in a unit cell in the reciprocal lattice by counting the j's which are in a primitive cell. Each j obeys:

$$j_n = 0, 1, 2, \ldots M - 1.$$

Thus there are $M^3 = N$ different j's in the unit cell. The volume per allowed $\mathbf{k}$ point is:

$$v_p = [(2\pi)^3/v_c]/N = [(2\pi)^3/\Omega]. \tag{3.22}$$

This proves the relationship because we can put:

$$\sum_{\mathbf{k}} g(\mathbf{k}) \approx \int \frac{d\mathbf{k}}{v_p} g(\mathbf{k}) = \frac{\Omega}{(2\pi)^3} \int d\mathbf{k}\, g(\mathbf{k}).$$

There is an important corallary to the proof: *there are exactly as many allowed $\mathbf{k}$'s in the Brillouin zone as unit cells in the crystal.*

If we combine Eq. (3.18), (3.19) and (3.20), we see that $H(\mathbf{k})$ is related to the conventional Fourier transform:

$$h(\mathbf{r}) = \frac{1}{2\pi^3} \int d\mathbf{k}\, \tilde{h}(\mathbf{k}) e^{i\mathbf{k}\cdot\mathbf{r}}$$
$$\tilde{h}(\mathbf{k}) = \int d\mathbf{r}\, h(\mathbf{r}) e^{-i\mathbf{k}\cdot\mathbf{r}}, \tag{3.23}$$

so that $\Omega H(\mathbf{k}) = \tilde{h}(\mathbf{k})$.

In the following we will have occasion to superimpose functions (such as atomic densities or potentials of a single ion) to make a periodic function; cf. Eq. (2.25). That is we put $h(\mathbf{r}) = \sum_{\mathbf{R}} h_1(\mathbf{r} - \mathbf{R})$. Using Eq. (3.19) we find:

$$H(\mathbf{k}) = \frac{1}{\Omega} \sum_{\mathbf{R}} \int d\mathbf{r}\, e^{-i\mathbf{k}\cdot\mathbf{r}} h_1(\mathbf{r} - \mathbf{R}) = \frac{1}{\Omega} \sum_{\mathbf{R}} e^{i\mathbf{k}\cdot\mathbf{R}} \tilde{h}_1(\mathbf{k}). \tag{3.24}$$

Now use Eq. (3.16) and put the result into Eq. (3.18). The result is:

$$h(\mathbf{r}) = \sum_{\mathbf{G}} \frac{1}{v_c} \tilde{h}_1(\mathbf{G})\, e^{i\mathbf{G}\cdot\mathbf{r}}. \tag{3.25}$$

This should be compared to Eq. (3.10).

3.3 Scattering

The observed macroscopic properties of crystals such as facets (see Figure 4.5) led to the suspicion that some solids are crystalline from early times. However, the first conclusive demonstration was done in 1912 by W. Friedrich, P. Knipping, and M. von Laue (Friedrich, Knipping & von Laue 1912). They scattered X-rays from a crystal and detected them on photographic film. The idea was von Laue's, and went as follows. It was already known that the distance between atoms in a solid was of the order of angstroms, and he suspected that the wavelength of X-rays was also very short. Thus the X-rays might diffract from the crystal, as light does from a diffraction grating whose spacing is of the order of a wavelength. A way to say this, due to the later work of W. L. Bragg and W. A. Bragg (Bragg & Bragg 1913, Bragg 1913), is to think of the crystal planes as partial reflectors. Then if the angles and wavelengths are correct, we get constructive interference, as shown in Figure 3.11. The condition is $2d \sin(\theta/2) = n\lambda$, where λ is the wavelength as in ordinary optics.

We mentioned above that each set of planes is associated with a reciprocal lattice vector, so for each reciprocal lattice vector we have constructive interference. In the case of the Friedrich *et al.* experiment we have what is now called a Laue pattern: the X-rays were a broad-band combination of wavelengths, so for several sets of planes which satisfied the diffraction condition there were spots on the film. These spots are now called Bragg spots or Bragg peaks.

Common scattering techniques used today involve not only X-rays but also neutrons and electrons. Since $d \sim 1$ Å, we need radiation with $\lambda \sim 1$ Å for all three cases. For electromagnetic radiation we need X-rays with energies of order 10^4 eV, while for neutrons, using $\lambda = h/mv$, we get 0.1 eV, thermal neutrons. Electrons turn out to be in the 100 eV range. Electrons are used mainly for surface studies because they are charged; electron beams of the proper energy do not penetrate far into condensed matter.

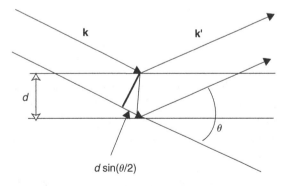

Fig. 3.11   **The Bragg construction for X-ray scattering. The angle θ is the scattering angle and λ the wavelength of the radiation. The total path length difference between the two beams is $2d \sin(\theta/2)$.**

3.3.1 Structure functions and scattering

It is useful to look at the scattering in a different way. Recall Huyghens principle: each atom in the solid which is struck by the X-ray beam is a secondary source of radiation which spreads out and interferes. For a random target like a gas, the interference is primarily destructive, except for the forward direction. The forward scattered beam combines with the incident beam, and we have the usual slowing down of light in matter. There is a weak diffuse background (e.g. the blue color of the sky) which arises from fluctuations in the medium. The situation in a crystal is quite different, as we have seen above: there can be large-angle scattering with large intensities in the Bragg peaks.

To sort this all out, we will show that for weak scattering the intensity as a function of angle is proportional to the absolute square of the Fourier transform of the density of scatterers. This is why scattering experiments are useful to deduce structures.

Scattering theory

Let us briefly review quantum mechanical elastic scattering theory. The cross-section for elastic scattering of a particle of mass M into a solid angle Θ is given by:

$$d\sigma/d\Theta = |f(\mathbf{k}, \mathbf{k}')|^2, \tag{3.26}$$

where f is the scattering amplitude from incident wavevector $\mathbf{k}$ to scattered wavevector $\mathbf{k}'$; see Figure 3.11. The scattering arises from a potential of interaction, $U(\mathbf{r})$ between the incident particle and the target. If the scattering is weak, we can use the Born approximation to write:

$$f(\mathbf{k}, \mathbf{k}') = f(\mathbf{k} - \mathbf{k}') = -\frac{M}{2\pi\hbar^2} \int d\mathbf{r} e^{-i\mathbf{k}'\cdot\mathbf{r}} U(\mathbf{r}) e^{i\mathbf{k}\cdot\mathbf{r}}. \tag{3.27}$$

The scattering is proportional to the matrix element of the potential between the initial and final states of the particle.

Neutrons

For neutrons the strongest scattering is with the nuclei, so U is the interaction with the nuclear density. For low energies the nuclear interaction leads to s-wave scattering, and the amplitude is equal to the scattering length, b. A useful way to express this is in terms of the Fermi pseudopotential which gives the interaction of a neutron at $\mathbf{r}$ with a nucleus at $\mathbf{R}$:

$$u(\mathbf{r}) = \frac{2\pi\hbar^2 b}{M}\delta(\mathbf{r} - \mathbf{R}). \tag{3.28}$$

This is to be used in the Born approximation (only), and gives $f = b$.

Define $\mathbf{q} = \mathbf{k} - \mathbf{k}'$, the wavevector transfer to the target. For a collection of nuclei at sites labeled by i we have:

$$U(\mathbf{r}) = \frac{2\pi\hbar^2}{M} \sum_i b_i \delta(\mathbf{r} - \mathbf{R}_i)$$

$$f(\mathbf{q}) = -\sum_i b_i e^{i\mathbf{q}\cdot\mathbf{R}_i}. \qquad (3.29)$$

The quantity $\sum_i b_i \delta(\mathbf{r} - \mathbf{R}_i)$ may be thought of as the density of nuclear matter weighted by the scattering length. The scattering amplitude is proportional to the Fourier component of the density at wave-vector $\mathbf{q} = \mathbf{k} - \mathbf{k}'$.

X-rays

For electromagnetic radiation, e.g. X-rays, the major scattering is with the electrons in the target. The nuclei are much heavier and have a negligible effect. For electrons the classical Thompson scattering expression (Jackson 1999) can be used: it simply results from allowing an incident electric field to accelerate an electron, and finding the radiated power from Larmor's formula. The result is;

$$d\sigma/d\Theta = r_o^2 \frac{1 + \cos^2\theta}{2}, \qquad (3.30)$$

for scattering from a single electron at the origin. Here $r_o = e^2/mc^2$ is the classical radius of the electron. For convenience we will simply ignore the angle dependent factor in what follows; this is by no means necessary, but it lightens the notation. Then the Thompson scattering amplitude is simply r_o. For atoms, we can treat the valence electrons as free provided the X-ray energy is large compared to the binding energy. This is valid for light elements. (The case where the X-ray energy is comparable to an atomic absorption is called dispersive or anomalous. It is interesting and useful, but we will not discuss it here.)

For an extended system scattered waves from different parts of the electron distribution will interfere, and we have to add up scattering amplitudes with the proper interference factors. The phase shift will involve a factor $e^{i\mathbf{k}\cdot\mathbf{r}}$ for exciting radiation, and $e^{-i\mathbf{k}'\cdot\mathbf{r}}$ for the scattered wave. Thus, for system of electrons with density $n(\mathbf{r})$ we get:

$$f(\mathbf{k}, \mathbf{k}') = r_o \int d\mathbf{r} e^{-i\mathbf{k}'\cdot\mathbf{r}} e^{i\mathbf{k}\cdot\mathbf{r}} n(\mathbf{r}). \qquad (3.31)$$

Note, just as above, that the scattering amplitude is proportional to the Fourier component of the electron density at wavevector $\mathbf{k} - \mathbf{k}'$.

The total electron density, in the simplest case, is the sum of the electron density of all the atoms in the target:

$$n(\mathbf{r}) = \sum_i n_i(\mathbf{r} - \mathbf{R}_i), \qquad (3.32)$$

where n_i is the atomic density for the atom at point $\mathbf{R}_i$.

We now can compute f.

$$f(\mathbf{k}, \mathbf{k}') = r_o \sum_i \int e^{-i\mathbf{k}' \cdot \mathbf{r}} n_i(\mathbf{r} - \mathbf{R}_i) e^{i\mathbf{k} \cdot \mathbf{r}} d\mathbf{r}$$

$$= r_o \sum_i \int e^{-i\mathbf{k}' \cdot [\mathbf{R} + \mathbf{R}_i]} n_i(\mathbf{R}) e^{i\mathbf{k} \cdot [\mathbf{R} + \mathbf{R}_i]} d\mathbf{R}. \tag{3.33}$$

We have set $\mathbf{R} = \mathbf{r} - \mathbf{R}_i$. Then:

$$f(\mathbf{q}) = \sum_i e^{i\mathbf{q} \cdot \mathbf{R}_i} r_o \int e^{i\mathbf{q} \cdot \mathbf{R}} n_i(\mathbf{R}) d\mathbf{R}$$

$$= \sum_i e^{i\mathbf{q} \cdot \mathbf{R}_i} f_i(\mathbf{q}). \tag{3.34}$$

Here $f_i(\mathbf{q})$ is the scattering amplitude of atom i.

We need $|f|^2$ to get the cross-section:

$$|f(\mathbf{q})|^2 = \sum_{j,l} e^{i\mathbf{q} \cdot \mathbf{R}_j} f_j(\mathbf{q}) e^{-i\mathbf{q} \cdot \mathbf{R}_l} f_l^*(\mathbf{q}). \tag{3.35}$$

Suppose now that all the atoms are identical and have scattering amplitude f_1.

$$d\sigma/d\Theta = |f_1|^2 I(\mathbf{q}) \tag{3.36}$$

$$I(\mathbf{q}) = \sum_{j,l} \langle e^{i\mathbf{q} \cdot [\mathbf{R}_j - \mathbf{R}_l]} \rangle. \tag{3.37}$$

This is an interesting formula: *the information about the atoms has been separated from the geometry.* In fact, the first factor is the scattering cross-section of a single atom, and the other is defined to be proportional to the (static) structure factor[†] of the target:

$$S(\mathbf{q}) = I(\mathbf{q})/N. \tag{3.38}$$

The $\langle \cdot \rangle$ in Eq. (3.37) indicates that we should take an average over any thermal motions that the atoms might undergo in the course of the experiment. Typically X-ray experiments are done slowly; thus we can replace the time average with an average over the equilibrium thermal ensemble.

[†] The conventional usage of the term "structure factor" is inconsistent. The quantity we call $S(q)$ involves the sum over all the atoms in the target. Later we will see another quantity which is also called by the same name, which is the sum over atoms in a single unit cell of a crystal. We denote it with a different symbol, $\mathcal{S}(q)$.

3.3.2 Scattering from a Bravais lattice of atoms

Suppose all the atoms are rigidly fixed at the points, $\mathbf{R}_j$, of a Bravais lattice. We need to evaluate:

$$I(\mathbf{q}) = \sum_{j,l} e^{i\mathbf{q}\cdot[\mathbf{R}_j - \mathbf{R}_l]} = N^2 \sum_{\mathbf{G}} \delta_{\mathbf{k},\mathbf{k}'+\mathbf{G}}. \tag{3.39}$$

We have used Eq. (3.16). This equation says that momentum transfer must be a reciprocal lattice vector for coherent scattering. Note that we have very strong, coherent, scattering at angles other than the forward direction. If $\mathbf{q} = \mathbf{G}$ each term in the sum is unity. From the geometry of the scattering in Figure 3.11 we see that $\sin(\theta/2) = |\mathbf{G}|/2|\mathbf{k}|$, where θ is the scattering angle.

X-rays can be produced as a monochromatic beam, with fixed $\mathbf{k}$. This occurs when electrons incident on the anode of the X-ray generator are energetic enough to excite electrons from a deep level in the atoms of the target. When the level is refilled, intense monochromatic X-rays are produced. (In the Friedrich–Knipping experiment the electrons were not energetic enough, and produced broad-band bremsstrahlung radiation).

For a monochromatic beam there is a construction due to P. Ewald to see when coherent scattering is possible. It goes as follows: we require $|\mathbf{k}| = |\mathbf{k}'|$ (elastic scattering). We draw $\mathbf{k}$ ending at the origin of the reciprocal lattice, and a sphere of radius $|\mathbf{k}|$ at its start. Where the sphere intersects a $\mathbf{G}$ we have an allowed scattering as shown in Figure 3.12. In general this does not happen, so the standard method with monochromatic beams is to rotate the crystal to find angles where the condition is met. Or, alternatively, one can grind up the crystal into a powder: the powder pattern is a series of rings on the film, each coming from a crystallite that happens to satisfy the Bragg condition. With broad-band radiation, there will, in general be reciprocal lattice vectors that are equal to $\mathbf{k} - \mathbf{k}'$; this is why Friedrich and Knipping succeeded.

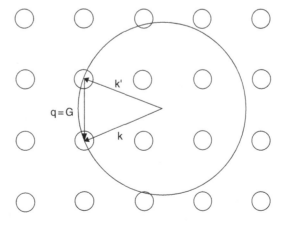

Fig. 3.12   **The Ewald construction.**

3.3.3 Crystals with a basis

If the crystal has a basis, we return to the above, and write the sum over positions as a sum over the Bravais lattice, $\mathbf{R}_j$, and a sum over the positions of the atoms in the basis, $\mathbf{s}_l$, so that the lth atom in the ith cell is at $\mathbf{R}_i + \mathbf{s}_l$. Then:

$$f(\mathbf{q}) = \sum_i \exp(i\mathbf{q} \cdot \mathbf{R}_i) \sum_l \exp(i\mathbf{q} \cdot \mathbf{s}_l) f_l(\mathbf{q}).$$

Thus:

$$|f(\mathbf{q})|^2 = \mathcal{S}(\mathbf{q}) \sum_{i,j} e^{i\mathbf{q} \cdot [\mathbf{R}_i - \mathbf{R}_j]}$$

$$= \mathcal{S}(\mathbf{q}) N^2 \sum_{\mathbf{G}} \delta_{\mathbf{k},\mathbf{k'}+\mathbf{G}}$$

$$\mathcal{S}(\mathbf{q}) = \left| \sum_l e^{i\mathbf{q} \cdot \mathbf{s}_l} f_l(\mathbf{q}) \right|^2. \tag{3.40}$$

The first factor is called the structure factor in this context (see footnote above). It affects the intensities of the Bragg peaks; the scattering angles are determined by the Bravais lattice alone.

Here is an interesting example of how this works out. Suppose we perversely regard the bcc lattice as a simple cubic lattice with a basis. Now the cube is bigger than the real unit cell, and the reciprocal lattice is smaller. Thus there are extra Bragg spots which should go away. Recall that the reciprocal lattice of bcc is fcc. To see how this sorts out, we compute $\mathcal{S}(\mathbf{G})$ for a sc lattice with a basis of identical atoms at $(0,0,0)(a/2,a/2,a/2)$, where a is the cube edge. For the sc lattice, the $\mathbf{G}$s are of the form $(2\pi/a)(n,l,m)$ where n,l,m are integers. Then:

$$\mathcal{S}(\mathbf{G}) = |f_1|^2 |1 + \exp(\pi i[n + l + m])|^2.$$

Now if $n + l + m$ is even $\mathcal{S}(\mathbf{G}) = 4|f_1|^2$, and otherwise $\mathcal{S}(\mathbf{G}) = 0$. That is, half the Bragg spots are missing. In particular, if

$$(n, l, m) = (1, 0, 0), (0, 1, 0), (0, 0, 1), (1, 1, 1)$$

there is no scattering. This is exactly what is needed to convert a simple cubic reciprocal lattice into an fcc one; see Figure 3.13.

3.3.4 The Debye–Waller factor

When the Friedrich–Knipping experiment was proposed, there was considerable doubt that any coherent scattering could be observed because it was known that the atoms in the lattice would vibrate away from their lattice positions at finite temperature by some amount $\mathbf{u}$. Any

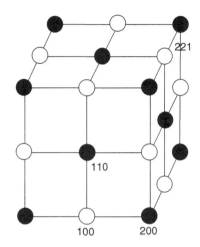

221

110

100   200

Fig. 3.13   **The missing reflections from a bcc lattice. The white circles have zero scattering.**

snapshot would see a disordered array, so there should be no effective diffraction grating. Friedrich and Knipping did the experiment anyway, with the results that we have seen.

The resolution of this paradox was worked out by P. Debye and I. Waller. Consider a Bravais lattice of identical atoms. For the vibrations we use the Einstein model discussed in Chapter 2. The atoms are effectively bound by springs so that each atom is a three-dimensional, isotropic, harmonic oscillator. Statistical physics says that if the atom deviates from its official position, $\mathbf{R}$, by x in a certain direction, the probability of the deviation is given by a Boltzmann factor:

$$P(x) \sim \exp(-M\omega_{\mathrm{E}}^2 x^2 / 2k_{\mathrm{B}}T). \tag{3.41}$$

We compute $I(\mathbf{q})$ again. Each atom is at $\mathbf{R}_i + \mathbf{u}_i$, so we have:

$$\sum_{i,j} e^{i\mathbf{q}\cdot(\mathbf{R}_i - \mathbf{R}_j)} \langle e^{i\mathbf{q}\cdot(\mathbf{u}_i - \mathbf{u}_j)} \rangle.$$

There are $N(N-1)$ terms in the sum with $i \neq j$, and only N with $i = j$, so with small error we can keep only the terms with $i \neq j$. The $\mathbf{u}$'s of different atoms vibrate independently, so we can write the term in brackets as:

$$\langle e^{i\mathbf{q}\cdot\mathbf{u}_i} \rangle \langle e^{-i\mathbf{q}\cdot\mathbf{u}_j} \rangle = |\langle e^{-i\mathbf{q}\cdot\mathbf{u}} \rangle|^2.$$

In the last equation we have used the fact that the thermal average is independent of i. Thus we have:

$$I(\mathbf{q}) = |\langle e^{i\mathbf{q}\cdot\mathbf{u}} \rangle|^2 N^2 \sum_{\mathbf{G}} \delta_{\mathbf{k},\mathbf{k}'+\mathbf{G}}.$$

To evaluate the average we need to do a simple Gaussian integral with the proper normalization:

$$\langle e^{i\mathbf{q}\cdot\mathbf{u}}\rangle = \frac{\int e^{iqx}e^{-M\omega_{\mathrm{E}}^2 x^2/2k_{\mathrm{B}}T}dx}{\int e^{-M\omega_{\mathrm{E}}^2 x^2/2k_{\mathrm{B}}T}dx}.$$

Here x is the component of $\mathbf{u}$ in the direction of $\mathbf{q}$. The result is:

$$I(\mathbf{q}) = e^{-2W}N^2\sum_{\mathbf{G}}\delta_{\mathbf{k},\mathbf{k}'+\mathbf{G}}; \quad W = q^2 k_{\mathrm{B}}T/M\omega_{\mathrm{E}}^2 = q^2\langle x^2\rangle. \tag{3.42}$$

W is called the Debye–Waller factor. The effect of temperature is to reduce the intensity of the Bragg scatterings by e^{-2W}, but not to wipe them out completely.

3.3.5 Neutron magnetic scattering

Thus far, the theories of X-ray and neutron elastic scattering are formally identical, with only a change in the overall cross-section. Of course, thermal neutrons are far harder to make since they require a nuclear reactor. However, neutron scattering is different from X-ray scattering in two important ways. One is that the neutron is sensitive to magnetic order, and, as we will see later, neutrons are useful for inelastic scattering.

The reason that the neutron sees magnetism is that it has a magnetic moment which interacts with the moments of the magnetic ions via the dipole-dipole interaction and a contact term. The cross-section due to this interaction is comparable to that arising from the direct nuclear interaction of the neutron with the nuclei.

The scattering amplitude now will be given by an equation like Eq. (3.29) with the density of nuclear matter replaced by the magnetization, $\mathbf{M}(\mathbf{r})$. For a ferromagnet $\mathbf{M}$ has the same periodicity as the crystal. Thus the magnetic scattering adds to the Bragg peaks. The situation is quite different for an antiferromagnet. In this case (see Figure 3.14), the magnetization has a different periodicity than the crystal. We expect new Bragg peaks to appear, see Figure 3.14. Another way to say this is to note that a wave that changes sign on the two sublattices will be sensitive to the staggered magnetization.

3.3.6 Electron scattering from surfaces: LEED and RHEED

Electron beams interact with both the electrons and the nuclei of solid targets via the Coulomb interaction. Because they scatter from surfaces and not the bulk, they are often used in surface science. The method is called Low Energy Electron Diffraction, LEED. The crystal lattice that is important is the two-dimensional lattice on the surface. The scattering condition applies only to the components of q in the plane of the surface. The third component can be anything since we don't get interference from deep in the crystal. The scattering condition is:

$$\mathbf{q} = (G_x, G_y, q_z)$$

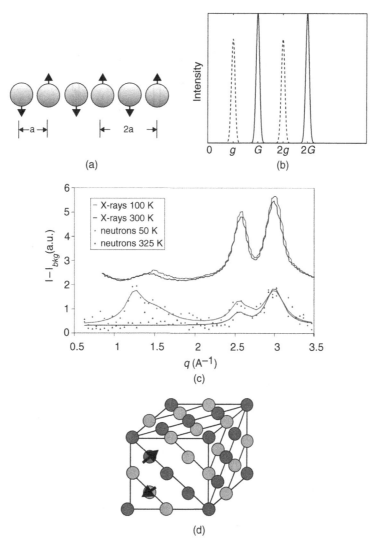

Fig. 3.14 (a) A one-dimensional antiferromagnet. Note that the periodicity of the magnetization is not a but $2a$. (b) Schematic representation of Bragg scattering as a function of q. For the crystal, or when the magnetic order is not present, there are peaks at multiples of $G = 2\pi/a$, solid lines. However, the magnetic scattering has peaks at multiples of $g = 2\pi/(2a)$, so that there are extra peaks, dotted lines. (c) An experimental realization of this effect in CoO. (In fact, these are Co core/CoO shell nanoparticles, but the small Co core does not contribute significantly.) Note that at low temperature extra lines occur in the magnetic scattering. Courtesy of S. Inderhees, G. Strycker, J. Borchers, and M. Aronson (d) The spin ordering of CoO.

where $G_{x,y}$ are the components of a two-dimensional reciprocal lattice vector of the surface and q_z is not restricted by diffraction, only by the scattering from individual atoms. Now we repeat the Ewald construction. Instead of intersecting the sphere with the points of the three-dimensional reciprocal lattice we need to intersect with rods which are spaced

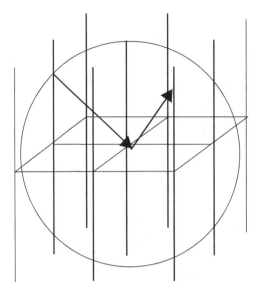

Fig. 3.15   **The two-dimensional version of the Ewald construction showing Bragg rods. Both k and k′ must point to places where a rod intersects the sphere of constant |k|.**

according to its two-dimensional counterpart; see Figure 3.15. Reflection High Energy Electron Diffraction (RHEED) is similar except that much higher energy electrons are used at glancing incidence (so that diffraction is still possible). In this case the Ewald sphere is very large since $|\mathbf{k}|$ is large, and the q_z's are small from the glancing incidence. The intersection of the Ewald sphere with the Bragg rods gives rise to a streaky pattern from many rods almost tangent to the sphere.

RHEED is also used in another mode: instead of looking at the diffracted beam, one can look at specularly reflected electrons. In this case, RHEED intensities have the interesting property of detecting partial occupancy of the top layer of atoms. For example, in crystal growth the intensity of the reflected beam oscillates in time with a period equal to the time to grow one layer. This is because step edges (boundaries between the region where the crystal is m and $m + 1$ layers high) increase diffuse scattering and deplete the specular beam. In a partially filled layer there are lots of such edges, but in complete layers there are few.

3.4 Correlation functions

We have seen that scattering amplitudes are given by the Fourier analysis of the density of scatterers. This implies that it is related to certain correlation functions. This is an interesting formal relationship, and is useful in dealing with scattering from non-periodic structures such as liquids, gases, and amorphous materials.

3.4.1 Density correlation functions

When a beam scatters, it sees the target in terms of the density of matter there. For example, neutrons see the nuclei. The extended arrangement of matter can be characterized by the spatial distribution of nuclei. We can write the effective density of scatterers as:

$$n(\mathbf{r}) = \sum_i b_i \delta(\mathbf{r} - \mathbf{R}_i), \tag{3.43}$$

where b_j is the scattering length of the jth nucleus. X-rays interact primarily with the electron clouds around the atoms; the relevant quantity is electron density. For a superposition of atomic densities:

$$n(\mathbf{r}) = \sum_i n_i(\mathbf{r} - \mathbf{R}_i). \tag{3.44}$$

The positions of the atoms do not lie on a lattice in a liquid, gas, or glass. In these cases the major scattering is for $\mathbf{q} = 0$, forward scattering. That is no longer true in a crystal where each Bragg peak has strong scattering. From the definition in the previous chapter, the intensity of the strong ($\propto N^2$) Bragg scattering is proportional to the square of the order parameter.

It is useful to define the density-density correlation function.

$$C(\mathbf{r}, \mathbf{s}) = \langle n(\mathbf{r})n(\mathbf{s}) \rangle = \sum_{i,j} \langle n_a(\mathbf{r} - \mathbf{r}_i)n_a(\mathbf{s} - \mathbf{r}_j) \rangle, \tag{3.45}$$

the last equality holds when the density is the sum of atomic densities.

Note that the double Fourier transform of C is exactly the scattering function: $C(\mathbf{q}, -\mathbf{q}) = \langle |n(\mathbf{q})|^2 \rangle$. Conversely, since $|f|^2 \propto n(\mathbf{q})n(-\mathbf{q})$ is a product, its Fourier transform is a convolution of densities:

$$\int e^{i\mathbf{q}\cdot\mathbf{r}} \langle |f_{tot}|^2 \rangle d\mathbf{q}/(2\pi)^3 \propto \int \langle n(\mathbf{r}+\mathbf{s})n(\mathbf{s}) \rangle d\mathbf{s} = \int C(\mathbf{r}+\mathbf{s}, \mathbf{s})d\mathbf{s}. \tag{3.46}$$

This function is called the Patterson function. It is useful in solving the "phase problem" of X-ray crystallography, see below.

It is useful to separate C into a short-ranged part and a long-ranged part to see the real correlations:

$$\bar{C}(\mathbf{r}, \mathbf{s}) = \langle (n(\mathbf{r}) - \langle n(\mathbf{r}) \rangle)(n(\mathbf{s}) - \langle n(\mathbf{s}) \rangle) \rangle = C(\mathbf{r}, \mathbf{s}) - \langle n(\mathbf{r}) \rangle \langle n(\mathbf{s}) \rangle. \tag{3.47}$$

That is:

$$|f_{tot}|^2 \propto |\langle n(\mathbf{q}) \rangle|^2 + \bar{C}(\mathbf{q}, -\mathbf{q}). \tag{3.48}$$

The first term in this equation is the coherent part: we expect it to be of order N^2 in a crystal if $\mathbf{q} = \mathbf{G}$ since $\langle n(\mathbf{r}) \rangle$ is periodic. The second term is an incoherent part. Since correlations are almost certainly short-ranged, the second term is of order N.

In macroscopically homogeneous systems such as amorphous solids, liquids, or gases, the average of the density is a constant, n, and the material is invariant under any translation. (A crystal is invariant only under a discrete set of translations.) Thus $\langle n(\mathbf{q}) \rangle = \Omega n \delta(\mathbf{q})$. Except for forward scattering, $\mathbf{q} = 0$, the short-range correlations are the whole signal. This means that for any translation $\mathbf{s}$:

$$C(\mathbf{r}, \mathbf{r}') = C(\mathbf{r} + \mathbf{s}, \mathbf{r}' + \mathbf{s}). \tag{3.49}$$

The result is that the density correlation function in Eq. (3.46) comes out of the integral. Scattering in translationally invariant materials measures the density correlation function directly.

3.4.2 Pair correlations

The most useful form of information about homogeneous systems comes from a slightly different function, the correlation of the position of a particle with *other* particles. For point particles located at $\mathbf{r}_i$:

$$p(\mathbf{r}, \mathbf{s}) = \sum_{i \neq j} \langle \delta(\mathbf{r} - \mathbf{r}_i) \delta(\mathbf{s} - \mathbf{r}_j) \rangle. \tag{3.50}$$

Note that the terms with $i = j$ are omitted compared to Eq. (3.45). The interpretation of this function is that $p(\mathbf{r}, \mathbf{s}) d\mathbf{r} d\mathbf{s}$ is the probability of finding a particle near $\mathbf{s}$ (within $d\mathbf{s}$) provided that there is one at $\mathbf{r}$ within $d\mathbf{r}$. If we take the Fourier transform we can relate p to the structure factor:

$$\int d\mathbf{r} \, d\mathbf{s} \, p(\mathbf{r}, \mathbf{s}) e^{i\mathbf{q} \cdot (\mathbf{r} - \mathbf{s})} = \sum_{i \neq j} \langle e^{i\mathbf{q} \cdot (\mathbf{r}_i - \mathbf{r}_j)} \rangle = N(S(\mathbf{q}) - 1). \tag{3.51}$$

The function p is used to define the *pair distribution function g*:

$$p(\mathbf{r}, \mathbf{s}) = \langle n(\mathbf{r}) \rangle g(\mathbf{r}, \mathbf{s}) \langle n(\mathbf{s}) \rangle. \tag{3.52}$$

The pair distribution function is unity if there is no correlation, for example if $|\mathbf{r} - \mathbf{s}|$ is very large. In homogeneous systems $\langle n(\mathbf{r}) \rangle = n$ and p and g depend only on the difference $\mathbf{r} - \mathbf{s}$. For this case we have the useful relation:

$$S(\mathbf{q}) = 1 + n \int d\mathbf{R} \, g(R) e^{i\mathbf{q} \cdot \mathbf{R}}, \tag{3.53}$$

which follows from Eq. (3.51). From this it is possible to measure g in a fluid, for example, by X-ray scattering experiments of the type described above. A sketch of what a typical result looks like is given in Figure 3.16. For a fluid there is a "hole" in g when two particles are close together.

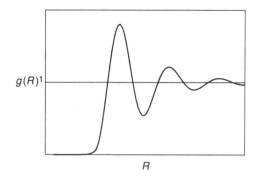

Fig. 3.16 **A sketch of the pair distribution function of a simple liquid. The peaks are at the positions of shells of nearest neighbors.**

If we integrate Eq. (3.50) with respect to $\mathbf{s}$ we find:

$$\int d\mathbf{s}\, p(\mathbf{r},\mathbf{s}) = (N-1)\langle n(\mathbf{r})\rangle,$$

$$-1 = n \int d\mathbf{R}(g(R)-1). \tag{3.54}$$

The last line is for a translationally invariant system. Both of these equations say the same thing, namely that the definition of p excludes one particle so that the hole near the origin in Figure 3.16 is big enough for a particle. We will use these ideas later in our discussion of interacting electrons.

For magnets we can go through the same discussion. The relevant correlation function is of the magnetization, namely

$$G(\mathbf{r},\mathbf{s}) = \langle \mathbf{m}(\mathbf{r})\mathbf{m}(\mathbf{s})\rangle. \tag{3.55}$$

If the magnet is ordered then if $|\mathbf{r}-\mathbf{s}| \to \infty$ we expect that:

$$G(\mathbf{r},\mathbf{s}) \to \langle \mathbf{m}(\mathbf{r})\rangle \langle \mathbf{m}(\mathbf{s})\rangle. \tag{3.56}$$

If this is so, the magnet is said to have *long-range order*. Disordered magnets, above the Curie temperature have short-range order, and G approaches zero, usually exponentially.

3.4.3 Structure determination and the phase problem

Modern crystallography methods are remarkable. The structure of very complex molecules such as proteins and DNA have been determined using the techniques pioneered by Friedrich, Knipping, von Laue, and the Braggs; see Drenth & Mesters (2007). For example, there is an on-line resource called the Protein Data Bank which contains huge numbers of protein structures with the locations of tens or hundreds of thousands of atoms.

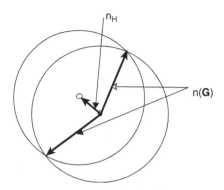

Fig. 3.17 **The Harker construction. The radii of the two circles are the intensities of the scattering before and after the substitution of the heavy atom. The complex scattering amplitude n_H of the heavy atom is assumed known. The two arrows are give two possibilities for the phase of the scattering for the original protein.**

This technique is not straightforward. As we pointed out in Eq. (3.46) the scattering cross-section, and the associated diffraction pattern does not completely determine the electron density, but only its correlation function. This is referred to as the "phase problem": we know the intensity of scattering for each $\mathbf{G}$, but we don't know its phase because the cross-section is an absolute square. If we did know the phase, we could find $n(\mathbf{r})$ directly.

There are several techniques which can overcome this problem. The one most used for proteins is called Multiple Isomorphous Replacement. In this method chemists introduce heavy atoms into the protein structure and do scattering again. In fact, two such substitutions must be made. The idea is the following: if we know the position in the unit cell of the heavy atom, $\mathbf{s}$, and the magnitude of its scattering amplitude, $|f_H| \equiv r_o|n_H|$ we can write, for the scattering intensity with the substitution:

$$|n'(\mathbf{G})|^2 = |n(\mathbf{G}) + |n_H|e^{i\mathbf{G}\cdot\mathbf{s}}|^2. \tag{3.57}$$

The solution to this equation is visualized in Figure 3.17. Note that there is a two-fold ambiguity in the solution. This is resolved by using another heavy atom, and repeating the process.

It remains to find the position of the heavy atom. One way to do this is to use the Patterson function, Eq. (3.46) of the substituted protein. Since the heavy atom scatters strongly compared to the light elements in the protein, there is only one position to find. This can be guessed from the density-density correlation function.

Suggested reading

Crystal structure and X-ray scattering is covered by all the general references. A collection of crystal structures is in:

Wyckoff (1963).

For neutron scattering see also:

Squires (1978).

Electron scattering from surfaces is reviewed in:

Zangwill (1988).

For structure determination in proteins and the phase problem:

Drenth & Mesters (2007).

Problems

1. Show that no two-dimensional Bravais lattice can have a 5- or 7-fold axis of rotation, i.e., no axis such that a rotation by $2\pi/5$ or $2\pi/7$ leaves that lattice invariant.

2. (a) Explicitly construct the reciprocal lattices for simple cubic, body-centered cubic, and fcc lattices. (b) Show that the reciprocal lattice of the reciprocal lattice is the direct lattice. (An easy way is the find the generators of the reciprocal of the reciprocal.)

3. (a) Prove the basic orthogonality relation:

$$\sum_{\mathbf{R}} e^{-i\mathbf{k}\cdot\mathbf{R}} e^{i\mathbf{k}'\cdot\mathbf{R}} = N \sum_{\mathbf{G}} \delta_{\mathbf{k}'-\mathbf{k},\mathbf{G}}$$

where $\mathbf{R}$ runs over the direct lattice, and $\mathbf{G}$ runs over the RL, and the delta function is the discrete (Kronecker) delta. This shows that when an X-ray of wavevector $\mathbf{k}$ scatters to $\mathbf{k}'$ then we must have $\mathbf{k}' - \mathbf{k} = \mathbf{G}$ (including $\mathbf{G} = 0$, of course) to give coherent scattering.

   Hint: if $\mathbf{k}' - \mathbf{k} = \mathbf{G}$ then all the exponentials are unity, obviously. This gives N when the Bragg condition is satisfied. The trick is to prove that if $\mathbf{k}' - \mathbf{k} \neq \mathbf{G}$ the sum is much smaller in the limit $N \to \infty$. To be precise, show that

$$\lim_{N\to\infty} \sum e^{-i\mathbf{q}\cdot\mathbf{R}}/N = 0$$

   except for $\mathbf{q}$ in a region near some $\mathbf{G}$ whose size goes to zero as $N \to \infty$.

4. We will often need to go back and forth between sums such as Eq. (3.18) and Eq. (3.23) so that we encounter Dirac delta functions and Knoecker delta functions. Show that:

$$\delta_{\mathbf{k},\mathbf{k}'} = \frac{(2\pi)^3}{\Omega}\delta(\mathbf{k}-\mathbf{k}').$$

5. Suppose there is scattering from a two-dimensional lattice, but the scattering is in three-dimensions. Show that there is no restriction on the component of $\mathbf{k}' - \mathbf{k}$ which is perpendicular to the plane of the lattice, but that the other two components must match some two-dimensional $\mathbf{G}$. This is the basis of the Bragg rod construction of Figure 3.15.

6. Show that the Laue condition $\mathbf{k} - \mathbf{k}' = \mathbf{G}$, for some reciprocal lattice vector is the same as the Bragg condition $2d \sin(\theta/2) = n\lambda$, where d is the distance between lattice planes, $\lambda = 2\pi/|\mathbf{k}|$, and θ is the scattering angle (the angle of deflection of the beam).

7. In this problem you will consider elastic scattering of thermal neutrons. (a) Model the potential between neutrons and nuclei by the Fermi pseudopotential: $v(r) = 2\pi\hbar^2 b\delta(\mathbf{r})/M$, where b is a constant (called the scattering length). Show that b^2 is the cross-section for elastic scattering from a single nucleus in the Born approximation. (b) Consider a crystal with more than one isotope so that the b's vary randomly. Show that, in the rigid crystal approximation (no thermal vibration, and elastic scattering):

$$d\sigma/d\Theta = N^2|\langle b\rangle|^2 \sum_{\mathbf{G}} \delta_{\mathbf{q},\mathbf{G}} + N(\langle b^2\rangle - |\langle b\rangle|^2).$$

Sketch as a function of scattering angle what the two terms do. Here $\langle b\rangle$ = average over the isotopes of the scattering length, and $\langle b^2\rangle$ = average over the isotopes of the square of the scattering length.

8. Consider the hcp crystal. It is made up of identical atoms. (a) Show that the structure factor, $S(\mathbf{G})$, takes on values $f(1 + e^{in\pi/3})$, where f is the scattering amplitude of the atoms, and $n = 1, \ldots, 6$ as $\mathbf{G}$ ranges over the RL for the hexagonal structure. (Hint: the two atoms in the unit cell are at 0 and at $\mathbf{r} = \frac{2}{3}\mathbf{a}_1 + \frac{1}{3}\mathbf{a}_2 + \frac{1}{2}\mathbf{a}_3$ where $\mathbf{a}_i$ are the primitive translation vectors.) (b) Show that all the points in the plane perpendicular to the c-axis that contains $\mathbf{G} = 0$ give non-zero Bragg scattering. (c) Show that points of zero S are found on alternate planes in the family of reciprocal lattice planes perpendicular to the c-axis. Show, for example, that the point above $\mathbf{G} = 0$ has zero form factor.

9. Derive Eq. (3.54).

4 Surfaces and crystal growth

In the previous chapter we considered perfect crystals which are periodic and thus infinite in extent. There is one deviation from perfect periodicity that is always present: all real crystals have surfaces. We argued above that the surface has a small effect on bulk properties. This may be true, but the *growth* of crystals is the process of adding surface layers. We will discuss a number of aspects of surfaces and growth here.

4.1 Observing surfaces: scanning tunneling microscopy

In the previous chapter we showed how that structure of a surface could be revealed by using LEED; this technique is useful for periodic surfaces. As we will see, crystal surfaces have interesting deviations from periodicity such as steps. An enormous advance in studying this aspect of surfaces was the development of direct atomic resolution microscopy.

A remarkable instrument of this type is the scanning tunneling microscope (STM) invented by G. Binnig and H. Rohrer (Binnig, Rohrer, Gerber & Weibel 1982). The idea of this device is that quantum mechanical tunneling is very sensitive to the distance through which the particles tunnel.

Suppose electrons tunnel through a classically forbidden region of length x. The current will be of order $j \sim e^{-x/a}$ where a is the attenuation length of the electron wavefunction in the forbidden region. The STM consists of a sharp metal tip which is brought into near contact with a surface by manipulating piezo-electric drivers. Electrons tunnel through the vacuum in the gap between the tip and the surface: the work function of the electrons in the tip (the energy to excite the electron out of the metal) is the barrier height. The current will depend very sensitively on the distance above the surface. If the tip is then scanned over the surface, and a feedback loop keeps the current constant by raising and lowering the tip, we can keep track of the height changes. This will, in many cases, translate into a map of the topography of the surface. Atomic resolution is not difficult to produce. See Figures 4.1, 4.2.

A similar gadget called the atomic force microscope (AFM) can be used for non-conducting surfaces; (Binnig, Quate & Gerber 1986). It measures the force between a tip and the surface: the instrument consists of a vibrating cantilever arm. A laser beam bouncing from the arm allows the resonance frequency of vibration to be measured. Shifts in the resonance occur due to loading from the van der Waals' forces of close-by surfaces.

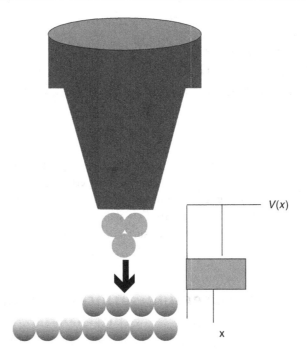

Fig. 4.1   **A schematic of a scanning tunneling microscope. The top section represents the piezoelectric transducers that drive the tip up and down and in a raster scan of the surface. The tunneling potential is shown: the difference of the level of the filled states of electrons is tuned by applying a voltage to the tip. If the bias is such that electrons flow into the sample, the empty states are probed. If current is drawn out of the sample, filled states are probed.**

The scanning is done as in the STM. The resolution of an AFM is usually lower than that of an STM.

4.2 Surfaces and surface tension

We now consider surfaces in more detail, starting with their thermodynamics.

4.2.1 Surface tension

If a sample of condensed matter is finite, its free energy will have a correction due to the finite surface area. In terms of chemical bonds, the existence of a surface always raises the energy since there are *dangling bonds* on the surface. From a macroscopic point of view the Gibbs' free energy of a sample with N particles and surface area A is:

$$G = \mu N + \gamma A. \tag{4.1}$$

The surface tension, γ, is the free energy per unit area. At low temperature γ approaches the surface energy per unit area.

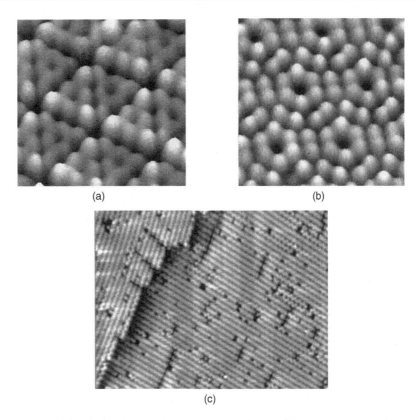

(a)   (b)

(c)

Fig. 4.2   **STM images: (a) Filled states on Si 111, atomic resolution. (b) Empty states on the same surface. (c) Larger scale picture of Si 100 showing a terrace and rows of atoms. Experimental pictures courtesy of B. G. Orr.**

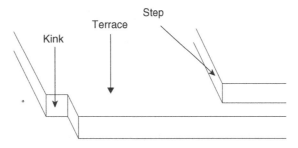

Fig. 4.3   **Steps, terraces, and kinks on a cubic crystal surface.**

To get a feeling for this, we introduce a very simple model, the Kossel crystal (after the work of W. Kossel). In this model the crystal is simple cubic, and there are nearest neighbor interactions. In the simplest, static view, it consists of terraces and steps, see Figure 4.3. At finite temperature there will be some extra atoms (adatoms) on the terraces which attach and detach from the steps. If an atom detaches from a step it creates two *kinks*. In general, if the step wanders, it will have many kinks.

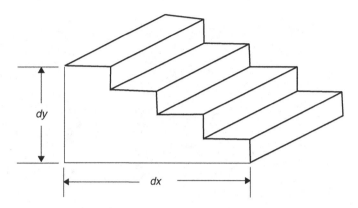

Fig. 4.4   **Steps in the Kossel crystal.**

We can find the surface energy by imagining the following experiment: we cut a crystal along a certain direction by doing work on all the bonds that cross the cutting surface. Suppose that breaking a bond costs work ϵ_b. We count all the bonds that are broken to get the energy; call the number $\mathcal{N}$ for area A. In the course of the procedure we have created two surfaces. Thus:

$$\gamma = \mathcal{N}\epsilon_b/2A. \tag{4.2}$$

For the Kossel crystal, suppose that the cutting surface is parallel to the z-axis. The cut rises dy in a length dx; see Figure 4.4:

$$\gamma = \frac{\epsilon_b}{2a^2} \frac{|dx| + |dy|}{\sqrt{dx^2 + dy^2}}. \tag{4.3}$$

We should note two things: the surface energy depends on the angle of the cutting surface with respect to the crystal axes, and this expression has a singular derivative near $dy = 0$: $\gamma(\theta) \propto |\theta|$.

The second statement may be put another way: the energy, β, to create a unit length of step is $\beta = \epsilon_b/2a$ because there is one extra nearest-neighbor bond for each unit of step. If dy is small so that we are near the [100] surface we have:

$$\gamma = \gamma_{100} + (\beta/a)|\theta|, \tag{4.4}$$

where θ is the angle between the surface normal and that of the [100] surface. A surface that is near a surface with a high density of bonds, such as [100], is called *vicinal*, and θ is called the *miscut*. The singular behavior of γ with miscut is typical of vicinal surfaces at low temperatures.

The energy of a miscut surface is higher than that of the singular surface. If we miscut, we need to cut the bonds on the surface and more (to make the steps). This is why, fundamentally, natural crystals are found with *facets*, macroscopic parts of the surface that are essentially atomically smooth, see Figure 4.5. However, there is much more to the story, as we will see below. Note that at finite T, β must be interpreted as the free energy of a step. Steps can wander and have entropy.

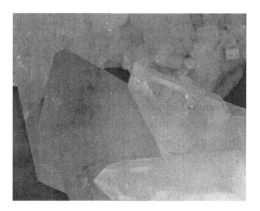

Fig. 4.5 **Large single crystals are smooth facets which occur at characteristic angles. Historically, naturally occurring single crystals were considered a rare curiosity – and they still are. This photo was taken in a rock shop; the price for the large quartz crystal was $196. It was only in the twentieth century that the techniques of the previous chapter revealed that many substances are crystalline, though the crystallites are usually small.**

In the foregoing discussion we talked about the crystal in contact with vacuum. If the crystal is in contact with a liquid or vapor, there will be energy of interaction of the two phases. The cost of a surface in this case is not the surface tension but the *interfacial tension* which includes the bonding of the liquid or vapor to the surface.

4.2.2 Reconstruction and surface stress

We have assumed that the structure of a surface is the same as that of the bulk, only terminated by the cutting plane. This is often not true. The dangling bonds on the surface often reconnect to lower the surface energy. For example, in a common reconstruction of the [100] surface of GaAs a surface Ga will connect with a surface As which is not its nearest neighbor. The resulting surface has rows of As dimers that are brought closer to each other breaking the symmetry of the surface. The Si surface in Figure 4.2(a,b) shows clear evidence of reconstruction.

A way to look at this effect is to note that the bonding on the surface need not be the same as in the bulk. That means that a surface, in general, will have an extra stress compared to the bulk: the surface could be under effective compression or tension. A way to measure the surface stress is to try to measure the work in stretching a surface of area A. (In a real experiment, the bulk will be stretched too, of course). The work will have two parts: $W/A = \gamma + \tau$. The first term is the work done in creating new surface, and the second is the extra work in stretching the surface bonds compared to the bulk bonds.

4.2.3 The Laplace law

The term "surface tension" is not an arbitrary name. There is an elementary experiment that shows that γ really is a tension: a loop of thread is made to float on the surface of water.

It assumes a limp shape. Now a surfactant (like soap or detergent) is put inside the loop. Surfactants decrease surface tension. The surface outside the loop now "pulls" harder than the surface inside, and the loop will become circular.

The result of this mechanical pull is that the inside of a small sample of liquid or solid has higher pressure than the outside because the surface is pulling it together. The difference in pressure is given by Laplace's law, which we now derive.

For a liquid, we consider a bit of surface of a drop. Suppose we add matter to the drop so that the surface grows by δA and the volume by δV. This will cost energy, so there is work associated with enlarging the liquid:

$$\delta W = (P_{in} - P_{out})\delta V = \gamma \delta A. \tag{4.5}$$

It is well known that at any point P on a surface we can define *principal radii of curvature*. That is, in two perpendicular planes the surface is locally the arc of a circle with radii R_1, R_2. We take the convention that if the center of the circle is in the liquid we take R to be positive. A concave surface has $R < 0$. It is easy to see that if s denotes arclength and θ the angle of the normal:

$$ds/d\theta = -R. \tag{4.6}$$

From Figure 4.6 we see that in one plane the change in length of the surface is $s\delta h/R_1$ where s is the arclength of the original surface and δh is the distance between the old and new surfaces. For both planes we get $\delta A = \delta h s^2(1/R_1 + 1/R_2)$ However $\delta V = s^2\delta h$. Combining this with Eq. (4.5) we arrive at the classical result of Laplace:

$$\delta W = (P_{in} - P_{out})s^2\delta h = \gamma \left(\frac{1}{R_1} + \frac{1}{R_2}\right)s^2\delta h; \tag{4.7}$$

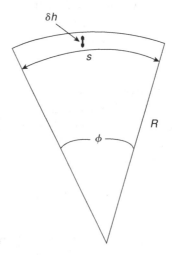

Fig. 4.6 **Increase in length of a surface when the volume inside is increased. The increase in arclength is $\delta s = \phi(R + \delta h) - \phi R = \phi \delta h$. However $s/R = \phi$ so that $\delta s = s\delta h/R$. Note that $s\delta h$ is the area between the two surfaces.**

$$P_{\text{in}} - P_{\text{out}} = \gamma \left(\frac{1}{R_1} + \frac{1}{R_2} \right). \tag{4.8}$$

For example, for a spherical droplet of radius R, $\delta P = 2\gamma/R$.

For a solid γ depends on angle. This gives an extra contribution to the work on the right-hand side of Eq. (4.7). For simplicity, consider the case of a cylindrical surface so that $R_2 \to \infty$. Note from Figure 4.6 that if δh, the displacement between the two surfaces, depends on the position on the surface the normal direction of the two surfaces is not the same, and θ depends on s. From simple geometry we see that $\delta\theta = d\delta h/ds$. Now there is a contribution to the change in the surface energy that is not present for a liquid: it is due to the relative rotation of the initial and final surfaces. Writing the energy and integrating by parts we have;

$$\delta W' = \int \delta\gamma dA = \int \frac{d\gamma}{d\theta}\delta\theta dA$$

$$= \int \frac{d\gamma}{d\theta}\frac{d\delta h}{ds}dA = -\int \frac{d^2\gamma}{d\theta ds}\delta h dA$$

$$= \int \frac{d^2\gamma}{d\theta^2}\frac{\delta h}{R}dA. \tag{4.9}$$

In the last step we used Eq. (4.6). If we consider a particular part of the surface with $dA = s^2$ we have, using Eq. (4.7):

$$(P_{\text{in}} - P_{\text{out}})s^2\delta h = \gamma s^2\delta h/R + \frac{d^2\gamma}{d\theta^2}s^2\delta h/R;$$

$$(P_{\text{in}} - P_{\text{out}}) = (\gamma + \gamma_{\theta\theta})/R. \tag{4.10}$$

The two terms are from stretching and rotating the surface. The quantity $\gamma + \gamma_{\theta\theta}$ is called the surface stiffness. The general formula, due to C. Herring, is:

$$P_{\text{in}} - P_{\text{out}} = (\gamma + \gamma_{\theta\theta})/R_1 + (\gamma + \gamma_{\phi\phi})/R_2. \tag{4.11}$$

It involves the stiffness in the two principal planes of curvature.

4.2.4 Gibbs–Thompson boundary condition

When a solid and a liquid are in equilibrium we are at the melting point. The melting temperature depends, in general, on the pressure. The same is true for a liquid and a gas at the boiling point. This dependence is determined by setting the chemical potentials of the two phases to be equal:

$$\mu_s(T_c, P) = \mu_l(T_c, P). \tag{4.12}$$

The assumption in this equation is that the pressure in the two phases is the same. However, if the pressures are not the same due to surface effects, the melting or boiling point must shift to $T_c + \delta T$.

To evaluate δT we recall that $d\mu = -\bar{s}dT + \bar{v}dP$ where $\bar{s}, \bar{v}$ are the entropy per particle and the volume per particle. Now we set the μ's equal taking Eq. (4.8) into account. We do this first for the liquid–gas case:

$$\mu_1(P + \delta P, T_c + \delta T) = \mu_g(P, T_c + \delta T)$$

$$\mu_1(T_c, P) - \bar{s}_1 \delta T + \bar{v}_1 \delta P = \mu_g(T_c, P) - \bar{s}_g \delta T$$

$$\delta T = \frac{-\bar{v}_1}{\bar{s}_g - \bar{s}_1} \delta P. \tag{4.13}$$

Now we use Eq. (4.8) and the fact that the latent heat per unit volume is given by $Q = T_c(\bar{s}_g - \bar{s}_1)/\bar{v}_1$. This gives for the depression of the boiling point of a small droplet:

$$\frac{\delta T}{T_c} = -\frac{\gamma}{Q}\left(\frac{1}{R_1} + \frac{1}{R_2}\right). \tag{4.14}$$

For a cylindrical solid which curves only in one plane we have the analog of Eq. (4.14) for the depression of the melting point:

$$\frac{\delta T}{T_c} = -\frac{\gamma + \gamma_{\theta\theta}}{Q}\frac{1}{R}. \tag{4.15}$$

We have discussed a crystal in equilibrium with its melt or vapor. Crystals can also grow from solution; the equilibrium is between atoms in the crystal and those in the solute. The analog to Eq. (4.15) is a shift in the equilibrium density of the solute: curved surfaces are in equilibrium with a larger density in solution.

4.3 Roughening

In the foregoing we have discussed only the energy part of the surface tension γ and the step free energy β. A free energy also contains entropy terms, and these are quite important. In fact, there is a remarkable phenomenon called the *roughening transition* which occurs when the entropy contribution to β overwhelms the energy, and the free energy to create a step is negative. In this case there is a proliferation of steps in equilibrium, and the static picture of smooth terraces separated by steps breaks down. The real crystal surface is a dynamic equilibrium of steps, islands surrounded by steps, and terraces. In the rough phase the cusp in $\gamma(\theta)$ is rounded off (though we will not demonstrate this) and the facet disappears.

To see how the transition occurs we use an approximate treatment due to Burton, Cabrera & Frank (1951) (BCF); we base our discussion on an argument due to Peierls. Suppose we consider a partly filled surface layer. Then, if energy dominates the formation of the structure, the extra atoms will group into a large island to minimize the number of dangling bonds. Breaking up this large island will increase the entropy. We estimate when this occurs.

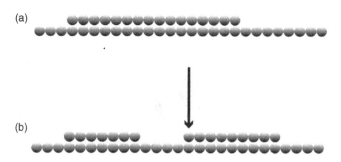

(a)

(b)

Fig. 4.7 Comparison of two states of with the same partial occupancy of the top layer in one dimension. (a) The state of lowest energy. (b) Breaking into two islands of the same total area increases the entropy because there is a different state for every position of the beginning of the new island (arrow).

In Figure 4.7 we consider the situation in one dimension, which is appropriate for the surface of the two-dimensional crystal (such as the edge of an island on a surface). The minimum energy state occurs when all the top-layer atoms are together. Separating the island into two costs energy ϵ_b. However, there is a gain in entropy: the number of ways to place the new island is N, the number of available sites, which we assume to be macroscopic. The entropy is $k_B \log(N)$ and the free energy for forming the new island is

$$F = E - TS = \epsilon_b - k_B T \log(N) < 0 \quad \text{as} \quad N \to \infty. \tag{4.16}$$

Since the free energy to make a new island is negative, in equilibrium (at any finite T) there will be a large number of such islands, and the surface will be *rough*, i.e., there will be no well-defined terrace-step structure in the thermal average surface.

In two dimensions the situation is different. Suppose, again, that there is an island that occupies a finite fraction of a macroscopic surface of N sites. The length of the edge measured in lattice constants is of order $N^{1/2}$. Now suppose we break the island up into two of comparable size. Each new island will have an edge length, $L \equiv ja$. The energy cost is of order $\beta_o L$, where β_o is the step *energy* per unit length. To estimate the entropy we imagine that we put the edge down on the free sites, whose number is $\mathcal{O}(N)$. For each starting point there will be z directions to choose to put the first link, of length a, of step. (For a surface of square symmetry, $z = 4$.) Then the next link will have $z - 1$ choices, and so on. We ignore, to get an estimate, the requirement that the new island be closed. Thus the entropy to make the new island is of order:

$$k_B \log(Nz(z-1)^{j-1}) \approx k_B \frac{L}{a} \left(\log(z-1) + \frac{1}{j} \log N \right). \tag{4.17}$$

Note that since $j = \mathcal{O}(N^{1/2})$ the last term is negligible for large N. Then the free energy per unit length is:

$$\frac{F}{L} = \frac{1}{a} \left(\beta_o a - k_B T \log(z-1) \right). \tag{4.18}$$

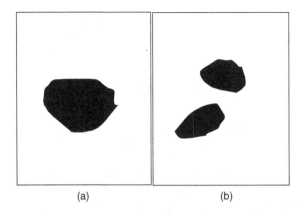

(a)   (b)

Fig. 4.8   **Comparison of two states of with the same partial occupancy of the top layer in two dimensions. (a) A state of low energy, a single compact island. (b) Breaking into two islands of the same total area increases the entropy because there is a different state for every position and shape of the new island. The energy increases by βL where L is the length of new edge created.**

Now for low T, F is positive, and for large T it is negative. The transition occurs at $k_{\mathrm{B}}T_{\mathrm{R}} \sim \log(z-1)/\beta_o a$. This is an estimate of the roughening transition temperature. BCF went further and pointed out that the transition could be mapped onto a two-dimensional Ising model: the mapping consists of identifying a top-surface atom with spin up, top surface vacancies with spin down. The famous Onsager solution to the Ising model then gives a better estimate of T_{R}. However, this will not do either. BCF realized that once the top-layer island broke up into smaller islands there was no guarantee that lower layers would be undisturbed. New islands could be formed by breaking up the terrace. The full solution of the model, including these effects is due to Chui & Weeks (1976). We will not explain it in detail, since it involves a complicated renormalization group calculation. For $T > T_{\mathrm{R}}$ the surface fluctuates thermally and any singularity in $\gamma(\theta)$ is smoothed out, and we have a rough surface.

Note that T_{R} depends on which crystal facet is being considered because the surface symmetry and the step energy depend on the facet. It is possible, and observed, that some crystal directions are rough while others are smooth. For example, ice is a crystal with a hexagonal structure. At low temperatures the equilibrium crystal consists of flat hexagonal prisms. Near the melting point of ice the base of the prism is still a facet, but the prism faces are rough. The nature of the coexistence of the two kinds of surface will be explained next when we consider macroscopic crystal shapes.

4.4 Equilibrium crystal shapes

For a cylindrical crystal in equilibrium Eq. (4.15) shows that the quantity $\bar{\gamma}/R \equiv (\gamma + \gamma_{\theta\theta})/R$ must be a constant on the surface. G. Wulff introduced a geometric construction (based on earlier work by J. W. Gibbs) that allows us to deduce the shape that keeps $\bar{\gamma}/R$ constant.

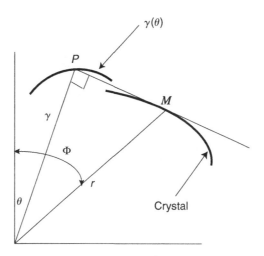

Fig. 4.9   **Construction of the pedal curve. The upper curve is the polar plot of $\gamma(\theta)$ and the other is the crystal surface. The perpendicular to the radius vector on the polar plot at P is tangent to the crystal surface at M.**

The method is based on a construction from geometry invented by Maclaurin, the *pedal curve*. (In fact, the curve we are about to describe is the inverse, or contrapedal of the usual definition). The idea is to make a polar plot of $\gamma(\theta)$ and then, at each point on the plot, construct the perpendicular to the radius vector. These perpendiculars include the tangents to the crystal shape: the crystal shape is the inner envelope of the perpendiculars: see Figure 4.9.

We can prove that this is the correct construction. We work in polar coordinates with r, θ being independent variables. Take r, Φ to be the polar coordinates of the point M on the crystal that is generated by the perpendicular from the point P on the polar plot of surface tension. Point P is at radius $\gamma(\theta)$ and polar angle θ. Then from Figure 4.9:

$$r = \frac{\gamma(\theta)}{\cos(\Phi - \theta)}$$

$$0 = \frac{\gamma_\theta}{\cos(\Phi - \theta)} - \frac{\gamma \sin(\Phi - \theta)}{\cos^2(\Phi - \theta)}. \tag{4.19}$$

The second line comes from differentiating the first, and noting that $dr/d\theta = 0$. Rearranging Eq. (4.19) we find:

$$r = \sqrt{\gamma^2 + \gamma_\theta^2}$$

$$\Phi = \theta + \tan^{-1}(\gamma_\theta/\gamma). \tag{4.20}$$

Now recall that:

$$R^2 = (ds/d\theta)^2 = (dr/d\theta)^2 + r^2(d\Phi/d\theta)^2 = (\gamma + \gamma_{\theta\theta})^2. \tag{4.21}$$

Thus $(\gamma + \gamma_{\theta\theta})/R$ is a constant, which is what we wanted to prove.

Fig. 4.10   **An example of the Wulff plot. The crystal shape is partly facetted and partly smoothly rounded.**

We give an example of the construction of a crystal shape in Figure 4.10. Note that the cusp in $\gamma(\theta)$ gives rise to a facet. When the roughening transition rounds off cusps, the facets in the corresponding crystal faces disappear.

4.5 Crystal growth

The equilibrium shapes that we described above are rarely observed in nature. Instead the shapes seen are the result of the growth history. The reason for this is that the time for a crystal to come to equilibrium with respect to its shape is very large, except for very tiny samples.

4.5.1 Dendrites

Crystal growth often gives rise to very complicated macroscopic shapes including the familiar snowflake. The growth occurs in the form of fingers or *dendrites* (from the Latin for root) that can be very long and complex. See Figure 4.11 for an example. This is a very old subject that was first considered in a systematic way by J. Kepler in *A New Years' Gift, or the Six-Cornered Snowflake* (1611). Kepler wanted to understand how the complex shapes of snowflakes arose, but admitted that he could not. This scientific problem has been solved, finally, within the last few decades, and is reviewed, for example, in Pelcé (2004). We will discuss some of the simple features of the solution.

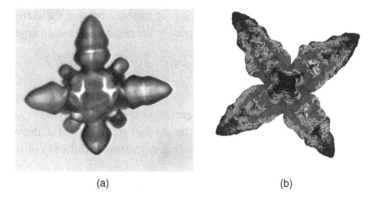

(a)   (b)

Fig. 4.11 (a) A crystal of NH$_4$Cl. The arms of the crystal are called dendrites. NH$_4$Cl is a cubic crystal, the macroscopic shape reveals the microscopic symmetry. Courtesy of A. Dougherty. (b) A numerical simulation of dendritic growth. The color coding indicates the temperature gradient at the surface. At the tips the gradients are large; see Section 4.5.2. Courtesy of T. Schulze.

On the microscopic level there has been a good deal of recent activity in understanding growth on smooth facets when new atoms are deposited in high vacuum. Growth typically proceeds by the motion of steps along the surface, cf. Figure 4.3. This approach was pioneered by BCF. We will not discuss this in detail.

4.5.2 Macroscopic growth laws

In order to understand growth we need to identify the *rate-limiting* step that controls the morphology. In this section we will be interested in a common sort of growth that gives rise to dendrites. It is the situation where the rate-limiting step is the arrival of matter by diffusion to the surface, diffusion-limited growth. If we have the opposite situation, where surface incorporation processes control growth, we speak of reaction-limited growth. In the latter case the shapes produced are usually less complicated.

Diffusion-limited growth and the Stefan problem

If a crystal grows in contact with its vapor (e.g. a snow crystal in a cloud) the rate-limiting step is the arrival of molecules at the surface. If it grows in solution, the same is true. If the crystal grows in contact with its liquid (ice in water, for example) there is no limitation due to the arrival of matter, but rather the rate-limiting step is the diffusion of latent heat away from the surface. All three cases can be described in a common way.

Let us review diffusion: we take the thermal case as an example. The fundamental principle in this case is Fourier's law:

$$\mathbf{j}_Q = -\kappa_T \nabla T. \tag{4.22}$$

Here $\mathbf{j}_Q$ is the heat current, i.e. the rate of transfer of energy from hot to cold per unit area, and κ_T is the thermal conductivity. We will estimate κ_T in Chapter 7.

The conservation law for the heat current says that the increase in internal energy per unit volume at some point is equal to the negative of the divergence of $\mathbf{j}_Q$

$$\partial q/\partial t = -\nabla \cdot \mathbf{j}_Q = \kappa_T \nabla^2 T$$
$$\partial T/\partial t = D_T \nabla^2 T. \tag{4.23}$$

In the second line (the heat equation) we have converted the change in q to a change in temperature using a heat capacity per unit volume, c_P. The thermal diffusivity, D_T, is thus given by κ_T/c_P. The corresponding statement for diffusion of molecular concentration, n, is based on Fick's law:

$$\mathbf{j}_n = -D\nabla n, \tag{4.24}$$

where D is the diffusion coefficient. This leads to an equation of identical form to Eq. (4.23), namely

$$\partial n/\partial t = D\nabla^2 n. \tag{4.25}$$

For the thermal case, suppose a crystal is growing from its melt. We will consider a simplified form of the problem (somewhat unphysical) where there is no heat conduction into the crystal, so that all the excess heat must flow away in the melt, and that the density of the two phases is equal. In time dt the surface advances by $v_n dt$ where v_n is the normal velocity of growth. This converts volume $Av_n dt$ where A is the area, and generates $QAv_n dt$ energy, where Q is the latent heat per unit volume of solid. On the other hand, Fourier's law says that the energy conducted away from the interface is proportional to the normal derivative of T: $-c_P D_T \partial T/\partial n$. This leads to the growth condition:

$$v_n = -\frac{c_P D_T}{Q}\frac{\partial T}{\partial n}. \tag{4.26}$$

In addition, we need boundary conditions. If we are to solidify at all, the ambient temperature must be below the melting point, T_c. And, at the surface, we have the Gibbs–Thompson condition. Thus:

$$T(\mathbf{r} \to \infty) = T_\infty < T_c \tag{4.27}$$
$$T(\text{surface}) = T_c(1 - (\gamma/Q)\kappa), \tag{4.28}$$

where κ is the sum of the inverse radii of curvature and we have neglected the difference between surface tension and surface stiffness.

In order to simplify notation we define a dimensionless diffusion field $u = c_P(T_c - T)/Q$ and a dimensionless undercooling, $\Delta = c_P(T_c - T_\infty)/Q$. Then Eq. (4.23), Eq. (4.27), Eq. (4.28), and Eq. (4.26) can be written:

$$\partial u/\partial t = D_T \nabla^2 u$$
$$u_\infty = \Delta$$
$$u_s = d_o\kappa$$
$$v_n = D_T \partial u/\partial n. \tag{4.29}$$

Here, $d_o = \gamma T_c c_P / Q^2$ is called the capillary length. It is usually a very small length, of order a few Å. The collection of equations Eq. (4.29) is a famous problem in applied mathematics called the Stefan problem. They take exactly the same form for the diffusion of molecules except for the definition of D_T, d_o, Δ.

Mullins–Sekerka instability

The key to understanding the complex shapes of diffusion-limited crystals is to realize that growth of a smooth, compact shape is unstable against the formation of bumps. This was shown in Mullins & Sekerka (1963). It is a general phenomenon in diffusion-limited processes. This effect arises as follows: if a smooth shape develops a bump, it will project into the cooler fluid, and grow faster. Bumps magnify, and round or flat crystals quickly develop fingers as in Figure 4.11.

To make this more quantitative, we follow Mullins and Sekerka and consider a small spherical crystal of radius R. To simplify matters (this is convenient, but not necessary) we take the limit of slow growth. Suppose the surface advances with typical velocity v in the x direction. We can estimate the relative size of the various terms in the diffusion equation. We go into the frame moving with the surface. Then we have $\partial u / \partial t = v \partial u / \partial x$. Thus:

$$\frac{1}{l_D} \frac{\partial u}{\partial x} = \frac{\partial^2 u}{\partial x^2}, \tag{4.30}$$

where $l_D = D/v$ is called the diffusion length. If v is small so that l_D is bigger than all the other scales of the problem (in this case, R) then the left hand side of Eq. (4.30) is small, and we can write:

$$\nabla^2 u \approx 0. \tag{4.31}$$

This is called the quasi-static approximation.

We are led to solve the Laplace equation, Eq. (4.31), with boundary conditions given in Eq. (4.29). Outside the sphere we have:

$$u = A + B/r.$$

Matching boundary conditions we find:

$$u = \Delta(1 - R/r) + 2d_0/r.$$

The growth boundary condition gives:

$$v_R = \dot{R} = \frac{D\Delta}{R}\left(1 - \frac{2d_0}{\Delta R}\right). \tag{4.32}$$

This is a differential equation for $R(t)$, and the growth approaches $R \propto \sqrt{t}$. If the spherical nucleus is smaller than $R_c = 2d_0/\Delta$ it will melt; otherwise it will grow.

Now suppose there are bumps on the surface, see Figure 4.12, and that $R > R_c$. For example, we can parameterize the bumps by:

$$r_s = R + \delta(t)Y_{lm}(\theta, \phi). \tag{4.33}$$

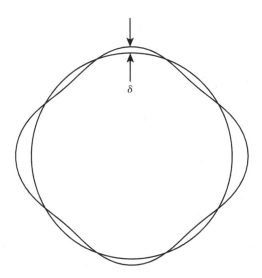

Fig. 4.12   **Cross-section of a slightly deformed sphere. The deformation is given by** Y_{30}.

The coefficient of the spherical harmonic gives the small deviation from sphericity. The essence of *linear stability analysis* is to see whether δ grows in time faster than R.

For simplicity we will neglect capillary effects, and set $d_0 = 0$ for the moment, and consider only $m = 0$. The solution to the Laplace equation is:

$$u = \Delta + B/r + CY_{l0}/r^{l+1}. \tag{4.34}$$

Matching the boundary condition at the surface means:

$$0 = \Delta + \frac{B}{r_s} + \frac{CY_{l0}}{r_s^{l+1}}. \tag{4.35}$$

We need to linearize in the sense that we keep only terms of order δ. Note that C is of the same order, so we can put $R^{-(l+1)}$ in the last term in Eq. (4.35). A bit of computation gives $C = -\delta\Delta R^l$ and $u = \Delta(1 - R/r) + CY_{l0}/r^{l+1}$. We need to look at the growth boundary condition. Setting $v = \dot{R} + \dot{\delta}Y_{l0}$ and matching terms with the same angular dependence gives:

$$\dot{R} = D\Delta/R \tag{4.36}$$

$$\frac{\dot{C}}{C} = \frac{\dot{\delta}}{\delta} = \frac{\Delta D(l-1)}{R^2}. \tag{4.37}$$

This means that if we neglect, in the second line, the rather slow growth of R, we have exponential growth of the instability. This is the source of complex crystal shapes. The $l - 1$ in front occurs because the lowest order deformation, $Y_{10} \propto z$ corresponds to moving the sphere. This motion is not amplified.

We can give an estimate of the effect of capillarity. Since Y_{l0} has $l - 1$ nodes, we can think of R/l as the wavelength of the deformation of the surface. We expect growth to stop if $R/l \sim d_0$. This means that there is a term like $(1 - ld_0/R)$ multiplying Eq. (4.37). In order to derive the full result we need an expression for the curvature in spherical coordinates, which is somewhat complicated. We will simply quote the result:

$$\frac{\dot{\delta}}{\delta} = \frac{(l-1)\dot{R}}{R} \left(1 - (2l+1)\frac{Dd_0(l+2)}{\dot{R}R^2} \right), \qquad (4.38)$$

where $\dot{R}$ is given by Eq. (4.36).

Dendritic growth

Given that a round crystal grows arms, we can ask about the long-time behavior of a growing crystal. There is an experimental observation that is crucial: if we focus on one arm of a growing crystal, the shape of the tip is preserved. That is, in a reference frame moving with the tip, the shape is constant. An enormous amount of sophisticated mathematics has been done to attempt to derive this simple fact.

The first step was to find an exact solution to the problem in the absence of surface tension. This was done by G. Ivantsov by expressing the problem in parabolic coordinates, and assuming that the crystal surface lay along one of the coordinate surfaces – i.e. that it is exactly a paraboloid of revolution; see Langer (1980). In the absence of surface tension the temperature is constant on the surface. The growth boundary condition implies that the product $\rho v = f(\Delta)$ where ρ is the radius of curvature of the tip. Thus the velocity of the tip is not specified, but only the product ρv.

There are two unsatisfactory features of this solution. One is that it is unstable. In fact, in the absence of surface tension a generic surface develops unphysical cusps in finite time. Also, the undercooling selects a unique velocity, not a continuum of v's. The two problems are related.

The *velocity selection* problem baffled the community until the 1980s when several groups showed that surface tension selected a velocity and a stable shape close to the Ivantsov solution, but not exactly the same. However, this is true only if anisotropy in γ is present. Otherwise there is no stable tip, and there is *tip-splitting*. The mathematical details of the solution are very complicated, and beyond the scope of this book; see Pelcé (2004).

Diffusion-limited aggregation

The dendritic solutions of the previous paragraph depend on there being a single crystal with a well-defined set of crystal axes so that there is a fixed anisotropy. However, diffusion-limited growth in a random environment, for example of a polycrystalline substance is also very common. A way to model this was devised by Witten & Sander (1981) in terms of a simple computer algorithm.

The algorithm is as follows: start with a nucleation center at the origin of coordinates. Then we need to add matter according to the diffusion equation. Instead, Witten and Sander solved the equation (actually the quasi-static version) by recalling that diffusion describes random walks (as A. Einstein showed in his famous 1905 paper). The algorithm involved

Fig. 4.13   A diffusion-limited aggregation cluster of a 50,000,000 particles.

Fig. 4.14   **A polycrystalline sample of GeSe$_2$ grown on a surface. An amorphous film was deposited and allowed to crystallize. (Radnoczi *et al.* 1987)**

launching a random walk from far away, and letting it diffuse until it came into contact with the center where it stuck irreversibly. Then another walker is launched, and allowed to proceed until it encountered the center or a previous particle. The result of many such events is shown in Figure 4.13.

There are a number of remarkable and interesting features of this pattern. It shows the Mullins–Sekerka instability in the growth of branches, but also has many tip-splittings. There is no surface tension in this model. The instability is controlled by the finite particle size. The pattern strongly resembles many shapes seen in nature, see Figure 4.14. And, most unexpectedly, the pattern is a *fractal* (Mandelbrot 1982), that is, it has scaling symmetry in the sense that each bit of a branch is statistically identical on scales larger than several particle sizes and less than the size of the whole cluster.

Suggested reading

The first two references are good overviews of the subject of this chapter.
Godreche (1991), especially Chapter 1 by P. Nozières.
Pelcé (2004)
Barabasi & Stanley (1995)
Langer (1980)
Zangwill (1988)
Sander (2000)
Meakin (1998)

Problems

1. (a) In a fog water droplets have a radius of about 1μm. Water at $20\,^{\circ}$C has a surface tension of 72.8 dynes/cm. What is the pressure inside the droplets in atmospheres? (b) A small Ge crystal has a radius of 10 Å. It is in contact with its melt. The interfacial tension is = 250 erg/cm^2 , the entropy of melting is 30.55 J/K mol, and $v_c = 2.3\ 10^{-23}$ cm^3. Find the depression of the normal melting point (1200K).

2. Consider a Kossel crystal with nearest neighbor interaction ϵ_1 and next nearest neighbor interaction ϵ_2 and lattice constant a. Show that:

$$\gamma_{100} = (\epsilon_1 + 4\epsilon_2)/(2a^2) \quad \gamma_{110} = (n\epsilon_1 + m\epsilon_2)/(2\sqrt{2}a^2)$$

   Here n, m are integers which you must find.

3. What is the ratio of the surface energy for a [100] and a [111] surface for a Kossel simple cubic crystal with nearest neighbor interactions?

4. Another form of the Wulff construction is available if the crystal has only facetted faces. Suppose the facets are labeled 1, 2,… n. The surface tensions are γ_n, and the distance of the facet from the center of mass of the crystal (the Wulff point) is h_n. Show that γ_n/h_n, is the same for each facet. That is, low energy facets are close to the origin.

   Hints: The crystal may be considered to be made up of pyramids whose base is the facet and whose altitude is h_n. Then the total volume of the crystal is $V = \sum_n h_n S_n/3$, where S_n

is the area of the nth facet. Use a work argument to show that $(P_{\text{in}} - P_{\text{out}})dV = \sum_n \gamma_n dS_n$. The volume of the crystal changes by $dV = \sum_n dh_n S_n$. Now use the fact that the volume is the sum of pyramids to get an expression for dV in terms of dS_n.

5. Consider a step which wanders thermally by detaching and attaching atoms. (a) Write the shape of the step as

$$x(y) = \sum_k x_k e^{iky}.$$

What k's are allowed if we use periodic boundary conditions for length L? Prove that

$$< |x_k|^2 >= k_B T / (L\beta k^2),$$

where β is the step energy /unit length. (b) Show that

$$< x^2 >= \sum_k < |x_k|^2 >= L k_B T / (12\beta).$$

Hints: The energy can be written in a coarse-grained way as $E = $ (energy/unit length)$\times$(arclength) $= \int \beta (1 + [dx/dy]^2)^{1/2} dy$. Expand the square root. If you prefer the more physical case of a step with fixed ends L apart, then for part (b) you should average over y. For (b) also note that the sum cannot be converted to an integral (Why?) The sum $\sum_n (1/n^2) = \pi^2/6$ may be useful.

Classical and quantum waves

So far we have considered only the ground state of some condensed matter systems. If we excite such objects many interesting things can happen. One of the simplest and most interesting is that we can excite wave-like motions of the atoms and molecules. If we hit a solid with a hammer, we know that sound waves will move through the body. These acoustic waves are part of a class of motions called *lattice vibrations* which also can be excited by thermal effects. If temperatures are low or the wave amplitude is small, quantum mechanics must be used. The quantized version of a lattice vibration is made up of *phonons*.

Less familiar is the fact that if we wiggle a spin in a magnet it will cause its neighbors to move, and the motion will be transmitted in a *spin wave*. Once more there is a quantum version, the *magnon*.

5.1 Lattice vibrations and phonons

We first look at the vibrations of atoms away from their equilibrium positions in the crystal. We do this in three ways. First we briefly allude to the macroscopic theory of elasticity which does not use microscopic details at all. Then we look at classical atoms bound together by forces. For vibrations near the equilibrium positions the forces become an effective set of springs that bind the atoms. Then we quantize the motions.

5.1.1 Theory of elasticity

From a macroscopic point of view, the vibrations of a solid are the subject of the theory of elasticity. There are many excellent books on the subject; my favorite is Landau, Lifshitz, Kosevich & Pitaevskii (1986). Elasticity refers to the fact that if a force is applied to a solid, it can deform reversibly. In most cases, if the external force is small, the deformation is proportional to the force and is given by Hooke's law, namely, force and displacement are proportional. For example, if a square rod of length L and area $A = l^2$ is stretched by force F, the change of its length, δL is given by:

$$F/A = Y\delta L/L. \tag{5.1}$$

The proportionality constant, Y, is called Young's modulus, and it depends on the material from which the rod is made. The rod acts like a spring. Further, the rod will shrink in the x

and y directions by δl so that:

$$\delta l/l = -\sigma \delta L/L. \tag{5.2}$$

The parameter σ is called Poisson's ratio, and it also depends on the material.

Static strain and stress

Note that in both Eq. (5.1) and Eq. (5.2) the important object is the relative deformation, $\delta L/L$. The classical theory is formulated in terms of this kind of object. We imagine that an elastic continuum is initially subject to no force, and we mark the positions of various points. Then in the presence of forces there will be displacements, **u** of the points. Now define the *strain* tensor in the material in terms of the components of **u** and the position **r**:

$$\epsilon_{ik} = \frac{1}{2}\left(\frac{\partial u_i}{\partial r_k} + \frac{\partial u_k}{\partial r_i}\right). \tag{5.3}$$

We also need to define how forces act. Suppose there is a force $\mathbf{f}d\Omega$ on each volume element of material. This generates internal forces in the material. Consider a small cubical parcel of matter. A force per unit area in the i direction is applied to the face whose normal is in the k direction by the neighboring bits of material. This is an element of the *stress tensor*, σ_{ik}. By Newton's third law the internal forces cancel in calculating the total force on a whole sample. The remaining unbalanced force is a surface integral:

$$\oint \sum_k \sigma_{ik} dS_k = \int \sum_k \frac{\partial \sigma_{ik}}{\partial r_k} d\Omega = \int f_i d\Omega. \tag{5.4}$$

We have used the divergence theorem. Thus the force is the divergence of the stress tensor:

$$f_i = \sum_k \partial \sigma_{ik}/\partial r_k. \tag{5.5}$$

In terms of these concepts we can rewrite the definition of Young's modulus for a rod in the x direction with no forces in the y or z direction: $\sigma_{xx} = Y\epsilon_{xx}$.

Dynamics

Since each part of a solid has mass if it is set in motion it will act like a set of coupled springs and support traveling sound waves. To see how this arises consider a one-dimensional solid rod of area A. We completely neglect changes in the y, z directions: this amounts to putting the Poisson ratio equal to zero. Now both the stress and strain tensors have only one interesting component.

The equation of motion of a parcel of material is just Newton's law. We get the force as the divergence of the stress tensor from Eq. (5.5):

$$\rho \frac{\partial^2 u_x}{\partial t^2} = f_x = \frac{\partial \sigma_{xx}}{\partial x} = Y\frac{\partial \epsilon_{xx}}{\partial x}. \tag{5.6}$$

From Eq. (5.3) $\epsilon_{xx} = \partial u_x / \partial x$. Inserting this we get:

$$\rho \frac{\partial^2 u_x}{\partial t^2} = Y \frac{\partial^2 u_x}{\partial x^2}. \tag{5.7}$$

This is the familiar wave equation. It has a solution which corresponds to a wave traveling with velocity $v_l = \sqrt{Y/\rho}$.

In three dimensions sound waves often come in three polarizations for a given wavelength, two transverse and one longitudinal. The sound velocity is given in terms of the elastic constants of the solid as in the one-dimensional case. For example, in an isotropic material the longitudinal and transverse waves have velocities:

$$v_l = \sqrt{\frac{Y(1-\sigma)}{\rho(1+\sigma)(1-2\sigma)}}; \quad v_t = \sqrt{\frac{Y}{2\rho(1+\sigma)}}. \tag{5.8}$$

Elasticity of crystals

In a crystal, as opposed to an isotropic solid, there are not just two elastic constants, Y and σ, but many more, depending on the symmetry of the crystal. For example, in a cubic crystal there are three independent elastic constants, while in a rhombohedral crystal there are six.

5.1.2 Lattice dynamics

We can take another approach and view the atoms or ions as discrete. We will work explicitly in the context of crystals. We will find (at long wavelength) modes like the ones described in Eq. (5.8), and other modes as well, not described by continuum theory; they correspond to vibrations within the unit cell of the crystal.

To begin we note that, as in the continuum case, when a crystal ion moves with respect to its equilibrium position there is, in general, a restoring force. As a result the ion vibrates around that position. Also, if the position of the atom in question is $\mathbf{R} = \mathbf{R}^o + \mathbf{u}$, where $\mathbf{u}$ is, again, the deviation from the equilibrium position, then we always have $|\mathbf{u}|/a << 1$ where a is the lattice constant. Even at the melting point this ratio is a few percent. This is a small parameter in the theory.

We consider the classical dynamics of the atoms or ions. Since they are classical, we write Newton's equation $\mathbf{F} = m\mathbf{a}$ for each one. This is a huge number of equations: there are $3NB$ equations, where N is the number of unit cells, and B the number of ions in the basis. Oddly enough, we will be able to reduce them to a manageable number by a series of tricks. In practice, as we will see, we will have $3B$ equations to solve.

We need to specify the forces that the atoms exert on one another. We should take the electronic degrees of freedom into account, since they mediate the forces. We will ignore this for the moment, and think of the ions as mass points which carry the electrons along rigidly. Even so, the form of the interaction is complicated, in general. For example, for an ionic crystal, we need to get the forces from a potential energy of the form:

$$V = (1/2) \sum_{i \neq j} \frac{\pm Ze^2}{|\mathbf{R_i} - \mathbf{R_j}|}. \tag{5.9}$$

In addition we need short-range repulsions. This is a very hard problem because of the long range of the Coulomb force. If we move a single ion, it creates an unbalanced dipole moment, and many other atoms are affected. We will return to some aspects of this problem below. See also Born & Huang (1985).

For the moment we should think of short-range forces such as the van der Waals' force. In many cases (but not in metals, for example) we can write:

$$V = (1/2) \sum_{i \neq j} \phi(\mathbf{R_i} - \mathbf{R_j}). \tag{5.10}$$

Here, ϕ is a pair potential which can be short ranged.

We now try to formulate the Newtonian dynamics of the BN atoms in our crystal using the small parameter above. We simply Taylor expand the potential in terms of the u's, and keep only the first two terms. We consider a crystal with a basis. We index the unit cells of the lattice with s (N values) and the elements of the basis with l (B values). In three dimensions we need to deal with $j = 1, 2, 3$ components. Thus the displacements from equilibrium will be labeled $u(slj)$. The total potential energy is a function of all of these quantities. We do a Taylor expansion.

$$V(\{u(slj)\}) = V(0) + \sum_{lsj} \frac{\partial V(0)}{\partial u(slj)} u(slj)$$

$$+ \frac{1}{2} \sum_{lsj; l's'j'} \frac{\partial^2 V(0)}{\partial u(slj) \partial u(s'l'j')} u(slj) u(s'l'j'). \tag{5.11}$$

The argument 0 is to indicate that the derivatives are to be evaluated at $u \equiv 0$. The second term is zero because the coefficient is the force on an atom in equilibrium, which vanishes. If we stop at this order the result is quadratic in the displacements, as in the case of a simple harmonic oscillator. This is called the harmonic approximation; it is the equivalent of Hooke's law in elasticity theory.

We define the *dynamical matrix*

$$G(slj; s'l'j') = \frac{\partial^2 V(0)}{\partial u(slj) \partial u(s'l'j')}. \tag{5.12}$$

It is the set of generalized spring constants of the problem. Note that because G is a second derivative it is symmetric when we interchange slj with $s'l'j'$.

For the special case of pair potentials there is a useful formula for G. From Eq. (5.10) we find (for simplicity we assume that $B = 1$ so the index l does not appear, and we work in one dimension):

$$V = (1/2) \sum_{i \neq j} \phi(R_i - R_j + u_i - u_j).$$

$$\partial V / \partial u_s = \sum_j \phi'(R_s - R_j + u_s - u_j).$$

$$G(s; s') = \delta_{ss'} \sum_j \phi''(R_s - R_j) - \phi''(R_s - R_{s'}). \tag{5.13}$$

Here ϕ', ϕ'' are the first and second derivatives of ϕ. In three dimensions we have, in like manner:

$$G(s, j; s'j') = \delta_{ss'} \sum_{s''} \frac{\partial^2 \phi(R_s - R_{s''})}{\partial r_j \partial r_{j'}} - \frac{\partial^2 \phi(R_s - R_{s''})}{\partial r_j \partial r_{j'}}. \tag{5.14}$$

Now we can write the equations of motion. Suppose there are a given set of u's. Then the force on the atom labeled by sl is the derivative of V:

$$-\frac{\partial V(\{u(slj)\})}{\partial u(slj)} = -\sum_{s'l'j'} G(slj; s'l'j') u(s'l'j'). \tag{5.15}$$

This quantity is the jth component of the force on atom sl due to the displacements of all the others. Each term in Eq. (5.15) has an interpretation: $-G(slj; s'l'j')$ is the jth component of the force on sl due to a unit displacement of $s'l'$ in the j' direction.

Since the cells of the crystal are equivalent, these forces, and thus G, can only depend on the difference $\mathbf{R}_s^o - \mathbf{R}_{s'}^o$. Sometimes we write G as a Cartesian tensor (a 3×3 matrix in the j, j') $\mathsf{G}(sl; s'l') = \mathsf{G}_{ll'}(\mathbf{R}_s^o - \mathbf{R}_{s'}^o)$.

Now we can write the $3NB$ equations of motion:

$$m_l \frac{d^2 u(slj)}{dt^2} = -\sum_{s'l'j'} G(slj; s'l'j') u(s'l'j'). \tag{5.16}$$

We can tame this huge set of equations by making a *normal mode* analysis. This amounts to writing a trial solution:

$$\mathbf{u}(sl) = \mathbf{U}_l(\mathbf{k}) \exp(i[\mathbf{k} \cdot \mathbf{R}_s^o - \omega t]). \tag{5.17}$$

We have expanded in running waves. $\mathbf{U}_l$ is the amplitude of the wave motion that is carried out by the lth atom in each cell; the direction of this vector gives the polarization of the wave.

The normal mode solution is a standard way to treat coupled linear equations. If we want to solve an initial-value problem, then at $t = 0$, a linear combination of solutions of the form of Eq. (5.17) is a Fourier series expansion. Then the development in time is determined by solving Eq. (5.16), which we will now do. Of course, the $\mathbf{k}$'s are the allowed vectors from periodic boundary conditions in the whole crystal.

Inserting the trial solution:

$$-\omega^2 m_l \mathbf{U}_l = -\sum_{s'l'} e^{i\mathbf{k}\cdot[\mathbf{R}^o_{s'}-\mathbf{R}^o_s]}\mathsf{G}(sl;s'l')\cdot\mathbf{U}_{l'}$$

$$= -\sum_{l'}\mathsf{G}_{ll'}(\mathbf{k})\cdot\mathbf{U}_{l'}. \tag{5.18}$$

The new quantity

$$\mathsf{G}_{ll'}(\mathbf{k}) = \sum_{\mathbf{R}} e^{-i\mathbf{k}\cdot\mathbf{R}}\mathsf{G}_{ll'}(\mathbf{R}) \tag{5.19}$$

is essentially the spatial Fourier transform of the dynamical matrix. It is a Hermitian matrix because interchanging all the subscripts changes $\mathbf{R}$ to $-\mathbf{R}$.

The equations of motion are now a much smaller set: Eq. (5.18) is a set of $3B$ equations governed by the $3B \times 3B$ matrix $\mathsf{G}_{ll'}(\mathbf{k})$. It is a generalized eigenvalue equation:

$$\sum_{l'}[\mathsf{G}_{ll'}(\mathbf{k}) - \omega^2 m_l \delta_{ll'}\mathsf{I}]\cdot\mathbf{U}_{l'} = 0 \tag{5.20}$$

with ω^2 playing the role of eigenvalue, and $\mathbf{U}_l$ the eigenvector. Here I is the unit matrix in three dimensions.

We should recall some facts about the generalized eigenvalue equation. It is a problem of the form:

$$\mathcal{G}\mathcal{U} = \lambda\mathcal{M}\mathcal{U}, \tag{5.21}$$

where $\mathcal{M}$ is a positive definite matrix. The application to our case is to form $\mathcal{U}$ as a $3B$ dimensional vector, and $\mathcal{M}$ as a $3B \times 3B$ diagonal matrix with the masses of the various atoms in the basis as diagonal elements. For example, if $b = 2$ then $\mathcal{M}$ is a 6×6 matrix with $m_1, m_1, m_1, m_2, m_2, m_2$ on the diagonal. If $\mathcal{G}$ is self-adjoint, as in our case, then eigenvectors U, V belonging to different eigenvalues, λ, μ are orthogonal in the sense that:

$$\mathcal{U}\cdot\mathcal{M}\mathcal{V} = 0, \tag{5.22}$$

where $\cdot$ is the usual Euclidean inner product.

Equation (5.20) is the basic equation of lattice dynamics. The $3B$ eigenvalues (or, rather, their square roots) are called the normal frequencies. The plot of $\omega(\mathbf{k})$ is called the dispersion relation.

Some simple examples

Consider the one-dimensional chain with motion along the chain and one atom per unit cell. We have:

$$G(k) - \omega^2 m = 0$$

$$\omega(k) = \sqrt{G(k)/m} \tag{5.23}$$

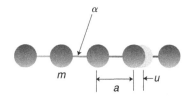

Fig. 5.1   **The one-dimensional chain.**

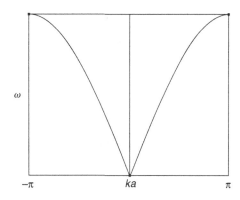

Fig. 5.2   **Dispersion relation in the 1d Brillouin zone.** ω is in units of $2\sqrt{\alpha/m}$.

The simplest explicit example is the chain with nearest neighbor interactions; see Figure 5.1. We may think of mass points connected by springs with spring constant α. Then:

$$V = \sum_s (\alpha/2)[u_s - u_{s+1}] = \sum_s \alpha[u_s^2 - u_s u_{s+1}]. \tag{5.24}$$

Therefore the only non-zero elements of G are:

$$G(0) = \partial^2 V/\partial u_s^2 = 2\alpha$$
$$G(\pm a) = \partial^2 V/\partial u_s \partial u_{s+1} = -\alpha. \tag{5.25}$$

And $G(k) = 2\alpha(1 - \cos(ka))$. This gives:

$$\omega(k) = 2\sqrt{\alpha/m}|\sin(ka/2)|. \tag{5.26}$$

See Figure 5.2. For small k the limiting behavior is $\omega \rightarrow vk, v = \sqrt{\alpha a^2/m}$, so that v is the velocity of sound.

Another easy one-dimensional example is of a lattice with a basis of two atoms. We take a chain where the masses alternate in the pattern $m_1, m_2, m_1 \ldots$; see Figure 5.3. Now there are two atoms in a cell. We can write the potential energy as before:

$$V = \sum_s (\alpha/2)[(u_{1,s+1} - u_{2,s})^2 + (u_{1,s} - u_{2,s})^2]. \tag{5.27}$$

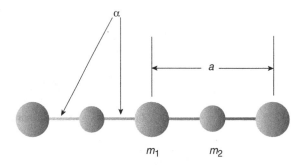

Fig. 5.3 **The 1d chain with a basis. The lattice constant is the distance between two atoms of type 1.**

The dynamical matrix follows as before:

$$G_{11}(a) = G_{22}(a) = 0$$
$$G_{11}(0) = G_{22}(0) = 2\alpha$$
$$G_{12}(0) = G_{21}(0) = -\alpha$$
$$G_{12}(-a) = G_{21}(a) = -\alpha. \tag{5.28}$$

The next step is to find $G(k)$.

$$G_{11}(k) = G_{22}(k) = 2\alpha$$
$$G_{12}(k) = -\alpha(1 + e^{-ika})$$
$$G_{21}(k) = -\alpha(1 + e^{ika}). \tag{5.29}$$

The result is:

$$(2\alpha - m_1\omega^2)U_1 - \alpha(1 + e^{-ika})U_2 = 0$$
$$-\alpha(1 + e^{ika})U_1 + (2\alpha - m_2\omega^2)U_2 = 0. \tag{5.30}$$

Setting the determinant of the coefficients to zero, and solving for the eigenvalues gives the following:

$$\frac{2\omega^2}{\alpha} = \frac{1}{m_1} + \frac{1}{m_2} \pm \left(\left[\frac{1}{m_1} + \frac{1}{m_2} \right]^2 - \frac{4}{m_1 m_2} \sin^2(ka/2) \right)^{1/2}. \tag{5.31}$$

There are now two modes for each k, as shown in the dispersion relation plotted in Figure 5.4. The upper curve is called the optical branch, and the lower the acoustic branch. By examining the eigenvectors, it is easy to show that for small k the two vibrations differ in the following way: for the acoustic branch both atoms in the cell move together, whereas for the optical they move oppositely. We can think of the optical mode for small k as being essentially an internal vibration of the molecule in the unit cell.

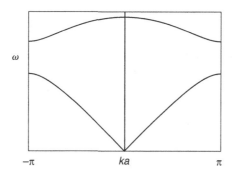

Fig. 5.4   **The acoustic and optical modes for two different atoms per unit cell.**

General features of the dispersion relation

In three dimensions lattice dynamics is much more challenging. Though many calculations have been done successfully, we should point out that it is *very seldom* the case that a nearest neighbor model is adequate. We could imagine using the coupling constants for many neighbors to fit the dispersion relation. This is not a fruitful approach because there are far too many parameters. We will see an explicit example of how to proceed for semiconductors in Chapter 9.

We can say some interesting general things about the dispersion relation from the structure of the dynamical equation, Eq. (5.18). It is clear that for each $\mathbf{k}$ there are $3B$ modes, because that is the dimension of $\mathsf{G}_{ll'}$, and the ω^2 are the eigenvalues of G. Thus we should label the eigenvalue for each mode $\omega_\lambda(\mathbf{k})$ where the mode index, λ takes on $3B$ values. For $\mathbf{k}$ directions of high symmetry the modes often correspond to two transverse and one longitudinal polarization.

The long wavelength (small k) acoustic modes should correspond to macroscopic sound as in Eq. (5.8). Thus we expect three modes, one longitudinal and two transverse, to behave like sound, and have $\omega \to 0$ as $k \to 0$. We can prove this.

Consider the effect of moving *each* atom the same amount in the same direction, $\mathbf{u}$. This amounts to picking up the crystal and moving it. Clearly there can be no restoring force. Recalling the expression for the net force on an atom, we find:

$$\sum_{sl} \mathsf{G}_{sl;s'l'} \cdot \mathbf{u} = 0. \tag{5.32}$$

in the case where $\mathbf{u}$ is the same for all s, l. Since $\mathbf{u}$ is arbitrary we have:

$$\sum_{sl} \mathsf{G}_{ll'}(\mathbf{R}_s^o - \mathbf{R}_{s'}^o) = 0 = \sum_{l'} \mathsf{G}_{ll'}(\mathbf{k} = 0). \tag{5.33}$$

The eigenvalue equation for $\mathbf{k} = 0$:

$$\sum_{l'} [\mathsf{G}_{ll'}(0) - \omega^2 M_l \delta_{ll'} \mathsf{I}] \cdot \mathbf{U}_{l'} = 0 \tag{5.34}$$

has the solution $\omega = 0$ if we take $\mathbf{U}$ independent of l'. These modes (three independent ones) are the limit of ordinary sound waves. Near, but not at $\mathbf{k} = 0$ we see that we expect the atoms in the basis to move together, since $\mathbf{U}$ is independent of l. We have the following rule: *For $\mathbf{k} = 0$ there are three acoustic modes with $\omega = 0$. The remaining $3B - 3$ roots do not go to 0: they are the optical modes.*

It is possible to go further and show that the acoustic modes approach 0 as $\omega_\beta \rightarrow v_\beta(\hat{\mathbf{k}})|\mathbf{k}|$ where the velocity of sound depends on the direction of $\mathbf{k}$. This is easy for the case of a Bravais lattice of atoms. To see this, return to Eq. (5.19) and expand the exponential.

$$\mathsf{G}(\mathbf{k}) \rightarrow \sum_{\mathbf{R}}[1 + i\mathbf{k} \cdot \mathbf{R} - (1/2)(\mathbf{k} \cdot \mathbf{R})^2 + ...]\mathsf{G}(\mathbf{R}) \tag{5.35}$$

The constant term vanishes because of Eq. (5.33), and the linear term vanishes because for a Bravais lattice there is inversion symmetry so that $\mathsf{G}(\mathbf{R}) = \mathsf{G}(-\mathbf{R})$. The leading term is quadratic in k_j. Since G is proportional to k^2, so are its eigenvalues ω_β^2. For a crystal with a basis the proof is more complicated because crystals do not necessarily have inversion symmetry; see Ashcroft & Mermin (1976).

In a previous chapter we introduced the Einstein model for which all the atoms vibrate *independently* with frequency ω_E. It is important to note that this model breaks the translational invariance which we have assumed. Thus there is no reason to have acoustic modes, a sound velocity, etc.

5.1.3 Bloch theorem

We introduced the normal mode form of solution as an *ad hoc* assumption, though it turned out to be very useful. We can take a deeper look at the matter using a theorem of F. Bloch. The equation of motion is of the form (again, simplify to a Bravais lattice, though the conclusion is general):

$$m\omega^2 \mathbf{u}(\mathbf{R}) = \sum_{\mathbf{R}'} \mathsf{G}(\mathbf{R} - \mathbf{R}') \cdot \mathbf{u}(\mathbf{R}'). \tag{5.36}$$

Think of this as a matrix operator equation in a space of $3N$ dimensions. What we will find is that the translational invariance of G, namely its dependence on $\mathbf{R} - \mathbf{R}'$, puts restrictions on its eigenvectors. Here is the result:

Bloch theorem For any eigenvalue equation of the form of Eq. (5.36) with a translationally invariant operator, the eigenvectors obey:

$$\mathbf{u}(\mathbf{R} + \mathbf{R}') = e^{i\mathbf{k} \cdot \mathbf{R}}\mathbf{u}(\mathbf{R}') \tag{5.37}$$

where $\mathbf{k}$ is an allowed k-vector. In particular, set $\mathbf{R}' = 0$ and $\mathbf{u}(0) = \mathbf{U}$. We recover the normal mode form of Eq. (5.17).

Proof Define a translation operator $\tau(\mathbf{R})$ by $\tau(\mathbf{R})\mathbf{u}(\mathbf{R}') = \mathbf{u}(\mathbf{R} + \mathbf{R}')$. Any two translation operators clearly commute:

$$\tau(\mathbf{R}_1)\tau(\mathbf{R}_2)\mathbf{u}(\mathbf{R}_3) = \mathbf{u}(\mathbf{R}_1 + \mathbf{R}_2 + \mathbf{R}_3) = \tau(\mathbf{R}_2)\tau(\mathbf{R}_1)\mathbf{u}(\mathbf{R}_3).$$

Also,

$$\tau(\mathbf{R})\mathsf{G} \cdot \mathbf{u} = \mathsf{G} \cdot \tau(\mathbf{R})\mathbf{u}.$$

Apply τ to the eigenvalue equation:

$$\tau(\mathbf{R}_1)\mathsf{G}\mathbf{u} = \tau(\mathbf{R}_1)\lambda\mathbf{u}$$

$$\mathsf{G}\mathbf{u}(\mathbf{R} + \mathbf{R}_1) = \lambda\mathbf{u}(\mathbf{R} + \mathbf{R}_1).$$

Therefore $\mathbf{u}(\mathbf{R} + \mathbf{R}_1)$ belongs to the same eigenvalue as $\mathbf{u}(\mathbf{R})$. There are two cases.

(i) The eigenvalue is non-degenerate. Then $\mathbf{u}(\mathbf{R}+\mathbf{R}_1)$ and $\mathbf{u}(\mathbf{R})$ differ by a constant factor, γ. Suppose the factors associated with the three generators of the lattice, $\mathbf{a}_j$ are called γ_j so that $\mathbf{u}(\mathbf{R} + \mathbf{a}_j) = \gamma_j\mathbf{u}(\mathbf{R})$.

Periodic boundary conditions require that $\tau(M\mathbf{a}_j) = 1$, the vibration pattern repeats when we go around the whole sample. This implies that $\gamma_j^M = 1$, so that the γ's are roots of unity. We must have

$$\gamma_j = \exp(2\pi i m_j/M),$$

for some set of three integers m_j. Now any translation $\mathbf{R}$ can be written $\sum_j n_j\mathbf{a}_j$ so that the factor associated with the translation is

$$\gamma_1^{n_1}\gamma_2^{n_2}\gamma_3^{n_3} = \exp(2\pi i[m_1 n_1 + m_2 n_2 + m_3 n_3]/M).$$

Define $\mathbf{k} = \sum_j m_j\mathbf{g}_j$; the $\mathbf{g}$'s were defined in Eq. (3.7). This gives:

$$\mathbf{u}(\mathbf{R} + \mathbf{R}') = e^{2\pi i[m_1 n_1 + m_2 n_2 + m_3 n_3]/M}\mathbf{u}(\mathbf{R}') = e^{i\mathbf{k}\cdot\mathbf{R}}\mathbf{u}(\mathbf{R}')$$

as required.

(ii) In the case where the eigenvalue is degenerate, the proof is essentially the same but more involved. Consider a two-fold degenerate state, $\mathbf{u}_i, i = 1, 2$. At worst τ will mix the two $\mathbf{u}$'s together:

$$\mathbf{u}_i(\mathbf{R} + \mathbf{R}') = \sum_j t_{ij}\mathbf{u}_j(\mathbf{R}'). \tag{5.38}$$

However, if we consider the t_{ij} to be elements of a matrix, we can diagonalize it by taking a basis of linear combinations of the $\mathbf{u}_i$. (It is possible to show that the matrix t is unitary, so it can be diagonalized; see below.) Any such linear combination is still an eigenvector with the same eigenvalue. Call the new eigenvectors $\mathbf{u}'_j$. We have, for the translation corresponding to $\mathbf{a}_1$:

$$\begin{pmatrix} \mathbf{u}'_1(\mathbf{a}_1) \\ \mathbf{u}'_2(\mathbf{a}_1) \end{pmatrix} = \begin{pmatrix} \gamma_1 & 0 \\ 0 & \gamma_2 \end{pmatrix} \begin{pmatrix} \mathbf{u}'_1(0) \\ \mathbf{u}'_2(0) \end{pmatrix}.$$

Now we repeat the argument above to show that the γ's are roots of unity. Furthermore, since the translations corresponding to $\mathbf{a}_{2,3}$ commute with this one, we may simultaneously diagonalize their t matricies. Now the argument goes through as before, and for the two $\mathbf{u}'$ we generate $\mathbf{k}$'s which are, in general, different.

5.1.4 Orthogonality of normal modes

We can think of the normal mode solution in a different way which is quite useful. Recall that the vector $\mathbf{u}_{\mathbf{k},\lambda}(\mathbf{R}) = \mathbf{U}_\lambda(\mathbf{k})e^{i\mathbf{k}\cdot\mathbf{R}}$ is a solution of the eigenvalue equation, Eq. (5.36). Think of $\mathbf{u}(\mathbf{R})$ as the components of a vector, $\mathcal{U}$, in a space of $3N$ dimensions (we are still taking $B = 1$.) Then Eq. (5.36) is a matrix eigenvalue equation of the form:

$$\lambda\mathcal{U} = \mathcal{G}\mathcal{U}. \tag{5.39}$$

In this space $\mathcal{G}$ is a real symmetric matrix, so its eigenvectors can be made orthonormal, if we define an inner product. That is we can arrange things so that for any two eigenfunctions:

$$\mathcal{U}_i \bullet \mathcal{U}_j = \delta_{i,j}. \tag{5.40}$$

We can define a Cartesian inner product:

$$\mathcal{U} \bullet \mathcal{V} = \frac{1}{N}\sum_{\mathbf{R}} \mathbf{u}^*(\mathbf{R}) \cdot \mathbf{v}(\mathbf{R}). \tag{5.41}$$

For the case of normal mode vectors in Bloch form we have an important result: we label the eigenvectors with $\mathbf{k}, \lambda$. Then

$$\mathcal{U}_{\mathbf{k},\lambda} \bullet \mathcal{U}_{\mathbf{k}',\lambda'} = \frac{1}{N}\sum_{\mathbf{R}} \mathbf{U}^*_{\mathbf{k},\lambda} \cdot \mathbf{U}_{\mathbf{k}',\lambda'}e^{i(\mathbf{k}-\mathbf{k}')\cdot\mathbf{R}}$$

$$= \mathbf{U}^*_{\mathbf{k},\lambda} \cdot \mathbf{U}_{\mathbf{k},\lambda'}\delta_{\mathbf{k},\mathbf{k}'}. \tag{5.42}$$

We conclude from Eq.(5.40) and Eq. (5.42) that:

$$\mathbf{U}^*_{\mathbf{k},\lambda} \cdot \mathbf{U}_{\mathbf{k},\lambda'} = \mathbf{U}_{-\mathbf{k},\lambda} \cdot \mathbf{U}_{\mathbf{k},\lambda'} = \delta_{\lambda,\lambda'}. \tag{5.43}$$

If $B > 1$ we must use the generalized orthogonality relation, Eq. (5.22).

We can also now prove that the matrix t in Eq. (5.38) is unitary. We are considering a case where two eigenfunctions do not have a definite $\mathbf{k}$, but do have the same $\lambda = \omega^2$. Then a translation will mix them. However:

$$\delta_{i,j} = \mathcal{U}_i \bullet \mathcal{U}_j = [\tau\mathcal{U}_i] \bullet [\tau\mathcal{U}_j]. \tag{5.44}$$

This must be true because τ just rearranges the terms in the sum in Eq. (5.41). However:

$$
\begin{aligned}
[\tau(\mathbf{R}')\mathcal{U}_i] \bullet [\tau(\mathbf{R}')\mathcal{U}_j] &= \frac{1}{N}\sum_{\mathbf{R}} \mathbf{u}_i^*(\mathbf{R}+\mathbf{R}') \cdot \mathbf{u}_j(\mathbf{R}+\mathbf{R}') \\
&= \sum_{k,l} \frac{1}{N}\sum_{\mathbf{R}} t_{ik}^* \mathbf{u}_k^*(\mathbf{R}) \cdot t_{jl}\mathbf{u}_l(\mathbf{R}) \\
&= \sum_{k,l} t_{ik}^* t_{jl} \mathcal{U}_k \bullet \mathcal{U}_j \\
&= \sum_{k} t_{ik}^* t_{jk} = [\mathsf{tt}^\dagger]_{j,i}.
\end{aligned}
\tag{5.45}
$$

Here, $\mathsf{t}^\dagger$ is the adjoint matrix to t. Comparing this to Eq. (5.44) we see that $\mathsf{t}^\dagger$ is the inverse of t. A matrix whose adjoint is its transpose is unitary.

5.1.5 Experimental results

The dispersion curves for lattice vibrations can be measured in a number of ways, the most convenient of which is neutron scattering, which we will treat below. The general nature of the result is given in the sketch in Figure 5.5.

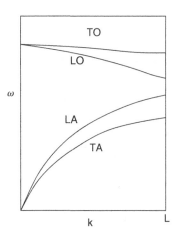

Fig. 5.5   A sketch of the phonon dispersion relation for Si. The modes are plotted from the center of the Brillouin zone to the middle of the hexagonal face (see Figure 3.10), the L point. LA is longitudinal acoustic, TA, transverse acoustic, LO, longitudinal optical, and TA transverse optical. Si has two atoms per unit cell, so that there are three acoustic modes and three optical. The transverse modes are degenerate for this direction. The LO mode intersects the zone edge at $\omega/2\pi \approx 10$ THz.

5.1.6 Origin of the potential

In all of the treatment above we have assumed that the potential of interaction between the ions is known and short-ranged. This is reasonable for solid Ar and similar Lennard-Jones solids, as in Problem 1.

In the general case things are not so simple. The forces that hold the solid together are mediated by the electrons. Consider, for example, a simple metal. We can think of charged ions as floating in a sea of electrons and interacting via the Coulomb force. As the ions move, the electrons move to retain charge neutrality. In a covalent material we bend and stretch the bonds and change the electron states. In both cases we need to understand the electrons, so we need to postpone calculating the forces.

Born–Oppenheimer approximation

Let us consider a (very bad) model for the ions in a metal, the Coulomb lattice. Suppose that the electrons, with density n are frozen in place and simply neutralize the average charge of the ions. Then when ions move they create electric fields which give rise to forces on the other ions. Consider a long wavelength longitudinal mode with $\mathbf{q}$ in the z direction. We can write $u_z(\mathbf{R}) = u_o e^{i(qz-\omega t)}$. Since a longitudinal wave gives rise to a periodic increase and decrease of the ion density there is an induced charge density $\delta\rho$ compared to the negative background. This is given by $\delta\rho = -nZe\nabla \cdot \mathbf{u} = -iqnZeu_z$ where Ze is the charge on the ion. (This follows from the interpretation of the divergence as the net flux out of a region).

Now the induced electric field is given by one of Maxwell's equations: $\nabla \cdot \mathbf{E} = 4\pi\rho$ which implies $E_z = -4\pi nZeu_z$. The equation of motion for the ions is:

$$M\frac{d^2 u_z}{dt^2} = ZeE_z = -4\pi nZ^2 e^2 u_z. \tag{5.46}$$

Thus the ions have a normal mode with natural frequency

$$\Omega_{\mathrm{p}} = \sqrt{4\pi nZ^2 e^2/M}. \tag{5.47}$$

This is a very peculiar result because Ω_{p} is *independent of* q. It is also wrong: longitudinal sound in metals has a dispersion relation just like that in insulators, $\omega = cq$. (Note that in this model transverse sound does have an ordinary dispersion relation.)

Where did we go wrong? This is clear: *we have frozen the electrons.* This is not physically correct: since electrons are much lighter than ions they move faster. In fact, they follow the ions and cancel out (screen) the long-range fields. In order to understand this in detail, we need to understand electron dynamics better. This is the subject of a later chapter: see Eq. (9.8).

However, there is a real-world example of this effect if we interchange the roles of electrons and ions. Since ions move sluggishly compared to electrons they serve, for many purposes, as a fixed neutralized background. The electrons do have a normal mode (the plasma oscillation) which has non-zero frequency as $q \rightarrow 0$, and whose value is $\omega_{\mathrm{p}} =$

$\sqrt{4\pi ne^2/m}$, namely the same as the formula above with the electron charge and mass; see Eq.(8.40).

There is another lesson of this example. When the ions move the electrons adjust, and since they are so light they adjust essentially instantly. Now consider, for example, the case of Si. As we move ions the covalent bonds are bent and stretched. This is the source of the restoring force in the lattice dynamics. We can get at the energetics of the restoring force by imagining that we have moved each ion by $\mathbf{u}(\mathbf{R})$ and frozen it in place. Then we figure out the total energy of the electrons (using techniques that we will develop later). Then we take another set of $\mathbf{u}$ and do it again. The change in the electron total energy gives rise to restoring forces. That is the electron energy in a frozen configuration of ions serves as the *potential energy* for the ions. In metals we need to figure out the total energy including screening.

This idea is the essence of the Born–Oppenheimer approximation. This method was developed to describe the vibrations of molecules; (Schiff 1968, Baym 1990). For example, in a diatomic molecule it is necessary to solve the wave equation as a function of the distance between the nuclei. Then the restoring force for vibrations is given by using the energy of the electrons as a potential energy. It is a practical method for doing lattice dynamics in solids – once we understand how to compute electron energies; see Section 9.6.1.

5.1.7 Long wavelength vibrations of polar crystals

For a polar crystal like NaCl or GaAs the forces between the ions are long range, as we remarked. We will not treat the special techniques necessary to handle this problem in general. See (Born & Huang 1985). However, we can find a number of interesting results using simple classical notions. In reality we need to consider the electron dynamics, as in the previous section. However, if we take the point of view that most of the physics captured in the transfer of one electron from the cation to the anion, we can get quite far.

As the ions move there are two sources for electric fields: there is a dipole field due to the relative motion of the positive and negative charges, and also the ions themselves can polarize, cf. Eq. (1.11). Just as above, these fields act on the ions and serve as the restoring force for the dynamics. We seek the normal modes which, in this case, may be thought of as waves of electronic and ionic polarization. Recall that the polarization of electromagnetic theory, $\mathbf{P}$, is the dipole moment per unit volume. We will distinguish between an ionic part, $\mathbf{P_i}$ and an electronic part $\mathbf{P_e}$. We will make the (crude) assumption that they add: $\mathbf{P} = \mathbf{P_i} + \mathbf{P_e}$.

For the reasons given above, the ions move more slowly than the electrons. In practice, for frequencies below the infra-red (for most solids) we need to consider the motion of both ions and electrons, and both sources of polarization are important. At the frequencies of visible radiation, we have only the electronic polarization. This is reflected in the frequency dependence of the dielectric constant, $\epsilon(\omega)$. We can write:

$$\mathbf{P_e} = n\alpha_e\mathbf{E}; \quad \mathbf{P_i} = n\alpha_i\mathbf{E}, \tag{5.48}$$

where n is the number of unit cells per unit volume, and the α's are the polarizabilities. We can suppose that α_e is the sum of the polarizabilities of the positive and negative ions in the unit cell. Now consider the usual relationship between the electric displacement, $\mathbf{D}$ and the field, $\mathbf{E}$: $\mathbf{D} = \mathbf{E} + 4\pi\mathbf{P} = \epsilon\mathbf{E}$. It follows that:

$$\epsilon = 1 + 4\pi n(\alpha_i + \alpha_e). \tag{5.49}$$

Since the ionic part will be frozen out for optical frequencies, we expect that $\epsilon_0 > \epsilon_\infty$. We use a standard convention and refer to optical frequencies as infinite (with respect to the ionic motions).

Local Fields and the Clausius–Mossotti relation

We have made an unjustified approximation in Eq. (5.48) above. The electric field, $\mathbf{E}$, in the equation is the macroscopic Maxwell field namely the spatial average of the microscopic electric fields. For a crystal this would be the average over a unit cell. However, the ions do not sit at a random point in the unit cell, but at very special points. The field there may be quite different from the average. It is referred to as the *local field*, $\mathbf{E}_{loc}$. We will work out $\mathbf{E}_{loc}$ for the case when the ions are at points of cubic symmetry (as in NaCl). See also Jackson (1999).

The way to proceed is due to H. Lorentz. Consider a crystal with uniform polarization $\mathbf{P}$ which arises from applying an external field. Now the field at any point arises from the external field plus the effect of the induced dipoles in all the unit cells. If we take the crystal to be a dielectric continuum everywhere, this gives $\mathbf{E}$. Now divide the crystal into two parts by introducing a sphere of radius R centered on the ion in question. We take R to be much larger than a lattice constant. In the distant region, outside of the sphere it is a good approximation to replace the discrete ions by a continuum. Inside the sphere it is not, if we are interested in the local field. Thus we can write:

$$\mathbf{E}_{loc} = \mathbf{E} - \mathbf{E}_{cont} + \mathbf{E}_{disc}. \tag{5.50}$$

Here $\mathbf{E}_{cont}$ is the field due to the dipoles inside the sphere treated as a continuum, and $\mathbf{E}_{disc}$ is the (correct) field if these same dipoles are counted as discrete.

The continuum approximation is easy to work out: if there is uniform polarization inside a sphere, then the effect of all the dipoles amounts to a surface charge $\sigma = \mathbf{P} \cdot \hat{\mathbf{n}}$ where $\hat{\mathbf{n}}$ is the normal to the surface of the sphere. Taking polar coordinates with z-axis along the polarization we have $\sigma = P\cos(\theta)$. The electric field at the origin is in the z direction by symmetry and is given by $-\cos(\theta)\sigma/R^2$. Integrating over the sphere we find:

$$\mathbf{E}_{cont} = -\int R^2 d\phi \sin(\theta)d\theta \cos^2(\theta)\mathbf{P}/R^2 = -4\pi\mathbf{P}/3. \tag{5.51}$$

The field due to discrete dipoles is found as follows: suppose we have dipoles of strength p at positions $\mathbf{r}_i$ where the vector is measured from the center of the sphere. The electric field due to these, say in the z-direction is

$$p \sum_i \frac{3z_i^2 - r_i^2}{r_i^5}. \tag{5.52}$$

If the environment of the ion has cubic symmetry we have:

$$\sum z_i^2/r_i^5 = \sum x_i^2/r_i^5 = \sum y_i^2/r_i^5. \tag{5.53}$$

Thus the sum in Eq. (5.52) vanishes. That is, $\mathbf{E}_{\text{disc}} = 0$. Putting all this together we find:

$$\mathbf{E}_{\text{loc}} = \mathbf{E} + 4\pi\mathbf{P}/3. \tag{5.54}$$

This result is called the Lorentz local field and it has an interesting consequence. Suppose we put, instead of Eq. (5.48), $\mathbf{P} = n\alpha\mathbf{E}_{\text{loc}}$, where α is the total polarizability. Then:

$$P = n\alpha(E + 4\pi P/3); \quad P/E = n\alpha/(1 - 4\pi n\alpha/3). \tag{5.55}$$

However we also know that $\epsilon = 1 + 4\pi P/E$. Combining these expressions we find:

$$\epsilon = \frac{1 + 8\pi n\alpha/3}{1 - 4\pi n\alpha/3}; \quad \frac{\epsilon - 1}{\epsilon + 2} = \frac{4\pi n\alpha}{3}. \tag{5.56}$$

This is the Clausius–Mossotti relation, discovered independently by O. F. Mossotti and R. Clausius. It was introduced in the context of optics by H. Lorentz and L. Lorenz in terms of the refractive index $\mathsf{n} = \sqrt{\epsilon}$.

Now we use the fact that at low frequencies the ions and the electrons can polarize, while at high frequencies the ions are frozen out. Thus:

$$\frac{\epsilon_0}{\epsilon_\infty} = \frac{1 + 8\pi n(\alpha_i + \alpha_e)/3}{1 - 4\pi n(\alpha_i + \alpha_e)/3} \frac{1 - 4\pi n\alpha_e/3}{1 + 8\pi n\alpha_e/3}. \tag{5.57}$$

Lyddane–Sachs–Teller relation

Now consider long-wavelength optical vibrations. Recall that the two ions in the unit cell will move opposite to one another in the optical mode. We can write an effective harmonic oscillator equation of motion for the difference variable $\eta = u_+ - u_-$ of the motions of the positive and negative ions in some direction:

$$\mu d^2\eta/dt^2 + \mu\omega_0^2\eta = eE_{\text{loc}}. \tag{5.58}$$

Here μ is the effective mass associated with the relative motion, and ω_0 is the natural frequency that would occur if the ions did not generate long-range electric fields. Note that in the absence of long-range fields we would expect the $q \to 0$ transverse and longitudinal modes to have the same frequency since they are both essentially uniform motions. With long-range fields this is no longer the case.

First consider the static case. We have $P = ne\eta = (ne^2/\mu\omega_0^2)E_{\text{loc}}$ so that the ionic part of the polarizability is $\alpha_{\text{i}} = e/\mu\omega_0^2$. For the free motion of the transverse mode, there is a local field generated because there is a non-zero polarization:

$$\mu d^2\eta/dt^2 + \mu\omega_0^2\eta = 4\pi eP/3. \tag{5.59}$$

However $P = P_{\text{i}} + P_{\text{e}} = ne\eta + n\alpha_{\text{e}}E_{\text{loc}}$. Thus $P = ne\eta/(1 - 4\pi n\alpha_{\text{e}}/3)$. Putting this into Eq. (5.59) we find the frequency of the transverse mode:

$$\omega_{\text{T}}^2/\omega_0^2 = 1 - \frac{4\pi n\alpha_{\text{i}}/3}{1 - 4\pi n\alpha_{\text{e}}/3} = \frac{1 - 4\pi n\alpha_{\text{i}}/3 - 4\pi n\alpha_{\text{e}}/3}{1 - 4\pi n\alpha_{\text{e}}/3}. \tag{5.60}$$

For a longitudinal mode there is another effect: charges accumulate at the end of the sample to give a charge density $\sigma = \mathbf{P} \cdot \hat{\mathbf{n}}$. This gives rise to a *depolarization* field $E_{\text{d}} = -4\pi P$. Now the total local field is $E_{\text{loc}} = (-4\pi + 4\pi/3)P = -(8\pi/3)P$. Repeating the steps above gives:

$$\omega_{\text{L}}^2/\omega_0^2 = 1 + \frac{8\pi n\alpha_{\text{i}}/3}{1 + 8\pi n\alpha_{\text{e}}} = \frac{1 + 8\pi n\alpha_{\text{i}}/3 + 8\pi n\alpha_{\text{e}}/3}{1 + 8\pi n\alpha_{\text{e}}/3}. \tag{5.61}$$

Take the ratio of ω_{L}^2 to ω_{T}^2, and use Eq. (5.57). The result is:

$$\frac{\omega_{\text{L}}^2}{\omega_{\text{T}}^2} = \frac{\epsilon_0}{\epsilon_\infty}. \tag{5.62}$$

This relation is due to Lyddane, Sachs & Teller (1941). The splitting of the longitudinal and transverse modes is due to the long-range electric fields generated by the charges. The Lyddane–Sachs–Teller (LST) relation has been tested by comparing measured vibration frequencies and dielectric constants. It is remarkably accurate.

Frequency dependence of ϵ and polaritons

Now return to Eq. (5.58) and suppose that there is an applied field, $\mathbf{E}$ at frequency ω which corresponds, for example to the (transverse) electric field of long-wavelength incident light. It is easy to see that the solution is now of the form:

$$\eta = \frac{A}{1 - \omega^2/\omega_{\text{T}}^2}E, \tag{5.63}$$

where A is some constant. Thus

$$\epsilon = 1 + 4\pi P_{\text{i}}/E + 4\pi P_{\text{e}}/E = \epsilon_\infty + B/(1 - \omega^2/\omega_{\text{T}}^2),$$

where $B = 4\pi neA$. By putting $\omega = 0$ we identify $B = \epsilon_0 - \epsilon_\infty$ so that

$$\epsilon(\omega) = \epsilon_\infty + \frac{\epsilon_0 - \epsilon_\infty}{1 - \omega^2/\omega_{\text{T}}^2}. \tag{5.64}$$

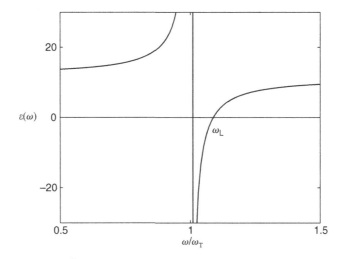

Fig. 5.6   **The dielectric function of Eq. (5.64) as a function of ω/ω_T. The parameters are appropriate for the polar semiconductor GaAs: $\epsilon_0 = 13.1, \epsilon_\infty = 11.1$. Experimental observations are pretty close to this behavior except for damping, which we have neglected; see Yu & Cardona (2001).**

Note that ϵ has a singularity at ω_T. Also, from the LST relation we see that $\epsilon(\omega_L) = 0$. These are general properties of dielectric functions. The dielectric function is plotted in Figure 5.6.

There is a region, $\omega_T < \omega < \omega_L$, for which $\epsilon < 0$, which has an interesting interpretation. A full discussion of crystal optics will be postponed to Section 8.5, below, but the main result that we need is simple to discuss. Suppose a light wave falls on a sample. Recall that the refractive index appears in the equation of a wave as follows:

$$E = E_0 \exp(i[k\mathsf{n}z - \omega t]). \qquad (5.65)$$

Now suppose that $\epsilon < 0$ so that $\mathsf{n} = i\kappa$ is imaginary. Then the wave does not propagate, but is attenuated: $|E| \propto \exp(-kz\kappa)$. In this case, as we will see later, the wave is *reflected*. That is, for a polar material, there will be strong reflection for $\omega_T < \omega < \omega_L$. This is observed. The reflection, which is usually in the infrared, is called the *Reststrahl* (residual ray).

We can go further: in a material medium electromagnetic theory says that the relation between frequency, ω and wavevector k is given by:

$$\omega^2/c^2 = k^2/\epsilon(\omega). \qquad (5.66)$$

If we put our expression for ϵ into this equation we get the result shown in Figure 5.7. The dispersion relation for light in the material is strongly affected by the optically active transverse optical modes. In fact, we should regard the electromagnetic wave as a mixed mode which becomes more and more like an optical vibration as $\omega \to \omega_T$. There is a forbidden region, $\omega_T < \omega < \omega_L$, for which there are no propagating waves. The mixed mode is called a polariton, and is readily observed.

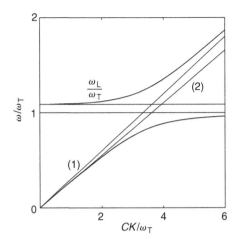

Fig. 5.7 **The dispersion relation for light in a polar crystal with optical properties of GaAs. The solution to Eq. (5.66) is shown. Also curve (1) corresponds to light with refractive index $n = \sqrt{\epsilon_\infty}$ and (2) is for $n = \sqrt{\epsilon_0}$. Note that they are tangent to the upper and lower sections of the curve when the mixing with the optical mode is small.**

5.1.8 Quantum mechanics of lattice vibrations

So far we have neglected quantum mechanics for the lattice vibrations. In quantum mechanics the $\mathbf{u}(\mathbf{R})$, like all position vectors, become operators along with their conjugate momenta, $\mathbf{p} = m\dot{\mathbf{u}}$. (From now on, we will put $B = 1$ for notational simplicity. Everything goes through for the general case.) The quantization rules are:

$$[\mathbf{u}(\mathbf{R}), \mathbf{u}(\mathbf{R}')] = [\mathbf{p}(\mathbf{R}), \mathbf{p}(\mathbf{R}')] = 0 \tag{5.67}$$

$$[p_j(\mathbf{R}), u_{j'}(\mathbf{R}')] = (\hbar/i)\delta_{j,j'}\delta_{\mathbf{R},\mathbf{R}'}.$$

We must write the Hamiltonian in terms of $\mathbf{p}, \mathbf{u}$. For the harmonic approximation the result is:

$$\hat{\mathcal{H}} = \sum_{\mathbf{R}} \frac{1}{2m}p^2(\mathbf{R}) + \sum_{\mathbf{R},\mathbf{R}'} \frac{1}{2}\mathbf{u}(\mathbf{R}) \cdot \mathbf{G}(\mathbf{R} - \mathbf{R}') \cdot \mathbf{u}(\mathbf{R}') \tag{5.68}$$

Einstein phonons

We first use the simplified model of Einstein. He considered each ion to be independently attached to its lattice point. In this case we have:

$$\hat{\mathcal{H}} = \sum_{\mathbf{R}} \left[\frac{1}{2m}p^2(\mathbf{R}) + \frac{m\omega_E^2}{2}u^2(\mathbf{R}) \right]. \tag{5.69}$$

This is the sum of $3N$ independent oscillators. For each one, we can introduce creation and annihilation operators for each direction j:

$$\hat{a}_j(\mathbf{R}) = \frac{1}{\sqrt{2m\hbar\omega_E}}(m\omega_E u_j(\mathbf{R}) + ip_j(\mathbf{R}))$$

$$\hat{a}_j^+(\mathbf{R}) = \frac{1}{\sqrt{2m\hbar\omega_E}}(m\omega_E u_j(\mathbf{R}) - ip_j(\mathbf{R}))$$

$$u_j(\mathbf{R}) = \sqrt{\frac{\hbar}{2m\omega_E}}(\hat{a}_j(\mathbf{R}) + \hat{a}_j^+(\mathbf{R}))$$

$$p_j(\mathbf{R}) = i\sqrt{\frac{m\hbar\omega_E}{2}}(\hat{a}_j(\mathbf{R}) - \hat{a}_j^+(\mathbf{R})). \tag{5.70}$$

Then the Hamiltonian is simple:

$$\hat{\mathcal{H}} = \sum_{\mathbf{R},j} \hbar\omega_E \left[\hat{a}_j^+(\mathbf{R})\hat{a}_j(\mathbf{R}) + 1/2\right]. \tag{5.71}$$

Furthermore the operator (the phonon number operator) $\hat{n}(\mathbf{R},j) = \hat{a}_j^+(\mathbf{R})\hat{a}_j(\mathbf{R})$ has spectrum $0, 1, 2 \ldots$. The eigenstates of $\hat{n}$ are the eigenstates of $\hat{\mathcal{H}}$ and can be labeled by a set of non-negative integers $n(\mathbf{R},j)$. The states can be written as:

$$|n(\mathbf{R}_1,j_1), n(\mathbf{R}_2,j_2), n(\mathbf{R}_3,j_3) \ldots\rangle. \tag{5.72}$$

We have:

$$\hat{n}(\mathbf{R}_k,j)|\ldots, n(\mathbf{R}_k,j) \ldots\rangle = n(\mathbf{R}_k,j)|\ldots, n(\mathbf{R}_k,j) \ldots\rangle$$

$$\hat{a}_j^+(\mathbf{R}_k)|\ldots, n(\mathbf{R}_k,j) \ldots\rangle = \sqrt{n(\mathbf{R}_k,j)+1}|\ldots, n(\mathbf{R}_k,j) \ldots\rangle$$

$$\hat{a}_j(\mathbf{R}_k)|\ldots, n(\mathbf{R}_k,j) \ldots\rangle = \sqrt{n(\mathbf{R}_k,j)}|\ldots, n(\mathbf{R}_k,j) \ldots\rangle \tag{5.73}$$

$$[\hat{a}_j(\mathbf{R}), \hat{a}_{j'}^+(\mathbf{R})] = \delta_{j,j'}\delta_{\mathbf{R},\mathbf{R}'}. \tag{5.74}$$

We interpret the equations by saying that $\hat{a}^+(\mathbf{R},j)$ creates a particle of type j at $\mathbf{R}$. This is called a (localized) phonon. Shortly we will do the same thing for the wave excitations for coupled atoms. In that case we will have phonons that have a dispersion relation identical to that of the classical vibrations.

$$E = \sum_{\mathbf{R},j} \hbar\omega_E [n(\mathbf{R},j) + 1/2]. \tag{5.75}$$

The Einstein model is often used to treat optical phonons. They can have rather flat dispersion relations so that a single frequency approximation is not bad. We can formally make running waves out of our localized phonons in the following way. We introduce new operators which depend on the **k**'s in the Brillouin zone:

$$u_j(\mathbf{R}) = \frac{1}{\sqrt{N}} \sum_{\mathbf{k}} u_j(\mathbf{k}) e^{i\mathbf{k}\cdot\mathbf{R}}$$

$$p_j(\mathbf{R}) = \frac{1}{\sqrt{N}} \sum_{\mathbf{k}} p_j(\mathbf{k}) e^{i\mathbf{k}\cdot\mathbf{R}}$$

$$\hat{a}_j(\mathbf{k}) = \frac{1}{\sqrt{N}} \sum_{\mathbf{k}} \hat{a}_j(\mathbf{R}) e^{i\mathbf{k}\cdot\mathbf{R}}$$

$$\hat{a}_j^+(\mathbf{k}) = \frac{1}{\sqrt{N}} \sum_{\mathbf{k}} \hat{a}_j^+(\mathbf{R}) e^{i\mathbf{k}\cdot\mathbf{R}}. \tag{5.76}$$

These operators create and destroy running waves. Then:

$$\hat{\mathcal{H}} = \sum_j \sum_{\mathbf{k},\mathbf{k}'} \left[\hbar\omega_{\mathrm{E}} \hat{a}_j^+(\mathbf{k}) \hat{a}_j(\mathbf{k}') (1/N) \sum_{\mathbf{R}} e^{i(\mathbf{k}-\mathbf{k}')\cdot\mathbf{R}} \right] \tag{5.77}$$

$$+ 3N\hbar\omega_{\mathrm{E}}/2$$

$$= \sum_{\mathbf{k},j} \hbar\omega_{\mathrm{E}} \left[\hat{a}_j^+(\mathbf{k}) \hat{a}_j(\mathbf{k}) + 1/2 \right].$$

Phonons in real solids

In the Einstein model, quantization was very simple: there is an independent quantized oscillator on each site. The Hamiltonian of Eq. (5.68) looks much more difficult to deal with because different **R**'s are coupled. However we can uncouple them with the normal mode transformation, We will end up with a sum of independent oscillators corresponding to running waves of oscillation.

The essential point will be to express u, p as linear combinations of normal modes, $\mathbf{U}_\lambda(\mathbf{k}) e^{i\mathbf{k}\cdot\mathbf{R}}$. We use these to define new operators for running waves;

$$\mathbf{u}(\mathbf{R}) = \sum_{\mathbf{k},\lambda} \sqrt{\frac{1}{N}} q_{\mathbf{k},\lambda} \mathbf{U}_\lambda(\mathbf{k}) e^{i\mathbf{k}\cdot\mathbf{R}}$$

$$\mathbf{p}(\mathbf{R}) = \sum_{\mathbf{k},\lambda} \sqrt{\frac{1}{N}} p_{\mathbf{k},\lambda} \mathbf{U}_\lambda(\mathbf{k}) e^{i\mathbf{k}\cdot\mathbf{R}} \tag{5.78}$$

The normal modes are a complete set, so this expansion is always possible. The new variables $q_{\mathbf{k},\lambda}, p_{\mathbf{k},\lambda}$ are the amplitudes of the normal modes.

We now express $\hat{\mathcal{H}}$ in terms of these new variables starting with the kinetic energy.

$$\sum_{\mathbf{R}} \frac{1}{2m} p^2(\mathbf{R}) = \sum_{\mathbf{R},\mathbf{k},\mathbf{k}',\lambda,\lambda'} \frac{1}{2mN} p_{\mathbf{k},\lambda} p_{\mathbf{k}',\lambda'} e^{i\mathbf{k}\cdot\mathbf{R}} e^{i\mathbf{k}'\cdot\mathbf{R}} \mathbf{U}_\lambda(\mathbf{k}) \cdot \mathbf{U}_{\lambda'}(\mathbf{k}')$$

$$= \sum_{\mathbf{k},\mathbf{k}',\lambda,\lambda'} \frac{1}{2m} p_{\mathbf{k},\lambda} p_{\mathbf{k}',\lambda'} \delta_{\mathbf{k},-\mathbf{k}'} \mathbf{U}_\lambda(\mathbf{k}) \cdot \mathbf{U}_{\lambda'}(\mathbf{k}')$$

$$= \sum_{\mathbf{k},\lambda} \frac{1}{2m} p_{\mathbf{k},\lambda} p_{-\mathbf{k},\lambda}. \tag{5.79}$$

We have used the orthogonality relation for the $\mathbf{U}$'s. The potential energy is similar.

$$V = \sum_{\mathbf{R},\mathbf{R}'} \frac{1}{2} \mathbf{u}(\mathbf{R}) \cdot \mathsf{G}(\mathbf{R} - \mathbf{R}') \cdot \mathbf{u}(\mathbf{R}')$$

$$= \sum_{\mathbf{R},\mathbf{R}',\mathbf{k},\mathbf{k}',\lambda,\lambda'} \frac{1}{2N} q_{\mathbf{k},\lambda} q_{\mathbf{k}',\lambda'} e^{i\mathbf{k}\cdot\mathbf{R}} e^{i\mathbf{k}'\cdot\mathbf{R}'}$$

$$\times \mathbf{U}_\lambda(\mathbf{k}) \cdot \mathsf{G}(\mathbf{R} - \mathbf{R}') \cdot \mathbf{U}_{\lambda'}(\mathbf{k}'). \tag{5.80}$$

We can reduce some of the inner sums as follows:

$$\sum_{\mathbf{R},\mathbf{R}'} \mathsf{G}(\mathbf{R} - \mathbf{R}') e^{i\mathbf{k}\cdot\mathbf{R}} e^{i\mathbf{k}'\cdot\mathbf{R}'}$$

$$= \sum_{\mathbf{R}} e^{i\mathbf{k}\cdot\mathbf{R}} e^{i\mathbf{k}'\cdot\mathbf{R}} \sum_{\mathbf{R}'} e^{-i\mathbf{k}'\cdot\mathbf{R}} e^{i\mathbf{k}'\cdot\mathbf{R}'} \mathsf{G}(\mathbf{R} - \mathbf{R}')$$

$$= N \delta_{\mathbf{k},-\mathbf{k}'} \mathsf{G}(\mathbf{k}'). \tag{5.81}$$

Thus:

$$V = \sum_{\mathbf{k},\lambda,\lambda'} \frac{1}{2} q_{\mathbf{k},\lambda} q_{-\mathbf{k},\lambda'} \mathbf{U}_\lambda(\mathbf{k}) \cdot \mathsf{G}(-\mathbf{k}) \cdot \mathbf{U}_{\lambda'}(-\mathbf{k})$$

$$= \sum_{\mathbf{k},\lambda,\lambda'} \frac{1}{2} q_{\mathbf{k},\lambda} q_{-\mathbf{k},\lambda'} m \omega_{\lambda'}^2(-\mathbf{k}) \mathbf{U}_\lambda(\mathbf{k}) \cdot \mathbf{U}_{\lambda'}(-\mathbf{k})$$

$$= \sum_{\mathbf{k},\lambda} \frac{m \omega_\lambda^2(\mathbf{k})}{2} q_{\mathbf{k},\lambda} q_{-\mathbf{k},\lambda}. \tag{5.82}$$

To get this result we have used the eigenvalue equation, Eq. (5.20), the orthogonality of the $\mathbf{U}$'s, and the fact (which follows from time-reversal invariance) that ω is even in $\mathbf{k}$.

Now the Hamiltonian is a simple sum of independent modes:

$$\hat{\mathcal{H}} = \sum_{\mathbf{k},\lambda} \left[\frac{1}{2m} p_{\mathbf{k},\lambda} p_{-\mathbf{k},\lambda} + \frac{m \omega_\lambda^2(\mathbf{k})}{2} q_{\mathbf{k},\lambda} q_{-\mathbf{k},\lambda} \right]. \tag{5.83}$$

The coupling of $\mathbf{k}$ and $-\mathbf{k}$ is a trivial matter which we will dispose of at once. It arises from the fact that $\mathbf{u}$ is real, so that $u_{\mathbf{k}} = u_{-\mathbf{k}}^*$ and likewise for $p_{\mathbf{k}}$.

Now we quantize. The commutation relations translate into:

$$[p_{\mathbf{k},\lambda}, q_{\mathbf{k}',\lambda'}] = (\hbar/i) \delta_{\mathbf{k},\mathbf{k}} \delta_{\lambda,\lambda'}. \tag{5.84}$$

We can define creation and annihilation operators:

$$q_{\mathbf{k},\lambda} = \sqrt{\frac{\hbar}{2m\omega_{\mathbf{k},\lambda}}}(\hat{a}_{\mathbf{k},\lambda} + \hat{a}^+_{-\mathbf{k},\lambda})$$

$$p_{\mathbf{k},\lambda} = i\sqrt{\frac{m\hbar\omega_{\mathbf{k},\lambda}}{2}}(\hat{a}_{\mathbf{k},\lambda} - \hat{a}^+_{-\mathbf{k},\lambda}). \tag{5.85}$$

This implies, for example:

$$\mathbf{u}(\mathbf{R}) = \sum_{\mathbf{k},\lambda}\sqrt{\frac{\hbar}{2m\omega_{\mathbf{k},\lambda}N}}(\hat{a}(\mathbf{k},\lambda) + \hat{a}^+(-\mathbf{k},\lambda))\mathbf{U}_\lambda(\mathbf{k})e^{i\mathbf{k}\cdot\mathbf{R}}$$

$$= \sum_{\mathbf{k},\lambda}\sqrt{\frac{\hbar}{2m\omega_{\mathbf{k},\lambda}N}}(\mathbf{U}_\lambda(\mathbf{k})e^{i\mathbf{k}\cdot\mathbf{R}}\hat{a}(\mathbf{k},\lambda) + \mathbf{U}^*_\lambda(\mathbf{k})e^{-i\mathbf{k}\cdot\mathbf{R}}\hat{a}^+(\mathbf{k},\lambda)). \tag{5.86}$$

The Hamiltonian is now diagonal:

$$\hat{\mathcal{H}} = \sum_{\mathbf{k},\lambda}\hbar\omega_{\mathbf{k},\lambda}(\hat{a}^+_{\mathbf{k},\lambda}\hat{a}_{\mathbf{k},\lambda} + 1/2). \tag{5.87}$$

This is the Hamiltonian for a set of independent simple harmonic oscillators.

Again, as in the Einstein case, the number operator $\hat{n}_{\mathbf{k},\lambda} \equiv \hat{a}^+_{\mathbf{k},\lambda}\hat{a}_{\mathbf{k},\lambda}$ has spectrum $0, 1, 2\ldots$. The eigenstates of $\hat{n}$ are the eigenstates of $\hat{\mathcal{H}}$ and can be labeled by a set of non-negative integers $n_{\mathbf{k},\lambda}$. The states can be written as:

$$\left| n_{\mathbf{k}1,\lambda1}, n_{\mathbf{k}2,\lambda2}, n_{\mathbf{k}3,\lambda3}, \ldots \right\rangle. \tag{5.88}$$

Furthermore:

$$\hat{n}_{\mathbf{k},\lambda}\left|\ldots, n_{\mathbf{k},\lambda}\ldots\right\rangle = n_{\mathbf{k},\lambda}\left|\ldots, n_{\mathbf{k},\lambda}\ldots\right\rangle$$

$$\hat{a}^+_{\mathbf{k},\lambda}\left|\ldots, n_{\mathbf{k},\lambda}\ldots\right\rangle = \sqrt{n_{\mathbf{k},\lambda}+1}\left|\ldots, n_{\mathbf{k},\lambda}\ldots\right\rangle$$

$$\hat{a}_{\mathbf{k},\lambda}\left|\ldots, n_{\mathbf{k},\lambda}\ldots\right\rangle = \sqrt{n_{\mathbf{k},\lambda}}\left|\ldots, n_{\mathbf{k},\lambda}\ldots\right\rangle \tag{5.89}$$

$$[\hat{a}_{\mathbf{k},\lambda}, \hat{a}^+_{\mathbf{k}',\lambda'}] = \delta_{\mathbf{k},\mathbf{k}'}\delta_{\lambda,\lambda'}. \tag{5.90}$$

We can interpret the equations by saying that $\hat{a}^+(\mathbf{R}, j)$ creates a particle of type j at $\mathbf{R}$. This is called a *phonon*. Phonons that have a dispersion relation identical to that of the classical vibrations. A phonon is *not* the same as a sound wave. What we are doing is quantizing the amplitude of the vibration. A sound wave is *many* phonons in the same state. This is possible because phonons are bosons, as the commutation relation in Eq.(5.90) shows. As above, the energy of any state is the sum of the energies of the phonons which are excited:

$$E = \sum_{\mathbf{k},\lambda}\hbar\omega_{\mathbf{k},\lambda}\left[n_{\mathbf{k},\lambda} + 1/2\right]. \tag{5.91}$$

The first term in the sum is the energy of the phonons. Each phonon has energy $\hbar\omega_{\mathbf{k},\lambda}$. The last term is the *zero-point* energy, i.e., the energy of the atoms in the crystal due to quantum fluctuations.

Beyond the harmonic approximation

In the harmonic approximation we have a set of independent bosons, i.e. phonons. Phonons do not interact. If we go to higher orders in $\mathbf{u}$ in the expansion of V we would get phonon-phonon interactions. Explicitly, if we do the normal mode transformation on the quadratic terms in V, and then express the higher order terms in terms of the $\hat{a}^+, \hat{a}$ we generate couplings. For example, in third order we would generate terms like

$$\hat{a}^+_{\mathbf{k}'',\lambda''}\hat{a}^+_{\mathbf{k}',\lambda'}\hat{a}_{\mathbf{k},\lambda}. \tag{5.92}$$

We can interpret this as a scattering term: a phonon in $\mathbf{k},\lambda$ is destroyed and two others in $\mathbf{k}',\lambda'$ and $\mathbf{k}'',\lambda''$ are created. This contributes to sound attenuation.

These are weak effects for small $\mathbf{k}$, though for zone-edge phonons there are lots of interactions. Many physical effects such as sound attenuation and thermal expansion require going beyond the harmonic approximation.

Crystal momentum

We can think of $\hbar\omega_{\mathbf{k},\lambda}$ as the energy of each phonon. We are tempted to think of $\hbar\mathbf{k}$ as the momentum of the particle. This is not really correct since the actual kinetic momentum of all the atoms in a crystal is zero if it is excited with wavevector $\mathbf{k}$. To see this, we set one of the $q_{\mathbf{k},\lambda}$ to be non-zero, and calculate the total momentum:

$$\sum_{\mathbf{R}} m\dot{\mathbf{u}}(\mathbf{R}, t) = \sqrt{\frac{1}{N}}\dot{q}_{\mathbf{k},\lambda}(t)\mathbf{U}_\lambda(\mathbf{k})\sum_{\mathbf{R}} e^{i\mathbf{k}\cdot\mathbf{R}} = 0. \tag{5.93}$$

The only mode that carries momentum is $\mathbf{k} = 0$.

Nevertheless, $\hbar\mathbf{k}$ is a significant quantity. It is called *crystal momentum* and it is a conserved quantity associated with the invariance of crystals under translations by lattice vectors, $\mathbf{R}$. In the basis of the states $|n\rangle$ the operator that measures the total $\mathbf{k}$ is:

$$\mathcal{K} = \sum \mathbf{k}\hat{n}(\mathbf{k}). \tag{5.94}$$

We now show that the translation operator in this basis is:

$$\hat{\tau}(\mathbf{R}) = \exp\left(-i\mathcal{K}\cdot\mathbf{R}\right). \tag{5.95}$$

To demonstrate this, we recall from quantum mechanics that if we prefer to transform operators rather than states we can look at $\hat{\mathcal{O}}_\tau = \hat{\tau}(\mathbf{R})\hat{\mathcal{O}}\hat{\tau}^{-1}(\mathbf{R})$, where $\hat{\mathcal{O}}$ is any operator.

Consider matrix elements of the creation operator:

$$\langle n_{\mathbf{k}} + 1 | \hat{a}_\tau^+(\mathbf{k}) | n_{\mathbf{k}} \rangle = \langle n_{\mathbf{k}} + 1 | \exp\left(-i\mathcal{K} \cdot \mathbf{R}\right) \hat{a}^+(\mathbf{k}) \exp\left(i\mathcal{K} \cdot \mathbf{R}\right) | n_{\mathbf{k}} \rangle$$
$$= \exp(-i\mathbf{k} \cdot \mathbf{R}) \langle n_{\mathbf{k}} + 1 | \hat{a}^+(\mathbf{k}) | n_{\mathbf{k}} \rangle. \tag{5.96}$$

Thus, using $\hat{\tau}$ to transform $\hat{a}^+$ is the same as multiplying by $\exp(-i\mathbf{k} \cdot \mathbf{R})$. With this in hand we can transform $\mathbf{u}(\mathbf{S})$:

$$\mathbf{u}_\tau(\mathbf{S}) = \sum_{\mathbf{k},\lambda} \sqrt{\frac{\hbar}{2m\omega_{\mathbf{k},\lambda}N}} (\hat{a}_\tau(\mathbf{k},\lambda) + \hat{a}_\tau^+(-\mathbf{k},\lambda)) \mathbf{U}_\lambda(\mathbf{k}) e^{i\mathbf{k}\cdot\mathbf{S}}$$

$$= \sum_{\mathbf{k},\lambda} \sqrt{\frac{\hbar}{2m\omega_{\mathbf{k},\lambda}N}} (\hat{a}(\mathbf{k},\lambda) + \hat{a}^+(-\mathbf{k},\lambda)) \mathbf{U}_\lambda(\mathbf{k}) e^{i\mathbf{k}\cdot[\mathbf{S}+\mathbf{R}]}$$

$$\equiv \mathbf{u}(\mathbf{S} + \mathbf{R}). \tag{5.97}$$

So $\hat{\tau}(\mathbf{R})$ really does translate by $\mathbf{R}$.

Now consider two states $|a\rangle, |b\rangle$ where $\mathcal{K}|a\rangle = \mathbf{k}_a|a\rangle$, $\mathcal{K}|b\rangle = \mathbf{k}_b|b\rangle$, and write down the matrix element of a periodic operator between them. Note that $\tau(\mathbf{R})$ commutes with any such operator.

$$\langle b | \hat{\mathcal{H}}_1 | a \rangle = \langle b | \tau(\mathbf{R}) \hat{\mathcal{H}}_1 \tau(-\mathbf{R}) | a \rangle$$
$$= \exp(i[\mathbf{k}_a - \mathbf{k}_b] \cdot \mathbf{R}) \langle b | \hat{\mathcal{H}}_1 | a \rangle. \tag{5.98}$$

If we think of this as a scattering matrix element for the process $a \to b$ we have two alternatives: either the matrix element is zero, or the phase factor is unity. This gives the important *conservation law for crystal momentum*:

$$\mathbf{k}_b = \mathbf{k}_a + \mathbf{G}. \tag{5.99}$$

For example suppose we go to cubic terms in the expansion of V. Since we have terms in u^3, this part of $\hat{\mathcal{H}}$ will contain terms like $\hat{a}_{\mathbf{k}_3}^+ \hat{a}_{\mathbf{k}_2}^+ \hat{a}_{\mathbf{k}_1}$. That is, a phonon with $\mathbf{k}_1$ can decay into two phonons with $\mathbf{k}_2$ and $\mathbf{k}_3$. However, the conservation law demands that

$$\mathbf{k}_1 = \mathbf{k}_2 + \mathbf{k}_3 + \mathbf{G}.$$

Processes with $\mathbf{G} = 0$ are called N-processes (Normal), and otherwise we have U-processes. U stands for the German word *Umklapp*, which means a sudden change in direction.

5.1.9 Thermal effects due to phonons

In thermal equilibrium at temperature T, each phonon state will have a probability to be excited. The spectrum of each phonon is that of a harmonic oscillator. Using this fact, and

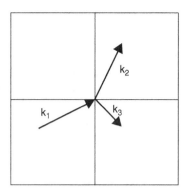
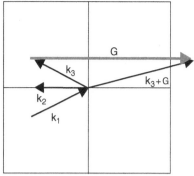

Fig. 5.8   **Left, N ($k_1 = k_2 + k_3$). Right, U ($k_1 = k_2 + k_3 + G$).**

elementary statistical mechanics we see that:

$$\langle \hat{n} \rangle_T = \frac{\sum_{n=0}^{\infty} n e^{-\beta \hbar \omega (n + 1/2)}}{\sum_{n=0}^{\infty} e^{-\beta \hbar \omega (n + 1/2)}}$$

$$= \frac{1}{e^{\beta \hbar \omega} - 1}. \tag{5.100}$$

Here, $\beta = 1/(k_B T)$.

When a solid is heated its internal energy, U, increases. For insulators most of the increase of U is due to the thermally excited phonon states. Thus:

$$U(T) = U_o + \sum_{\mathbf{k}, \lambda} \hbar \omega_{\mathbf{k}, \lambda} \langle \hat{n}_{\mathbf{k}, \lambda} \rangle_T. \tag{5.101}$$

U_o includes chemical energy, etc. Using Eq. (5.100) we have:

$$U(T) = U(T = 0) + \sum_{\mathbf{k}, \lambda} \frac{\hbar \omega_{\mathbf{k}, \lambda}}{e^{\beta \hbar \omega_{\mathbf{k}, \lambda}} - 1}. \tag{5.102}$$

The heat capacity, $C_V = \partial U / \partial T |_V$ is easy to measure. It had great historical significance in sorting out quantum effects in solids. For temperatures such that $k_B T \gg \hbar \omega_{max}$ where ω_{max} is the largest frequency present, we can expand the exponential and arrive at simple classical formula (the law of P. Dulong and A. Petit):

$$U - U(0) \rightarrow k_B T \sum_{\mathbf{k}, \lambda} 1 = 3NB k_B T; \quad C_V \rightarrow 3NB k_B. \tag{5.103}$$

Recall that $3NB$, is the total number of modes (three times the total number of atoms) because there are N values of $\mathbf{k}$ in the Brillouin zone.

Low-temperature heat capacity

In the Einstein model the low-temperature heat capacity is small because there is an energy gap, $\hbar \omega_E$, between the ground state and the first excited state. C_V is dominated by the

Boltzmann factor required to make an excitation on any site: $C_V \propto \exp(-\hbar\omega_E/k_B T)$. In fact, $\lim_{T\to 0} C_V = 0$ is experimentally correct, and this is in violent contrast with the classical theory above. However, for real solids we have sound waves whose frequency vanishes for long wavelength. Correspondingly, the energy of small $\mathbf{k}$ phonons goes to 0, and there is no energy gap. Nevertheless, the heat capacity vanishes, though the functional form is different.

At low T the sum in Eq. (5.102) will be dominated by the terms associated with low frequency modes. There are three such modes for each $\mathbf{k}$, and their energy is $\hbar v_\lambda k$, where, as above, λ is the mode index. For small T we can approximate the sum by:

$$\sum_{\mathbf{k},\lambda} \frac{\hbar\omega_{\lambda,\mathbf{k}}}{e^{\beta\hbar\omega_{\lambda,\mathbf{k}}-1}} = \frac{\Omega}{(2\pi^3)} \sum_\lambda \int_{BZ} \frac{\hbar\omega_{\lambda,\mathbf{k}}}{e^{\beta\hbar\omega_{\lambda,\mathbf{k}}-1}} d^3k \qquad (5.104)$$

$$\approx \frac{\Omega}{(2\pi^3)} \sum_\lambda \int \frac{\hbar v_\lambda k}{e^{\beta\hbar v_\lambda k} - 1} d^3k$$

$$\propto \sum_\lambda v_\lambda \int_0^\infty \frac{k}{e^{\beta\hbar v_\lambda k} - 1} k^2 dk$$

$$\propto T^4 \sum_\lambda v_\lambda^{-3} \int_0^\infty \frac{y^3}{e^y - 1} dy. \qquad (5.105)$$

In the last step we have made the change of variables $y = \beta\hbar v_\lambda k$. We have, in general, $\lim_{T\to 0} U(T) = U(0) + AT^4$, and $\lim_{T\to 0} C_V = 4AT^3$ where A is a constant, in agreement with experiment.

The Debye model

For arbitrary temperature we should evaluate the integral over the Brillouin zone in Eq. (5.104) numerically. Long before this was possible in a practical way, Debye found a useful interpolation formula which gives a good account for all T. He noted that for large T, the details of the spectrum do not matter: equipartition tells us that the important thing to get right is the total number of modes. At low T we are dominated by the acoustic modes. So Debye made two simplifications:

(i) Replace the Brillouin zone by a sphere with the same volume. This conserves the number of modes. From Eq. (3.12) we see that if k_D is the radius of the Debye sphere:

$$4\pi k_D^3/3 = (2\pi)^3/v_c. \qquad (5.106)$$

(ii) Replace the real spectrum by an isotropic spectrum with a mean sound speed:

$$\omega_{\mathbf{k},\lambda} \to \bar{v}k. \qquad (5.107)$$

From the last line of Eq. (5.105) it is evident that we need to choose $3\bar{v}^{-3} = \sum_\lambda v_\lambda^{-3}$.

This model is often parameterized by the maximum energy of the acoustic modes at the Debye sphere, converted to temperature units. This defines the *Debye temperature*, Θ as follows:

$$\hbar \bar{v} k_D = k_B \Theta. \tag{5.108}$$

Using the Debye approximation in Eq. (5.104) we find, after some algebra:

$$U = 9Nk_B T \left(\frac{T}{\Theta}\right)^3 \int_0^{\Theta/T} \frac{y^3 dy}{e^y - 1}. \tag{5.109}$$

This is the Debye formula. For the heat capacity we need to differentiate with respect to T:

$$C_V = 9Nk_B \left(\frac{T}{\Theta}\right)^3 \int_0^{\Theta/T} \frac{y^4 e^y dy}{(e^y - 1)^2}. \tag{5.110}$$

It is easy to check that we get the low temperature limit for $T << \Theta$, i.e., $C_V \propto T^3$. For $T >> \Theta$ we recover the classical value.

In practice, Eqs. (5.109) and (5.110) are used as fitting functions. There is one parameter, the Debye temperature, Θ. This one parameter fit is quite accurate for most cases. For many materials $\Theta \approx 300K$. For crystals with optical modes we can add modes with a constant frequency (the Einstein model). For a typical crystal optical modes are frozen out except at quite large T.

The phonon density of states

There is another way to look at thermal properties which is useful in a general context. Consider the internal energy, from above:

$$U(T) - U(0) = \sum_{\mathbf{k},\lambda} \frac{\hbar \omega_{\lambda,\mathbf{k}}}{e^{\beta \hbar \omega_{\lambda,\mathbf{k}}} - 1}$$

$$= \frac{\Omega}{(2\pi^3)} \sum_\lambda \int_{BZ} \frac{\hbar \omega_{\lambda,\mathbf{k}}}{e^{\beta \hbar \omega_{\lambda,\mathbf{k}}} - 1} d^3k. \tag{5.111}$$

Note that the integrand is a function of the phonon energy, $\hbar \omega_{\lambda,\mathbf{k}}$ alone. This leads us to change variables so that the integral will be over energy. To this end we define the *density of states* by:

$$\mathcal{D}(E) = \sum_{\mathbf{k},\lambda} \delta(E - \hbar \omega_{\lambda,\mathbf{k}}). \tag{5.112}$$

Now it is clear that:

$$U(T) - U(0) = \int_0^\infty \mathcal{D}(E) E \frac{1}{e^{\beta E} - 1} dE. \tag{5.113}$$

Once $\mathcal{D}(E)$ is calculated for a given spectrum, any thermal average can be performed.

For small E it is possible to find a general formula for $\mathcal{D}(E)$ because the low energy excitations are acoustic phonons. We have, in d dimensions:

$$\mathcal{D}(E) \propto \sum_{\lambda} \int \delta(E - \hbar v_{\lambda} k) dE \propto E^{d-1}. \tag{5.114}$$

As an example we give the density of states for the Debye model:

$$\mathcal{D}(E) = 3 \frac{\Omega}{2\pi^2} \frac{1}{\hbar \bar{v}} \left(\frac{E}{\hbar \bar{v}} \right)^2. \tag{5.115}$$

Some authors define the density of states in frequency, with $\delta(\omega - \omega_{\lambda,\mathbf{k}})$ in the equation above. In this notation, for the Debye model we get

$$\mathcal{D}(\omega) = 3\Omega\omega^2 / (2\pi^2 \bar{v}^3).$$

5.2 Spin waves and magnons

Magnets also have wave-like motion of the magetization. In this section we will give some elementary ideas about this subject. Our goal here is to compare these modes to the phonons we have just studied. A number of effects that are important in real applications to magnets such as anisotropy, demagnetization factors, etc. will be ignored. See White (1970) for these details.

5.2.1 Spin waves

It is easy to give a simple treatment for spin waves in the classical domain. Consider a one-dimensional chain of spins coupled ferromagnetically along the z-axis. As in the case of lattice vibrations we look for small deviations from the ground state. In the ground state the spins have only a z-component. We will look for states with small, but non-zero x and y components.

The Heisenberg Hamiltonian, in this case, looks like:

$$\hat{\mathcal{H}} = -2J \sum_{l} \mathbf{s}_l \cdot \mathbf{s}_{l+1}. \tag{5.116}$$

In the spirit of molecular field theory, we think of each spin as seeing an effective field due to exchange. The term involving spin j can be written:

$$-2J\mathbf{s}_j \cdot (\mathbf{s}_{j-1} + \mathbf{s}_{j+1}) = -\gamma\hbar\mathbf{s}_j \cdot (2J/(\gamma\hbar))(\mathbf{s}_{j-1} + \mathbf{s}_{j+1}). \tag{5.117}$$

By comparison with Eq. (2.6) we see that the exchange field on spin j is

$$\mathbf{H}_j = (2J/(\gamma\hbar))(\mathbf{s}_{j-1} + \mathbf{s}_{j+1}). \tag{5.118}$$

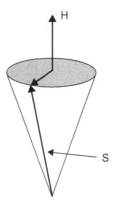

Fig. 5.9 **Precession of a classical spin in a magnetic field. The spin lies on a cone, and rotates around the field with frequency ω_L.**

The classical equation of motion is that the rate of change of the spin angular momentum is the torque, i.e.

$$\hbar d\mathbf{s}/dt = \gamma \hbar \mathbf{s} \times \mathbf{H}. \tag{5.119}$$

For a single spin in a magnetic field the solution of this equation corresponds to precession around the direction of $\mathbf{H}$ (see Figure 5.9). This is easy to see. For $\mathbf{H}$ in the z-direction we have:

$$
\begin{aligned}
ds^z/dt &= 0 \\
ds^x/dt &= -\gamma H s^y \\
ds^y/dt &= \gamma H s^x.
\end{aligned}
\tag{5.120}
$$

Set $s^+ = s^x + is^y$, and $\omega_L = |\gamma| H$. Then

$$ds^+/dt = i\omega_L s^+; \quad s^+(t) = s^+(0)\exp(i\omega_L t). \tag{5.121}$$

That is, the x and y components of the vector lie on a circle and rotate around the z-axis with frequency ω_L, the Larmor frequency.

For our coupled chain we find;

$$\frac{d\mathbf{s}_j}{dt} = \gamma \mathbf{s}_j \times (2J/\gamma\hbar)(\mathbf{s}_{j-1} + \mathbf{s}_{j+1}) = \frac{2J}{\hbar}(\mathbf{s}_j \times \mathbf{s}_{j-1} + \mathbf{s}_j \times \mathbf{s}_{j+1}). \tag{5.122}$$

We now linearize Eq. (5.122) by using the fact that $s \approx s^z \gg s^x, s^y$. Writing the equation in terms of components and keeping terms to first order in s^x, s^y, and setting $\omega_0 = 2Js/\hbar$

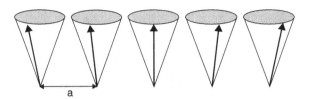

Fig. 5.10 **Spin wave in one dimension. Each spin precesses around the direction of the magnetization with frequency $\omega(k)$ and with a phase that advances by ka for each step along the chain.**

we get:

$$ds^z/dt = 0$$
$$ds_j^x/dt = -\omega_0(s_{j-1}^y + s_{j+1}^y - 2s_j^y)$$
$$ds_j^y/dt = \omega_0(s_{j-1}^x + s_{j+1}^x - 2s_j^x). \tag{5.123}$$

We seek a solution in the same way we solved Eq. (5.120), only for coupled spins. Set $s_j^+ = s_j^x + is_j^y$. Then:

$$ds_j^+/dt = i\omega_0(s_{j-1}^+ + s_{j+1}^+ - 2s_j^+). \tag{5.124}$$

Now, as in the lattice vibration case, we do a normal mode analysis:

$$s_j^+(t) = s^+(0)\exp(i[kx_j - \omega(k)t]).$$

This solution corresponds to a wave of precession as in Figure 5.10. Inserting this trial expression into Eq. (5.124) we find that we have a solution if:

$$\omega(k) = 2\omega_0(1 - \cos(ka)). \tag{5.125}$$

This is the dispersion relation for spin waves in a one-dimensional chain. Note that for $k \to 0$, long wavelengths, $\omega(k) \propto k^2$, in contrast to phonons.

For a three-dimensional lattice, a similar calculation gives:

$$\omega(\mathbf{k}) = \frac{2Js}{\hbar}\sum_\delta[1 - \cos(\mathbf{k}\cdot\delta)], \tag{5.126}$$

where, as usual, the δ's are the vectors that point to the nearest neigbors.

For an antiferromagnetic chain the calculation is not much more difficult. The result is that the spin wave frequency is linear in k at long wavelengths.

5.2.2 Magnons

The classical theory of spin waves is, if anything, simpler than that of lattice vibrations. However, the quantum theory is more complicated because the spin operators are complicated quantum objects. Nevertheless, a relatively straightforward theory exists, which we now present. It will be an expansion around the classical case (large spin).

Let us recall the quantum mechanics of spin operators. The spin vector is an operator which obeys $\hat{\mathbf{s}} \cdot \hat{\mathbf{s}} = s(s+1)$ with s an integer or half integer. It is useful to use the raising and lowering operators: $\hat{s}^{\pm} = \hat{s}_x \pm i\hat{s}_y$. Recall that $\hat{s}^-$ decreases m_z, the projection along the z-axis by 1. In the classical treatment above we found that spin waves correspond to tilting the spin vector away from the ordering direction by a small amount. Thus we expect that $\hat{s}^-$ will be involved in creating a quantized spin wave, a magnon.

To make this precise, recall the commutation relation for these operators:

$$[\hat{s}^+, \hat{s}^-] = 2\hat{s}^z. \tag{5.127}$$

If we want to make boson operators to make quantized spin waves we should have $\hat{a} = \hat{s}^+/\sqrt{2\hat{s}^z}, \hat{a}^+ = \hat{s}^-/\sqrt{2\hat{s}^z}$, because then we have the expected commutation relation, Eq. (5.90).

To accomplish this we introduce the transformation due to T. Holstein and H. Primakoff: we introduce boson operators $\hat{a}, \hat{a}^+$ and put:

$$\hat{s}^+/\sqrt{2s} = (1 - \hat{a}^+\hat{a}/2s)^{1/2}\hat{a}$$
$$\hat{s}^-/\sqrt{2s} = \hat{a}^+(1 - \hat{a}^+\hat{a}/2s)^{1/2}. \tag{5.128}$$

It is easy to see $\hat{s}^{\pm}$ obey Eq. (5.127). Further:

$$\hat{s}^z = s - \hat{a}^+\hat{a}. \tag{5.129}$$

That is, the number of bosons gives the spin deviation from the z-direction. In what follows we will linearize for the case of large spin. (In practice, spin 5/2 is already pretty large). This amounts to replacing the square roots on the right side of Eq. (5.128) by unity.

All of this is for a single spin. Now consider a Bravais lattice of spins of N spins labeled by j. We expect a running wave of spin deviation to be the normal mode. Define:

$$\hat{b}_{\mathbf{k}} = \frac{1}{\sqrt{N}} \sum_j e^{i\mathbf{k}\cdot\mathbf{r}_j}\hat{a}_j; \quad \hat{b}_{\mathbf{k}}^+ = \frac{1}{\sqrt{N}} \sum_j e^{-i\mathbf{k}\cdot\mathbf{r}_j}\hat{a}_j^+. \tag{5.130}$$

It is easy to show that these are independent bosons:

$$[\hat{b}_{\mathbf{k}}, \hat{b}_{\mathbf{q}}] = 0 = [\hat{b}_{\mathbf{k}}^+, \hat{b}_{\mathbf{q}}^+]; \quad [\hat{b}_{\mathbf{k}}, \hat{b}_{\mathbf{q}}^+] = \delta_{\mathbf{k},\mathbf{q}}. \tag{5.131}$$

In the linear approximation:

$$\hat{s}_j^+ = \sqrt{2s/N} \sum_{\mathbf{k}} e^{-i\mathbf{k}\cdot\mathbf{r}_j} \hat{b}_{\mathbf{k}}$$

$$\hat{s}_j^- = \sqrt{2s/N} \sum_{\mathbf{k}} e^{i\mathbf{k}\cdot\mathbf{r}_j} \hat{b}_{\mathbf{k}}^+$$

$$\hat{s}_j^z = s - (1/N) \sum_{\mathbf{k},\mathbf{q}} e^{i(\mathbf{k}-\mathbf{q})\cdot\mathbf{r}_j} \hat{b}_{\mathbf{k}}^+ \hat{b}_{\mathbf{q}}. \tag{5.132}$$

We will need the z-component of the total spin which we get by summing the last equation above:

$$\hat{S}^z = \sum \hat{s}_j^z = Ns - \sum \hat{b}_{\mathbf{k}}^+ \hat{b}_{\mathbf{k}}. \tag{5.133}$$

The Heisenberg Hamiltonian is given by Eq.(2.3); it involves the quantity:

$$\hat{\mathbf{s}}_j \cdot \hat{\mathbf{s}}_{j+\delta} = \hat{s}_j^z \hat{s}_{j+\delta}^z + \frac{1}{2}(\hat{s}_j^+ \hat{s}_{j+\delta}^- + \hat{s}_j^- \hat{s}_{j+\delta}^+). \tag{5.134}$$

If we substitute from Eq. (5.132) and keep only terms that are bilinear in the $\hat{b}$'s we get:

$$\hat{\mathcal{H}} = -JNzs^2 + 2Jzs \sum_{\mathbf{k}} [1 - \gamma_{\mathbf{k}}] \hat{b}_{\mathbf{k}}^+ \hat{b}_{\mathbf{k}},$$

$$\gamma_{\mathbf{k}} = \frac{1}{z} \sum_{\delta} e^{i\mathbf{k}\cdot\delta} = \frac{1}{z} \sum_{\delta} \cos(\mathbf{k} \cdot \delta). \tag{5.135}$$

We have used the fact that a Bravais lattice is inversion symmetric.

Eq. (5.135) is precisely analogous to Eq. (5.87). There are quantized spin waves and their dispersion relation is:

$$\hbar\omega_{\mathbf{k}} = 2Jzs(1 - \gamma_{\mathbf{k}}), \tag{5.136}$$

the same as the classical expression, Eq. (5.126).

The effects that we have left out by dropping the square root in Eq. (5.128) and the four-operator terms, above, give rise to magnon-magnon interactions.

The theory of magnons in antiferromagnets is complicated further by the fact that a simple transcription of the classical Néel state is not an eigenstate of the Hamiltonian. This is easy to see: for a one-dimensional chain the state $|s, -s, s, \ldots\rangle$ is mixed with $|s - 1, -s + 1, \ldots\rangle$ by the term $\hat{s}_1^- \hat{s}_2^+$. For a treatment of this point and the full theory the student is referred to the literature, e.g., Kittel (1963).

5.2.3 Thermal magnons

As in the case of phonons, low-temperature thermal properties of magnets show evidence of excitations of the normal modes. We will give two examples.

There is a magnon contribution to the heat capacity. If we consider a ferromagnet and set $\omega_{\mathbf{k}} = Dk^2$ then we have, for the magnetic part of the internal energy:

$$U_M(T) - U_M(0) = \sum_{\mathbf{k}} \hbar\omega_{\mathbf{k}} \langle \hat{b}_{\mathbf{k}}^+ \hat{b}_{\mathbf{k}} \rangle = \frac{\Omega}{2\pi^3} \int d^3k\, \hbar Dk^2 / (e^{\beta\hbar Dk^2} - 1). \qquad (5.137)$$

As usual, for low temperature we can extend the limit on the integral to infinity and make a change of variables, $y^2 = \beta\hbar Dk^2$. A simple computation gives:

$$U(T) - U_M(0) = \frac{\Omega}{2\pi^2} \frac{(k_B T)^{5/2}}{(\hbar D)^{3/2}} \int_0^\infty \frac{y^4 dy}{e^{y^2} - 1}; \quad C_M \propto T^{3/2}. \qquad (5.138)$$

This has been experimentally verified.

Each excited magnon decreases the magnetization. Thus, near $T = 0$ we can write, using Eq. (5.133):

$$M_z = \frac{\gamma\hbar}{\Omega} \langle \hat{S}^z \rangle = \frac{\gamma\hbar}{\Omega} \left(Ns - \sum_{\mathbf{k}} \langle \hat{b}_{\mathbf{k}}^+ \hat{b}_{\mathbf{k}} \rangle \right). \qquad (5.139)$$

With the same change of variables as before:

$$M_z(0) - M_z(T) = \frac{\gamma\hbar}{2\pi^3} \int \frac{d^3k}{e^{\beta\hbar Dk^2} - 1} = \frac{\gamma\hbar}{2\pi^2} \left(\frac{k_B T}{\hbar D} \right)^{3/2} \int_0^\infty \frac{y^2 dy}{e^{y^2} - 1}. \qquad (5.140)$$

This is the *Bloch $T^{3/2}$ law* which is confirmed by experiment. Note that the magnetization decreases from its saturation value much faster than predicted by the molecular field theory of Eq. (2.10).

5.3 Neutron scattering

In an earlier chapter we discussed Bragg scattering, but when we treated the vibration of the atoms we used the classical Einstein approximation. Now we can give a treatment of the Debye–Waller factor using real phonons. We will do this in the context of neutron scattering.

Neutrons with energies on the order of $k_B T$ (where T is room temperature) have a de Broglie wavelength of the order of the lattice spacing of solids. They can be, and are, used for structure determination in the same way as X-rays. However, neutrons have a significant advantage over X-rays. A typical phonon energy, in temperature units, is Θ, is also of the order of room temperature. Thus, *inelastic* neutron scattering has a special role: we can imagine scattering from a crystal and emitting a phonon. The energy shift of the scattered neutron is of the order of the incident energy, and thus easy to detect. This is quite unlike the case of X-rays where the energy shift is a very small fraction of the incident energy. For a magnetic crystal the magnetic scattering of neutrons allows magnons to be detected in the same way.

We start by writing down the general expression for the neutron scattering cross-section in the Born approximation (cf. Eq. (3.37)):

$$\frac{d^2\sigma}{d\Theta d\omega} = \left[\frac{d^2\sigma}{d\Omega d\epsilon}\right]_1 \frac{1}{\hbar}\sum_{i,f} p_i \sum_{\mathbf{R},\mathbf{S}} \langle i|e^{i\mathbf{q}\cdot\mathbf{S}}|f\rangle\langle f|e^{-i\mathbf{q}\cdot\mathbf{R}}|i\rangle\delta(\omega + (E_f - E_i)/\hbar)$$

$$\equiv \left[\frac{d^2\sigma}{d\Omega d\epsilon}\right]_1 \frac{N}{\hbar}S(\mathbf{q},\omega). \tag{5.141}$$

We have summed over final states and averaged over initial states with a probability p_i, i.e. a Boltzmann factor for systems in thermal equilibrium. The quantity $S(\mathbf{q},\omega)$ is called the dynamic structure factor. The energy transfer to or from the crystal is $\hbar\omega$. It is a generalization of $S(\mathbf{q})$ in Eq. (3.38): $\int d\omega\, S(\mathbf{q},\omega) = S(\mathbf{q})$.

We can put S in a more appealing form by replacing the δ function with its integral representation:

$$\delta(x) = \int_{-\infty}^{\infty} e^{ixt}dt/(2\pi). \tag{5.142}$$

Further, we recall the Heisenberg picture of quantum mechanics which associates a time dependence with operators:

$$e^{i(E_f - E_i)t}\langle f|\hat{O}|i\rangle = \langle f|\hat{O}(t)|i\rangle, \tag{5.143}$$

where $\hat{O}$ is any operator. Putting this into the definition of S we get:

$$S(\mathbf{q},\omega) = \int_{-\infty}^{\infty} e^{i\omega t}\frac{dt}{2\pi N}$$

$$\times \sum_{i,f} p_i \sum_{\mathbf{R},\mathbf{S}} \langle i|e^{i\mathbf{q}\cdot\mathbf{u}(\mathbf{S})}|f\rangle\langle f|e^{-i\mathbf{q}\cdot\mathbf{u}(\mathbf{R},t)}|i\rangle e^{i\mathbf{q}\cdot(\mathbf{S}^\circ - \mathbf{R}^\circ)}$$

$$= \int_{-\infty}^{\infty} e^{i\omega t}\frac{dt}{2\pi N}$$

$$\times \sum_{\mathbf{R},\mathbf{S}} \langle e^{i\mathbf{q}\cdot\mathbf{u}(\mathbf{S})}e^{-i\mathbf{q}\cdot\mathbf{u}(\mathbf{R},t)}\rangle_{\mathrm{T}} e^{i\mathbf{q}\cdot(\mathbf{S}^\circ - \mathbf{R}^\circ)}. \tag{5.144}$$

In order to get to this expression we have used the completeness relation, $\sum_f |f\rangle\langle f| = 1$. As usual, $\mathbf{R} = \mathbf{R}^\circ + \mathbf{u}(\mathbf{R})$. The expression $\langle\hat{O}\rangle_{\mathrm{T}} = \sum_i p_i\langle i|\hat{O}|i\rangle$ is the thermal expectation value of operator $\hat{O}$.

We need to deal with the exponential operators in this equation. To get an idea of what is happening, we expand the exponentials to second order. Note that $\mathbf{u}$ is linear in $\hat{a},\hat{a}^+$ (c.f. Eq. (5.86)) so that the first-order terms have zero expectation value.

$$\langle e^{i\mathbf{q}\cdot\mathbf{u}(\mathbf{S})}e^{-i\mathbf{q}\cdot\mathbf{u}(\mathbf{R},t)}\rangle_{\mathrm{T}} \approx 1 - \frac{1}{2}(\langle(\mathbf{q}\cdot\mathbf{u}(\mathbf{S}))^2\rangle_{\mathrm{T}} + \langle(\mathbf{q}\cdot\mathbf{u}(\mathbf{R},t))^2\rangle_{\mathrm{T}})$$

$$+ \langle[\mathbf{q}\cdot\mathbf{u}(\mathbf{S})][\mathbf{q}\cdot\mathbf{u}(\mathbf{R},t))]\rangle_{\mathrm{T}} + \cdots \tag{5.145}$$

The expression $\langle (\mathbf{q} \cdot \mathbf{u}(\mathbf{R}, t))^2 \rangle_T$ should be independent of position and time – it is just q times the average of the vibration of any atom in the $\mathbf{q}$ direction. Thus, to the same order we can write this expression as:

$$\langle e^{i\mathbf{q}\cdot\mathbf{u}(S)} e^{-i\mathbf{q}\cdot\mathbf{u}(\mathbf{R},t)} \rangle_T = e^{-\langle(\mathbf{q}\cdot\mathbf{u})^2\rangle_T} e^{\langle[\mathbf{q}\cdot\mathbf{u}(S)][\mathbf{q}\cdot\mathbf{u}(\mathbf{R},t)]\rangle_T}. \qquad (5.146)$$

Remarkably, this expression is not an approximation, but exact (Mermin 1966). In fact, for any operators $\hat{A}, \hat{B}$ that are linear in $\hat{a}, \hat{a}^+$ for a harmonic oscillator Hamiltonian it is true that (Messiah 1968):

$$\langle e^{\hat{A}} e^{\hat{B}} \rangle_T = e^{\langle \hat{A}^2 + \hat{B}^2 + 2\hat{A}\hat{B}\rangle_T/2}, \qquad (5.147)$$

from which Eq. (5.146) follows.

5.3.1 Elastic scattering revisited

All of the time dependence in Eq. (5.146) is in the second factor. We now expand it in a power series and keep only the first term so that $e^{\langle[\mathbf{q}\cdot\mathbf{u}(S)][\mathbf{q}\cdot\mathbf{u}(\mathbf{R},t)]\rangle_T}$ is unity. Then substitute in Eq. (5.144).

$$S(\mathbf{q}, \omega) = e^{-\langle(\mathbf{q}\cdot\mathbf{u})^2\rangle_T} \int_{-\infty}^{\infty} e^{i\omega t} \frac{dt}{2\pi N} \sum_{\mathbf{R},\mathbf{S}} e^{i\mathbf{q}\cdot(\mathbf{S}^\circ - \mathbf{R}^\circ)}$$

$$= e^{-2W} N \sum_{\mathbf{G}} \delta_{\mathbf{q},\mathbf{G}} \delta(\omega). \qquad (5.148)$$

The factor $\delta(\omega)$ means that this is the elastic scattering part: there is no energy transfer to or from the crystal. The Debye–Waller factor is:

$$2W = \langle(\mathbf{q}\cdot\mathbf{u})^2\rangle_T = q^2 \langle u^2 \rangle_T/3, \qquad (5.149)$$

compare Eq. (3.42). The last expression is correct if the modes are isotropic. This expression for the Debye–Waller factor is correct for X-ray scattering too.

Take the neutron mass to be M_n, the incident wavevector $\mathbf{k}_1$ and the scattered wavevector $\mathbf{k}_2$. The conservation laws are:

$$\hbar^2 k_1^2/2M_n = \hbar^2 k_2^2/2M_n \quad \mathbf{q} = \mathbf{k}_1 - \mathbf{k}_2 = \mathbf{G}. \qquad (5.150)$$

Our new expression for the Debye–Waller factor is in terms of the exact phonon spectrum. Using Eq. (5.86) we have:

$$W = \frac{\hbar}{2mN} \sum_j \frac{(\mathbf{q} \cdot \mathbf{U}_j)^2}{\omega_j} [\langle \hat{n}_j \rangle_T + 1/2]. \qquad (5.151)$$

It is not hard to write down a formula for W in the Debye model. For low temperatures:

$$W(T) = \frac{3\hbar^2 q^2}{8mk_B\Theta}\left[1 + \frac{2\pi^2}{3}\left(\frac{T}{\Theta}\right)^2 + \cdots\right] \tag{5.152}$$

5.3.2 Inelastic scattering

In elastic scattering, discussed above, the final state of the crystal is the same as the initial one. It gives the intense Bragg peaks. In addition, there is a weaker (but very interesting) part of the scattering where there is energy transfer: either the neutron leaves behind energy (creates a phonon with energy $\hbar\omega_{k,\lambda}$), or takes some away (absorbs a phonon). The conservation laws are;

$$\hbar^2 k_1^2/2M_n = \hbar^2 k_2^2/2M_n \pm \hbar\omega_{k,\lambda} \quad q = k_1 - k_2 = G \pm k. \tag{5.153}$$

The $+$ corresponds to phonon creation, and $-$ to absorption.

To get expressions like this we use the next term in the expansion of $e^{\langle[q\cdot u(S)][q\cdot u(R,t)]\rangle_T}$. Since u is linear in the creation and annihilation operators the possible final states differ from the initial by one phonon created or destroyed. Following this through gives the conservation rules above.

In practice, neutron scattering consists in looking for diffuse scattering near Bragg peaks. When the diffuse scattering is energy analyzed, peaks are found at $\Delta E = \pm\hbar\omega_{k,\lambda}$, and displaced from the Bragg peak by an angle corresponding to $k_1 - k_2 = G \pm k$.

5.4 Mössbauer effect

There is a peculiar and useful effect related to the considerations above, that of "recoilless emission" in solids. There are radioactive nuclei, the most useful of which is Fe-57, which emit soft γ-rays. In gases there is a large Doppler shift from the recoil of the nucleus, and Doppler broadening due to the center of mass thermal motion of the nucleus. What Mössbauer discovered is that in solids at sufficiently low temperatures there is a very sharp emission line, basically the natural linewidth of the transition. Recoil and thermal broadening are absent; see Frauenfelder (1962), Kittel (1963). This came as a shock to nuclear physicists studying the phenomenon, and led to much misunderstanding.

The effect is very useful. Since the emission line (and also related absorption lines) are extremely narrow, precision measurements are possible. For example, hyperfine interactions which depend on the environment of the nucleus can be detected because they detune the resonant absorption. This is called Mössbauer spectroscopy.

In view of the previous sections the effect is not mysterious at all. The reason that the emitting nucleus doesn't recoil is that the entire crystal takes up the momentum of the γ-ray. The probability to do this is reduced by a Debye–Waller factor so that the intensity of the effect decreases as T increases.

To see how this goes, we use the standard expression from quantum mechanics for the emission of radiation of wavevector $\mathbf{k}$, frequency ω, and polarization $\mathbf{e}$. The rate will be:

$$w = C\omega |\langle f | e^{i\mathbf{k}\cdot(\mathbf{R}+\mathbf{r})} \mathbf{e} \cdot \mathbf{p} | i \rangle|^2 d\Omega. \qquad (5.154)$$

Here, C is a constant, $\mathbf{R}$ is the center-of-mass coordinate of the nucleus, $\mathbf{r}$ the coordinate with respect to the center of mass, and $\mathbf{p}$ the operator coupling the internal states. In free space we would take the expectation value over internal coordinates and also the final center of mass wavefunction of the nucleus. Ordinary conservation of momentum comes from the fact that the center of mass must recoil with $-\mathbf{k}$ or the matrix element is zero.

If the nucleus is in a solid, we must take the matrix element over the crystal states rather than plane wave states for the center of mass. Suppose that we want recoilless emission. This just means that the final state of the crystal is the same as the initial, so that ω is equal to $E_o/\hbar$, the unperturbed transition energy. We work at $T = 0$ for simplicity. Then the rate will be:

$$w = w_{\text{free}} |\langle 0 | e^{i\mathbf{k}\cdot(\mathbf{R}^o+\mathbf{u})} | 0 \rangle|^2. \qquad (5.155)$$

That is, the rate is multiplied by the Debye–Waller factor, e^{-2W}. The amount of recoilless emission is given by this factor and the rest of the oscillator strength of the transition goes into diffuse background involving the emission of one or more phonons.

5.5 Two dimensions

Let us return to Eq. (5.151) and try to evaluate the Debye–Waller factor for low but non-zero temperature in two dimensions. We have, using Eq. (5.114):

$$W \propto \sum_{k,\lambda} \frac{1}{\omega_{k,\lambda}} \frac{k_B T}{\omega_{k,\lambda}} \propto \int_0 \frac{\mathcal{D}(E)}{E^2} dE \propto \int_0 \frac{dE}{E}. \qquad (5.156)$$

The integral diverges. This seems to indicate that thermal fluctuations wipe out crystalline order in two dimensions, in the sense that the intensities of the Bragg peaks are zero. Going beyond this heuristic observation, Mermin (1968) proved that there is no ordinary crystalline order in two dimensions. See Problem 8 for the case of magnetic order.

What actually happens in two dimensions is that at low enough temperatures there is a state with *quasi long-range order*; (Nelson & Halperin 1979, Young 1979). The Bragg peaks are no longer δ functions, but rather power-law divergences, but the orientational order is long-range. And, most remarkably, melting consists of two transitions. Between the isotropic liquid and the solid there is another phase, the hexatic phase, with orientational order, but no positional order. This state of affairs was first predicted theoretically. Analogous transitions have been observed in certain liquid crystals. Two-dimensional colloidal crystals seem to show this behavior too (Murray & Grier 1996).

Suggested reading

The classic on lattice dynamics is:
  Born & Huang (1985)
See also:
  Ziman (1979)
and the general references. Magnons are in:
  White (1970)
  Mattis (1988).

Problems

1. Consider a linear chain of atoms of alternating mass interacting via the Lennard-Jones nearest neighbor potential. (a) Solve for the dispersion relation and exhibit the optical modes. (b) Show that at long wavelength the optical modes correspond to atoms in the unit cell moving opposite to one another, but that for the acoustic modes they move together.

2. Consider a linear chain of atoms interacting with each other via springs of spring constant κ and relaxed length a, but also with *fixed* lattice points with lattice constant a, with another spring constant K. Find and plot the dispersion relation. You should find that $\omega(k)$ does not go to zero as k goes to 0. Explain.

3. Consider a bcc crystal with nearest neighbor central force interactions given by $v(r)$ (a) Find $G^{jj'}(R), j = x, y, z$ for all non-zero components in terms of the second partial derivative of v at the nearest neighbor distance a. Partial answer: $G^{xx}(0) = 8v''(a)/3$. (b) Show that $G(\mathbf{k})$ is a diagonal matrix for $\mathbf{k}$ along the x, y or z axes. (c) Find the sound velocities for $\mathbf{k} \parallel (100)$ and (111). (d) Find $\omega(\mathbf{k})$ at the zone edge for the two cases in (c). (e) Verify that $\sum_{\mathbf{R}} G(\mathbf{R}) = 0 = \sum_{\mathbf{R}} \mathbf{R}G(\mathbf{R})$.

4. Suppose that the potential energy of interaction is given by the sum of pair potentials:

$$V = \frac{1}{2} \sum_{i \neq j} v(\mathbf{R}_i - \mathbf{R}_j).$$

Take one atom per unit cell, and suppose we know the Fourier transform of the pair potential $v(\mathbf{r}) = \sum_{\mathbf{k}} v(\mathbf{k})e^{i\mathbf{k}\cdot\mathbf{r}}$. (a) Show that the dynamical matrix can be written:

$$G(\mathbf{k}) = N \sum_{\mathbf{K}} [(\mathbf{K} + \mathbf{k})(\mathbf{K} + \mathbf{k}) \, v(\mathbf{K} + \mathbf{k}) - \mathbf{K}\mathbf{K} \, v(\mathbf{K})],$$

where $\mathbf{K}$ runs over the reciprocal lattice. (b) Show that the following sum rule holds:

$$\sum_{\lambda} M\omega_{\lambda}^2(\mathbf{k}) = \mathrm{Trace}(G(\mathbf{k})),$$

where λ runs over the three acoustic modes. (d) Apply the foregoing to the unscreened Coulomb lattice of Section 5.1.6. Take

$$v(r) = \frac{e^2}{r} e^{-\Lambda r}.$$

The exponential factor makes it possible to calculate $v(k)$. At the end of the calculation take $\Lambda = 0$. You should find $v(k) = 4\pi e^2/(\Omega k^2)$. Now show that:

$$\sum_\lambda \omega_\lambda(\mathbf{k})^2 = \Omega_p^2.$$

Compare Eq. (5.46). This shows that the transverse mode frequencies of the Coulomb lattice approach zero as $k \to 0$.

5. (a) Show that the low-T lattice heat capacity of a d-dimensional crystal behaves as T^d. Is this result exact in the harmonic theory, or does it depend on the Debye approximation? (b) Magnons have dispersion relation $\omega_k = Dk^2$. Show that for low T the magnon heat capacity goes as $T^{d/2}$ in d-dimensions.

6. (a) Verify Eq. (5.152). (b) Show that for $T \ll \Theta$ the Debye–Waller factor is linear in T and find the coefficient.

7. Consider a Bravais lattice of atoms with one impurity that differs from the others only by its mass, so that all the atoms have mass M, but the one at the origin has mass M_0. (a) Show that the perturbation of the lattice Hamiltonian is of the form: $\delta\hat{\mathcal{H}} = \left[\frac{1}{M_0} - \frac{1}{M}\right]\hat{F}$ where the operator $\hat{F}$ is bilinear in $\hat{a}^+(\mathbf{k}), \hat{a}(\mathbf{k})$. b.) Show that $\delta\hat{\mathcal{H}}$ gives rise to phonon scattering. (c) Show that the cross-section for acoustic phonon scattering from such an impurity is proportional to k^4, i.e. Rayleigh scattering. (Use the Fermi Golden rule of quantum mechanics).

8. Try to work out the decrease in magnetization for a two-dimensional ferromagnet at finite temperature. Do you think that there is ferromagnetic order in two dimensions? The Ising model is very different. Why?

6   The non-interacting electron model

When atoms are assembled into a condensed state, it is often the case that the outer valence electrons become delocalized, and are no longer associated with a given atom. The most obvious such case is a metal where the electrons are free to move, and can conduct electricity. An ionized classical plasma is another system of this sort, but, as we have seen, the electrons in solids must be treated with quantum theory. That is the subject of this chapter. Metallic liquids and glasses exist, but we will concentrate on metallic crystals. As we will see, the theory that we will develop will also apply to the valence electrons in semiconductors and insulators.

Since electrons in metals are free to move it is natural to think of them as a gas; the term *electron gas* is often used. The most extreme version of this idea is surprisingly useful, namely the idea of the *free* electron gas. In this idealization the electrons don't see the ions that they were detached from except in an average way, to neutralize their charge. Also, in this model the electrons are non-interacting, and act as if their Coulomb repulsion is not present.

We should say from the outset that both these assumptions appear to be totally unreasonable. The strength of the electron-electron interaction in Cu was estimated in Chapter 1. It turned out to be about 3 eV, which is the same order of magnitude as the energies that we will find in the next section. As for the potential of interaction of the atoms with the electrons, it also must be of the same order: the 1s electron in Cu will see a charge of $|e|$ on the ions left behind, giving about 3 eV again. Nevertheless, the non-interacting model often gives an excellent description of metals.

This success should be regarded as a mystery. We will try to penetrate part of the mystery later in this chapter when we explain why the electron-ion interaction appears weak: this is basically the notion of pseudo-potentials. In a later chapter we investigate the electron-electron interaction, and Landau's Fermi liquid theory, which explains, in some measure, why this interaction appears to be weak. Of course, there are some materials for which the model is a bad approximation, as we will see later.

6.1 Sommerfeld model

The free electron gas model (which is due to A. Sommerfeld) is that of a collection of N electrons without charge in a box of volume Ω. The wave function of each individual particle is a plane wave:

$$\phi_{\mathbf{k},\sigma}(\mathbf{r}) = \frac{e^{i\mathbf{k}\cdot\mathbf{r}}}{\sqrt{\Omega}}\chi. \qquad (6.1)$$

Here χ is a spinor for spin-1/2: the electron can be in spin-up or spin-down states. We label these by $\sigma = \pm 1$. The $\mathbf{k}$'s are those allowed by the periodic boundary conditions that we enforce on the electrons. We will briefly denote this function as $\phi_j(\mathbf{r})$, and j is a shorthand for $\mathbf{k}, \sigma$. The ϕ's are called *orbitals*.

The only energy in the Sommerfeld model is the kinetic energy of the electron which is given by;

$$\hat{\mathcal{H}}\phi_{\mathbf{k}}(\mathbf{r}) = \frac{\hbar^2 k^2}{2m}\phi_{\mathbf{k}}(\mathbf{r}), \qquad (6.2)$$

where m is the mass of the electron. In general we will denote the energy associated with an orbital (which could be more complicated than a plane wave) by ϵ_j. For the Sommerfeld model $\epsilon_j = \hbar^2 k^2/2m$. We will often use the abbreviation $\epsilon_o(\mathbf{k})$ for this quantity.

6.1.1 Many electron wavefunctions

The wavefunction of a combination of independent, indistinguishable, particles cannot be taken to be a product of ϕ's because identical fermions need to have an antisymmetric wavefunction. That is, if we have N electrons who populate states with N labels we need to take sum of all the antisymmetric products of the ϕ's. The sum is over all the permutations, P, of the set of j's.

$$\Psi_{\{j\}}(\{\mathbf{r}_j\}) = \frac{1}{\sqrt{N}}\sum_P (-1)^P [\phi_{P(j1)}(\mathbf{r_1})\phi_{P(j2)}(\mathbf{r_2})\cdots\phi_{P(jN)}(\mathbf{r_N})]$$

$$= \frac{1}{\sqrt{N!}}\mathrm{Det}\begin{pmatrix} \phi_{j1}(\mathbf{r_1}) & \phi_{j2}(\mathbf{r_1}) & \cdots & \phi_{jN}(\mathbf{r_1}) \\ \phi_{j1}(\mathbf{r_2}) & \phi_{j2}(\mathbf{r_2}) & \cdots & \phi_{jN}(\mathbf{r_2}) \\ \cdots & \cdots & & \\ \cdots & \cdots & & \\ \phi_{j1}(\mathbf{r_N}) & \phi_{j2}(\mathbf{r_N}) & \cdots & \phi_{jN}(\mathbf{r_N}) \end{pmatrix}. \qquad (6.3)$$

Here p is the number of pair interchanges to produce the permutation in question from $j1, j2, \ldots, jN$. The last form is called a Slater determinant, after J. Slater. For example for two electrons:

$$\Psi_{k,l}(\mathbf{r}_1, \mathbf{r}_2) = \frac{1}{\sqrt{2}}(\phi_k(\mathbf{r}_1)\phi_l(\mathbf{r}_2) - \phi_l(\mathbf{r}_1)\phi_k(\mathbf{r}_2)). \qquad (6.4)$$

We need not use plane waves to make Slater determinants. Any basis set of one particle wavefunctions will do. Note that Ψ vanishes if two labels are the same – two columns of the determinant will be the same. This is the Pauli principle: no orbital can be doubly occupied. In what follows we will assume that some definite order has been taken for the orbital labels. The determinant with the labels interchanged in Eq. (6.4) gives $-\Psi_{j_1,j_2}$.

The orbitals live in the Hilbert space of one-particle wavefunctions, and the Ψ_N in the Hilbert space of N-particle functions. In what follows we will be using the Grand Canonical ensemble for the statistical mechanics. This means that we are considering systems with variable numbers of particles. To represent all the wavefunctions we need to construct *Fock space*, the direct product space of wavefunctions of any N. The basis for this space is the set (in Dirac notation):

$$|0\rangle \leftrightarrow \text{vacuum \quad state}$$

$$|j\rangle \leftrightarrow \phi_j(\mathbf{r})$$

$$|j,k\rangle \leftrightarrow \frac{1}{\sqrt{2}}(\phi_j(\mathbf{r_1})\phi_k(\mathbf{r_2}) - \phi_k(\mathbf{r_1})\phi_j(\mathbf{r_2}))$$

$$\cdots$$

$$|j_1,j_2,\ldots,j_N\rangle \leftrightarrow \Psi_N(\{\mathbf{r}_j\}) \tag{6.5}$$

We are hijacking the notation of quantum field theory in which, for entirely different reasons, you have different numbers of excitations of a quantum field. To make the analogy complete, we take the notation where the *occupancy* of a state j is $n_j = 0, 1$. An N particle Slater determinant can be denoted by $|n_1, n_2, \ldots n_j \cdots\rangle$. The sum of the n's is the number of particles present.

We can introduce operators which link different parts of the Fock space. The *creation operator* is defined by:

$$\hat{c}_j^+|n_1, n_2, \ldots\rangle = (-1)^{\nu}|n_1, n_2, \ldots, n_j + 1, \ldots\rangle, \tag{6.6}$$

where $\nu = \sum_{k=1}^{j-1} n_j$, i.e. the sum of occupancies in the list before j. Clearly $\hat{c}_j^+|n_j = 0\rangle = \pm|n_j = 1\rangle, \hat{c}^+|n_j = 1\rangle = 0, (\hat{c}_j^+)^2 = 0$. We also define the adjoint operator to $\hat{c}_j^+$, the *annihilation operator*, by:

$$\hat{c}_j|n_1, n_2, \ldots\rangle = (-1)^{\nu}|n_1, n_2, \ldots, n_j - 1, \ldots\rangle. \tag{6.7}$$

Clearly $\hat{c}_j|n_j = 1\rangle = \pm|n_j = 0\rangle, \hat{c}_j|n_j = 0\rangle = 0, (\hat{c}_j)^2 = 0$. Putting these relations together we find that $\hat{c}, \hat{c}^+$ anticommute:

$$\hat{c}\hat{c}^+ + \hat{c}^+\hat{c} \equiv \{\hat{c}, \hat{c}^+\} = 1. \tag{6.8}$$

It is easy to see that operators with different labels also anticommute. For example consider the combinations $\hat{c}_j^+\hat{c}_k^+$ and $\hat{c}_k^+\hat{c}_j^+$ applied to an N-particle state, $|A\rangle$. Suppose that j is earlier in the list of labels. A simple calculation shows that the phase factor, ν, associated with $\hat{c}_k^+\hat{c}_j^+|A\rangle$ is $\sum_{j+1}^{k-1} n_l + 1$. For $\hat{c}_k^+\hat{c}_j^+|A\rangle$ we get $\sum_{j+1}^{k-1} n_l$. Thus:

$$\hat{c}_j^+\hat{c}_k^+ + \hat{c}_k^+\hat{c}_j^+ = 0. \tag{6.9}$$

Thus: $\{\hat{c}_j^+, \hat{c}_k^+\} = \{\hat{c}_j, \hat{c}_k\} = 0, \{\hat{c}_j, \hat{c}_k^+\} = \delta_{j,k}$.

Any N-particle state composed of a single determinant can be written in terms of the vacuum state as:

$$\hat{c}_{j1}^+ \hat{c}_{j2}^+ \cdots \hat{c}_{jN}^+ |0\rangle. \tag{6.10}$$

Any N-particle state can be expanded in terms of these basis states. To completely define these basis states we need to agree on some order of the labels ji. Otherwise we may have differences of sign because:

$$\hat{c}_k^+ \hat{c}_l^+ |\Psi\rangle = -\hat{c}_l^+ \hat{c}_k^+ |\Psi\rangle, \tag{6.11}$$

from the commutation relations.

It is sometimes useful to change the basis orbitals, say, from the ϕ_j's to Φ_J. The expansion of the new orbitals in terms of the old is given by the standard expression:

$$\Phi_J = \sum_j \phi_j \langle j | J \rangle. \tag{6.12}$$

These same matrix elements allow us to find creation and annihilation operators with respect to the new basis:

$$\hat{c}_J^+ = \sum_j \hat{c}_j^+ \langle j | J \rangle. \tag{6.13}$$

This can be checked by applying both operators to $|0\rangle$, and comparing to Eq. (6.12).

As in the case of phonons, it is useful to define an operator that counts the occupancy of and orbital. It is:

$$\hat{n}_j = \hat{c}_j^+ \hat{c}_j. \tag{6.14}$$

Operators such as the Hamiltonian can be written in terms of the $\hat{c}^+, \hat{c}$. For example, consider any single-particle operator such as kinetic energy, or an external potential. Since the electrons are indistinguishable, the only physical operators must have the form:

$$V_1(\{\mathbf{r}\}) = \sum_{l=1}^N v(\mathbf{r}_l). \tag{6.15}$$

This operator is completely defined by its matrix elements in the basis of the Ψ's. By considering such matrix elements, it is easy to show that the only non-zero matrix elements are between states that differ by, at most, having the occupancy in one of the states, j different. And the permutation must be the same on both sides of the matrix element since the orbitals are orthogonal. Thus:

$$V_1 \leftrightarrow \sum_{jk} \langle j | v | k \rangle \hat{c}_j^+ \hat{c}_k$$

$$\langle j | v | k \rangle = \int d\mathbf{r} \phi_j(\mathbf{r})^* v(\mathbf{r}) \phi_k(\mathbf{r}). \tag{6.16}$$

For example, for a single particle, spin-independent, Hamiltonian, in the plane wave basis we have:

$$\hat{\mathcal{H}} = \sum_{l,\sigma} [\hat{p}^2/2m + v(\mathbf{r}_l)]$$

$$\leftrightarrow \sum_{\mathbf{k},\sigma} \frac{\hbar^2 k^2}{2m} \hat{c}^+_{\mathbf{k},\sigma} \hat{c}_{\mathbf{k},\sigma} + \sum_{\mathbf{k},\mathbf{q},\sigma} \langle \mathbf{k}|v|\mathbf{q}\rangle \hat{c}^+_{\mathbf{k},\sigma} \hat{c}_{\mathbf{q},\sigma}. \tag{6.17}$$

We will also deal with interactions between electrons. The most common form considered is a two-particle interaction (such as the Coulomb interaction):

$$V_2 = \frac{1}{2} \sum_{l \neq m} v(\mathbf{r}_l, \mathbf{r}_m). \tag{6.18}$$

For indistinguishable particles this is the only form that is acceptable. Further, v must be symmetric in its arguments.

The matrix elements of an operator like this can involve the change of two labels. It is a bit tedious, but not hard, to verify that the correct expression is:

$$V_2 \leftrightarrow \frac{1}{2} \sum_{ijkl} \langle ij|v|kl\rangle \hat{c}^+_i \hat{c}^+_j \hat{c}_l \hat{c}_k \tag{6.19}$$

$$\langle ij|v|kl\rangle = \int d\mathbf{r}_1 d\mathbf{r}_2 \phi_i(\mathbf{r}_1)^* \phi_j(\mathbf{r}_2)^* v(\mathbf{r}_1, \mathbf{r}_2) \phi_k(\mathbf{r}_1) \phi_l(\mathbf{r}_2). \tag{6.20}$$

Note the order of the labels in Eq. (6.19) which is not a misprint.

This notation is called *second quantization*. It is very useful for the theory of interacting electrons.

6.1.2 Ground state of the Sommerfeld model

We now return to the Sommerfeld model, and focus on the occupancies of the plane-wave orbitals. The ground state will be made by occupying the lowest energy orbital first, then the next, and so on, until the electrons are used up. Since the kinetic energy is proportional to k^2, the electrons must be put as close to the origin in k-space as possible. That is, the occupied orbitals form a sphere in k-space, known as the Fermi sphere. Call the radius of the Fermi sphere k_F. Thus, for each spin state:

$$\langle \hat{n}_{\mathbf{k},\sigma}\rangle = 1 \quad |\mathbf{k}| < k_F$$

$$= 0 \quad |\mathbf{k}| > k_F. \tag{6.21}$$

We can find k_F by counting the orbitals inside the Fermi sphere:

$$N = \sum_{\mathbf{k},\sigma} \langle \hat{n}_{\mathbf{k},\sigma} \rangle = 2[\Omega/(2\pi)^3] \int_{|\mathbf{k}|<k_F} d^3k \qquad (6.22)$$

$$k_F = (3\pi^2 N/\Omega)^{1/3}. \qquad (6.23)$$

The factor of 2 accounts for spin. The radius, k_F depends on the electron density, $n_e = N/\Omega$. This implies that in the ground state there are electrons with a large (zero-point) kinetic energy. The maximum energy in the ground state is called the Fermi energy;

$$E_F = \frac{\hbar^2 k_F^2}{2m} = \frac{\hbar^2}{2m}(3\pi^2 n_e)^{2/3}. \qquad (6.24)$$

For most metals E_F is a large energy, on the order of 3 eV. For almost all materials $k_B T \ll E_F$ right up to the melting point. For $\epsilon < E_F$ orbitals are occupied, and for $E > E_F$ they are empty in the ground state.

It is useful to define the density of states for electrons, as we did for phonons.

$$\mathcal{D}(E) = \sum_j \delta(E - \epsilon_j). \qquad (6.25)$$

As in the case of phonons, this function allows us to convert a sum over states of any function of ϵ alone into an integral over energy:

$$\sum_j g(\epsilon_j) = \int g(E)\mathcal{D}(E)dE. \qquad (6.26)$$

It is a simple calculation to show that for free electrons:

$$\mathcal{D}(E) = \Omega \frac{2^{1/2} m^{3/2}}{\pi^2 \hbar^2} E^{1/2} \qquad (6.27)$$

and that $\mathcal{D}(E_F) = 3N/2E_F$.

Now consider the case where ϵ is more complicated than $\hbar^2 k^2/2m$, but is some function which could, for example, depend on the direction of $\mathbf{k}$. Recall from classical physics that the group velocity of a wave is the gradient of the frequency: $\mathbf{v}_g = \partial\omega/\partial\mathbf{k}$. In our case we can put $\mathbf{v}_g = (1/\hbar)\nabla_{\mathbf{k}}\epsilon(\mathbf{k})$. Now we write:

$$\mathcal{D}(E) = 2\sum_{\mathbf{k}} \delta(E - \epsilon(\mathbf{k}))$$

$$= \frac{2\Omega}{(2\pi)^3} \int d\mathbf{k}\delta(E - \epsilon(\mathbf{k})). \qquad (6.28)$$

The volume element $d\mathbf{k}$ can be thought of as an element of area, dS on a surface $E = \epsilon(\mathbf{k})$ multiplied by the perpendicular distance between the surface at E and the one at $E + dE$, see

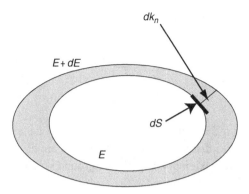

Fig. 6.1   **Integrating over constant energy surfaces. The volume element $d\mathbf{k}$ is the cylinder whose base is dS and height is dk_n.**

Figure 6.1. The gradient, $\nabla_{\mathbf{k}}\epsilon(\mathbf{k})$ is also perpendicular to the surface, and $dE = |\nabla_{\mathbf{k}}\epsilon(\mathbf{k})|dk_n$. Putting this together we find:

$$
\begin{aligned}
\mathcal{D}(E) &= \frac{\Omega}{4\pi^3}\int dS\,\frac{dE}{|\nabla_{\mathbf{k}}\epsilon(\mathbf{k})|}\delta(E-\epsilon(\mathbf{k})) \\
&= \frac{\Omega}{4\pi^3\hbar}\int \frac{dS}{|\mathbf{v}_g|}.
\end{aligned}
\tag{6.29}
$$

For the simplest case, the surface is a sphere, and $\mathbf{v}_g = \hbar\mathbf{k}/m$. This expression shows that if there are places where $\mathbf{v}_g$ is small there are large contributions to the density of states.

We can use $\mathcal{D}$ to find the total kinetic energy in the ground state:

$$
E_0 = \int_0^{E_F} E\mathcal{D}(E)dE = \frac{3}{5}NE_F.
\tag{6.30}
$$

The average kinetic energy per electron is $(3/5)E_F$.

6.2 Thermally excited states and heat capacity

The statistical physics of independent Fermions at finite temperature is characterized by the Fermi–Dirac distribution, $f(\epsilon)$, namely that the average thermal occupancy of orbitals at temperature T is:

$$
\langle\hat{n}_j\rangle_T = f(\epsilon_j) = \frac{1}{e^{\beta(\epsilon_j-\mu)}+1}.
\tag{6.31}
$$

Here μ is the thermodynamic chemical potential, and ϵ_j is the energy of orbital j, assuming that they are eigenstates of $\hat{\mathcal{H}}$.

The chemical potential is implicitly determined by the finite temperature analog of Eq. (6.22):

$$N = \sum_j f(\epsilon_j). \tag{6.32}$$

Note that as $T \to 0, f(\epsilon)$ approaches a step function,

$$f = 1, \quad \epsilon < \mu, \quad f = 0, \quad \epsilon > \mu.$$

That is the mean occupancy goes to zero if $\epsilon > \mu$. However, for $T = 0$, the ground state, we have already seen that orbitals are unoccupied for $\epsilon > E_F$. This leads to the identification: $\lim_{T \to 0} \mu(T) = E_F$. For classical gases, $\mu < 0$ since entropy dominates. .

A good deal of attention was paid to the internal energy and heat capacity of metals at the beginning of the twentieth century. This was because of a classical paradox, namely the prediction that metals would have a heat capacity per atom at least 50% higher than that of insulators. This is not the case.

However, the reasoning is simple, and correct in classical physics. The idea is that electrons in a metal are free, and form a gas. This is the basis of the classical model of P. Drude which had great success in describing electrical conductivity; see Chapter 7. The heat capacity follows from equipartition. For the ions we have (as we have seen), $3N$ degrees of freedom from the kinetic energy, and $3N$ from the potential. Since each degree of freedom acquires mean thermal energy $k_B T/2$, we have, for insulators the law of Dulong and Petit, $U = 3Nk_B T, C_V = 3Nk_B$. For a monatomic metal we should have another $3N$ degrees of kinetic freedom for the electrons so that the total internal energy and heat capacity would be: $U = 3Nk_B T + 3Nk_B T/2, C_V = 9Nk_B/2$. However, in practice metals and insulators have around the same heat capacity.

The resolution of this paradox is one of the great successes of the Sommerfeld model. We can see qualitatively what is happening by the following estimate. Consider the behavior of $f(E)$ for $k_B T \ll E_F$, see Figure 6.2. For a band of energies near E_F electrons can be excited, and behave more-or-less classically. The rest of the orbitals are deep within the "Fermi sea,"

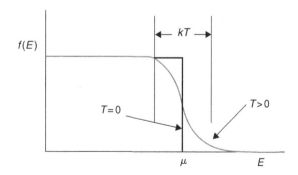

Fig. 6.2   **The Fermi distribution function for two different temperatures.**

and cannot be excited. So we estimate that the effective number of electrons is:

$$N_{\text{eff}} \approx k_B T \mathcal{D}(E_F) = 3Nk_B T/(2E_F). \tag{6.33}$$

Thus the electronic part of the internal energy is about $(3k_B T/2)N_{\text{eff}}$, so that $C_V^{\text{el}} \approx (9Nk_B/2)(k_B T/E_F)$. We conclude that the electronic specific heat is linear in T and much smaller than Dulong and Petit because of the small factor $k_B T/E_F$. A more careful calculation requires calculating the leading temperature dependences of U and μ.

There is a formula which gives the leading temperature dependence for the quantities we are calculating, known as the Sommerfeld expansion:

$$\int_0^\infty g(E)f(E)dE \to \int_0^\mu g(E)dE + \frac{\pi^2}{6}(k_B T)^2 g'(\mu) + \cdots \tag{6.34}$$

Here $g(E)$ is any smooth function of energy. The derivation (see problems) involves integrating by parts and noticing that $\partial f/\partial E$ is zero almost everywhere except near $E = \mu$ for $k_B T \ll \mu$, see Figure 6.2. Note for future reference that this means that $-\partial f/\partial E \approx \delta(E - \mu)$ for low temperatures.

Now we apply this to $U(T)$ and rewrite Eq. (6.32):

$$U(T) = \int_0^\infty E\mathcal{D}(E)f(E)dE$$

$$N = \int_0^\infty \mathcal{D}(E)f(E)dE. \tag{6.35}$$

Using Eq. (6.34) and eliminating μ gives the standard result:

$$C_V \to \frac{\pi^2}{3}\mathcal{D}(E_F)k_B^2 T = \frac{\pi^2 Nk_B}{2}\frac{k_B T}{E_F}, \tag{6.36}$$

which is pretty close to our guess. The details are left for a problem.

There are many thermodynamic quantities that are very well predicted by the Sommerfeld theory. For example, the low-temperature spin susceptibility in paramagnetic metals is temperature independent (Pauli). The Sommerfeld theory is also the basis for the theory of electrical transport in metals, see Chapter 7.

The successes of the Sommerfeld free electron model are disturbing. We need to understand why the strong forces of the ions on the electrons are somehow not important, and why the strong forces that electrons exert on each other are, also, somehow not important.

6.3 Band theory

In this section we will look at the forces of the ions on the electrons. The simplest way to do this is to pose a (fictitious) one electron problem:

$$\hat{\mathcal{H}}_1\psi = [\hat{p}^2/2m + v(\mathbf{r})]\psi = \epsilon\psi, \tag{6.37}$$

where the potential, v, is periodic in the lattice, and investigate the nature of the solutions. Then we use the orbitals which are the solutions to Eq. (6.37) to construct Slater determinants.

It is immediately obvious that v is a problematic object. For example, suppose we take all but one electron out of the solid, and look at the one-particle problem for that electron and all the ions. This is a bad starting point for understanding because the ions have charge $+NZe$ (for atomic number Z) which is balanced only by the charge of the one electron, $-e$. The potential is enormous, and quite unlike what an electron in a real solid feels. In the physical situation the other electrons neutralize (*screen*) the charge of the ions almost completely. Also, charged electrons interact.

The problem we really need to solve involves, necessarily, all the electrons. The best we can do is to interpret v as an effective potential which comes from the interacting, many-electron problem. We will return to this point below and see how an approximate effective potential is constructed using density functional theory. For the moment we will simply guess what v is or use its Fourier components as fitting parameters to experimental data.

6.3.1 Bloch theorem and bands

With the foregoing in mind we ask about the nature of the one-electron orbitals that are solutions to Eq. (6.37). We already have the answer from the last chapter: since $\hat{\mathcal{H}}_1$ is invariant under lattice translations, we can follow through the reasoning that we used for lattice vibrations. The orbitals must be *Bloch functions* labeled by allowed $\mathbf{k}$'s which satisfy:

$$\psi_{\mathbf{k}}(\mathbf{r} + \mathbf{R}) = e^{i\mathbf{k}\cdot\mathbf{R}}\psi_{\mathbf{k}}(\mathbf{r}). \tag{6.38}$$

Note that $e^{-i\mathbf{k}\cdot\mathbf{r}}\psi_{\mathbf{k}}(\mathbf{r})$ is periodic. This means that we can represent:

$$\psi_{\mathbf{k}}(\mathbf{r}) = e^{i\mathbf{k}\cdot\mathbf{r}}u_{\mathbf{k}}(\mathbf{r}), \tag{6.39}$$

where u is periodic in the unit cell.

Return to Eq. (6.37) and apply $\hat{\mathcal{H}}_1$ to $e^{i\mathbf{k}\cdot\mathbf{r}}u$. This gives the Schrödinger equation:

$$\left[\frac{\hbar^2}{2m}\left(\frac{\nabla}{i} + \mathbf{k}\right)^2 + v(\mathbf{r}) \right] u_{\mathbf{k}}(\mathbf{r}) = \epsilon(\mathbf{k})u_{\mathbf{k}}(\mathbf{r}). \tag{6.40}$$

The boundary conditions can be applied, for example on the surface of the Wigner–Seitz cell, namely $u_{\mathbf{k}}(\mathbf{r}) = u_{\mathbf{k}}(\mathbf{r} + \mathbf{R})$. This is an eigenvalue problem in *one unit cell*. Just as in the case of lattice vibrations, we are not obliged to solve throughout the crystal, but in a restricted volume.

The nature of the solutions is clear from Eq. (6.40), a Schrödinger equation in a finite volume. The eigenvalues will be discrete, and we label them with an index n, the band index. As we change $\mathbf{k}$, the eigenvalue $\epsilon(\mathbf{k})$ will change. The form of the equation is such that a small change in $\mathbf{k}$ will give a small change in ϵ. So as we move $\mathbf{k}$ in some direction through the Brillouin zone the energy levels will trace out a curve called a *band*. For each

n there will be a band, and it will be continuous except at the boundaries of the zone. The collection of bands is called the band structure. The actual calculation of band structures has been pursued by a large number of people over a period of decades.

There is an instructive soluble example, a version of the Kronig–Penney model (Kronig & Penney 1931). It involves a delta-function potential repeated periodically in one dimension. It is possible to construct the bands for this case analytically; see Problem 5.

There is something to note: the Bloch functions are eigenfunctions of the Hamiltonian. Thus *Bloch electrons do not scatter from ions.* They travel freely through the lattice even in the presence of and arbitrarily strong v because the diffraction conditions are not met. This is completely unlike the situation in a plasma where the presence of interaction with the *random* potential centers due to the ions leads to strong scattering.

6.3.2 Nearly free electron model

A first approach to the subject is to assume that the lattice potential is weak and smooth as first done by R. Peierls. If v is small, then it is reasonable to assume that a good starting point is the free electron orbitals of Sommerfeld theory, $\phi = e^{i\mathbf{q}\cdot\mathbf{r}}/\sqrt{\Omega}$. However, these are not of the Bloch form, because $\mathbf{q}$ can take on any value and u is absent. We can repair this by mapping $\mathbf{q}$ into the Brillouin zone, by putting $\mathbf{q} = \mathbf{k} + \mathbf{G}$, i.e. by subtracting some reciprocal lattice vector. Then we put:

$$\phi_{\mathbf{G},\mathbf{k}}(\mathbf{r}) = u_{\mathbf{G}}e^{i\mathbf{k}\cdot\mathbf{r}}; \quad u_{\mathbf{G}} = e^{i\mathbf{G}\cdot\mathbf{r}}/\sqrt{\Omega}. \tag{6.41}$$

This is an 'empty lattice' Bloch function; by inspection, u is periodic in the lattice.

There is a remarkable fact: if we interpret the free-electron energies as bands by "folding" the parabola into the zone (i.e., replacing $\mathbf{q}$ by $\mathbf{k}$), we get a reasonable account of the experimental band structure for simple metals (except near the zone boundaries)! The folding process is called going to the "reduced zone scheme." The original, free-electron case, where $\mathbf{q}$ is not restricted to the Brillouin zone is called the "extended zone scheme." There is also the "periodic zone scheme" where the Brilloun zone is repeated periodically to emphasize that the unit cell is arbitrary.

Now we turn on the potential, and, at first, don't assume it is weak. We can rewrite the Schrödinger equation in a useful way by expanding all the periodic functions in reciprocal lattice vectors:

$$u(\mathbf{r}) = \sum_{\mathbf{G}} u(\mathbf{G})e^{i\mathbf{G}\cdot\mathbf{r}}; \quad v(\mathbf{r}) = \sum_{\mathbf{G}} v(\mathbf{G})e^{i\mathbf{G}\cdot\mathbf{r}}. \tag{6.42}$$

Putting these into Eq. (6.40) and identifying the coefficient of $\mathbf{G}$ we find:

$$(\epsilon_\circ(\mathbf{k} + \mathbf{G}) - \epsilon(\mathbf{k}))u(\mathbf{G}) + \sum_{\mathbf{G}'} v(\mathbf{G} - \mathbf{G}')u(\mathbf{G}') = 0. \tag{6.43}$$

Here $\epsilon_\circ(\mathbf{k}) = (\hbar^2 k^2/2m)$ is the free-electron band energy. This equation turns the Schrödinger equation into a matrix eigenvalue equation for the $u(\mathbf{G})$, considered as a vector with one component for each point in reciprocal space. In practice, for a smooth potential, we can truncate the equation by considering only a few $\mathbf{G}$'s that are close to the origin.

If the $v(\mathbf{G})$ are small, the off-diagonal parts of the matrix are small, and can be treated by perturbation theory. In this case we can go back to the original Schrödinger equation and use standard quantum-mechanical perturbation theory. We use as zero-order states the free electron states from the empty lattice case whose energies are $\epsilon_\circ(\mathbf{k} + \mathbf{G})$. We use v as a perturbation. We need some matrix elements:

$$\langle \mathbf{G}', \mathbf{k}' | v(\mathbf{r}) | \mathbf{G}, \mathbf{k} \rangle = \sum_{\mathbf{G}''} v(\mathbf{G}'') \frac{1}{\Omega} \int d^3 r e^{i(\mathbf{G}+\mathbf{G}''-\mathbf{G}')\cdot\mathbf{r}} e^{i(\mathbf{k}-\mathbf{k}')\cdot\mathbf{r}}$$

$$= v(\mathbf{G}' - \mathbf{G})\delta_{\mathbf{k},\mathbf{k}'}. \qquad (6.44)$$

Note that the only diagonal matrix element is $v(\mathbf{G} = 0)$. We can set this to be zero by changing the origin of energies.

Now, to second order, for the band labeled by $\mathbf{G}$:

$$\epsilon(\mathbf{k}) = \epsilon_\circ(\mathbf{k} + \mathbf{G}) + \sum_{\mathbf{G}' \neq \mathbf{G}} \frac{|v(\mathbf{G}' - \mathbf{G})|^2}{\epsilon_\circ(\mathbf{k} + \mathbf{G}) - \epsilon_\circ(\mathbf{k} + \mathbf{G}')}. \qquad (6.45)$$

As is usual in quantum mechanics, we need another formula when

$$\epsilon_\circ(\mathbf{k} + \mathbf{G}) = \epsilon_\circ(\mathbf{k} + \mathbf{G}'), \qquad (6.46)$$

i.e. when two free-electron bands touch and the energy denominator is zero. It is clear that this occurs generically when $|\mathbf{k} + \mathbf{G}| = |\mathbf{k} + \mathbf{G}'|$. If we go back to the extended zone scheme, we have $|\mathbf{q}| = |\mathbf{q}'|$; $\quad \mathbf{q}' = \mathbf{q} + \mathbf{G} - \mathbf{G}'$. This is just the Bragg condition, and it generically occurs at the edge of the zone.

Suppose two bands labeled by $\mathbf{q}$ and $\mathbf{q} + \mathbf{G}$ touch at some zone edge. Quantum theory says that we must find the eigenvalues of the matrix:

$$\begin{pmatrix} \epsilon_\circ(\mathbf{q}) & v(\mathbf{G}) \\ v^*(\mathbf{G}) & \epsilon_\circ(\mathbf{q} + \mathbf{G}) \end{pmatrix}. \qquad (6.47)$$

The eigenvalues of the matrix give two bands:

$$\epsilon(\mathbf{q}) = \frac{(\epsilon_\circ(\mathbf{q}) + \epsilon_\circ(\mathbf{q} + \mathbf{G}))}{2}$$

$$\pm \left(\left[\frac{(\epsilon_\circ(\mathbf{q}) - \epsilon_\circ(\mathbf{q} + \mathbf{G}))}{2} \right]^2 + |v(\mathbf{G}|^2 \right)^{1/2}. \qquad (6.48)$$

Exactly at the zone edge we have Eq. (6.46). Thus:

$$\epsilon_\pm = \epsilon_\circ(\mathbf{q}) \pm |v(\mathbf{G})|. \qquad (6.49)$$

The bands split and form a *band gap* at the zone edge.

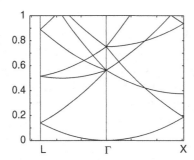 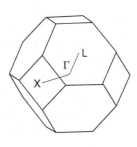

Fig. 6.3   Free electron bands in a fcc crystal. The units are atomic units, $\hbar = m = e = 1$. The lattice constant a is appropriate for Si, namely 10.26. The energy is plotted starting at the L point in the Brillouin zone, to the center, and then out to the X point, as shown on the right.

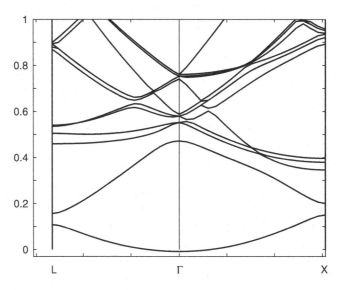

Fig. 6.4   The same situation as in Figure 6.3 except that a potential whose Fourier components are $0.03 \exp(\ G^2/2)$ has been turned on.

A numerical solution of Eq. (6.43) is not difficult in simple cases. In Figure 6.4 we show the result of a Matlab solution for the same case as in Figure 6.3 above, but with a potential of the form:

$$v(\mathbf{G}) = -v_0 \exp(-\alpha G^2). \tag{6.50}$$

The gaps that have opened up where the free-electron bands crossed are clear.

6.3.3 Tight binding or LCAO

The opposite point of view from the foregoing is to assume that the atomic potentials are so strong that the wavefunctions in the lattice are closely related to the wavefunctions of the isolated atoms. This idea has substantial conceptual value, and can be used in special cases in real solids.

Suppose we are interested in atoms with bound state wavefunctions, $\phi_\gamma(\mathbf{r})$ where γ labels the atomic levels. These are not an orthogonal set because of *overlaps* of the tails of the wavefunctions on adjacent sites. This method works best when the spacing between atoms is large, and overlaps small. We make this assumption in what follows, and try to build band wavefunctions from the ϕ's located on the different sites. We still must satisfy the Bloch condition, $\psi(\mathbf{r} + \mathbf{R}) = e^{i\mathbf{k}\cdot\mathbf{R}}\psi(\mathbf{r})$. A candidate is the following:

$$\psi_{\mathbf{k}}(\mathbf{r}) = \frac{1}{\sqrt{N}}\sum_{\mathbf{S},\gamma} b_\gamma e^{i\mathbf{k}\cdot\mathbf{S}}\phi_\gamma(\mathbf{r} - \mathbf{S}). \tag{6.51}$$

The coefficients b_γ are to be determined. This is called a tight binding or Linear Combination of Atomic Orbitals (LCAO) wavefunction. It is obvious that this function satisfies the Bloch condition: put $\mathbf{r} \to \mathbf{r} + \mathbf{R}$ and relabel the terms in the sum. We can think of this as a variational problem: we need to minimize the expectation of $\hat{\mathcal{H}}$ with respect to the b's.

For the simplest case, with only one ϕ, this reduces to finding the expectation value of $\hat{\mathcal{H}}$. Let $\hat{\mathcal{H}}_{\mathbf{R}}$ be the Hamiltonian for the isolated atom at $\mathbf{R}$ which satisfies $\hat{\mathcal{H}}_{\mathbf{R}}|\mathbf{R}\rangle = \epsilon_\gamma|\mathbf{R}\rangle$. Then the variational estimate of the energy is:

$$\langle\mathbf{k}|\hat{\mathcal{H}}|\mathbf{k}\rangle = \frac{1}{N}\sum_{\mathbf{S},\mathbf{S}'} e^{i\mathbf{k}\cdot(\mathbf{S}-\mathbf{S}')}\langle\mathbf{S}|[v - v_a(\mathbf{r} - \mathbf{S})] + \hat{\mathcal{H}}_{\mathbf{S}}|\mathbf{S}\rangle$$

$$= \epsilon_\gamma + \sum_\delta e^{i\mathbf{k}\cdot\mathbf{R}_1}\Gamma. \tag{6.52}$$

Here we have taken only overlaps with nearest neighbors, situated at

$$\mathbf{S}' - \mathbf{S} = \mathbf{R}_1,$$

and dropped smaller terms. The overlap integral, Γ, is defined by:

$$\Gamma = \int d\mathbf{r}[v - v_a(\mathbf{r} - \mathbf{R}_1)]\phi_\gamma^*(\mathbf{r} - \mathbf{R}_1)\phi_\gamma(\mathbf{r}). \tag{6.53}$$

For a simple cubic lattice with spherically symmetric ϕ's the result for the band energy is:

$$\epsilon(\mathbf{k}) = \epsilon_\gamma + 2\Gamma[\cos(k_x a) + \cos(k_y a) + \cos(k_z a)]. \tag{6.54}$$

The band width is 4Γ. This gets bigger as atoms are brought closer together; see Figure 6.5.

In practice the tight-binding method is used for tightly bound states. For example, we might describe the sp-electrons with plane waves, and the d- and f-electrons with tight binding.

6.3.4 Pseudopotentials

Real band structure calculations rarely use the two methods above, but rather build in a feature of the band states that we have neglected so far. In the course of the discussion

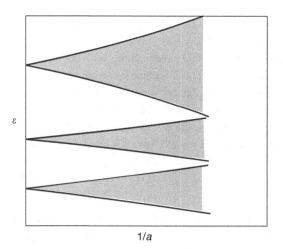

1/a

Fig. 6.5   **Band edges for the tight binding model for three atomic states. The shaded areas are the bands. When the bands start to overlap we should minimize with respect to the b_γ in Eq. (6.51).**

we will get a clue to the mysterious fact that the potentials in real solids look weak. Two points need to be made: the first is that electrons cover up (screen) the Coulomb potential of the ions. This will be treated later, but we can take it into account approximately by letting a band electron in Na, for example, see charge $Z = 1$ on the nucleus rather than the full atomic number. However, this potential is still too strong. The other point, which has been realized since the 1930s is that the band states may look like plane waves between the atoms, but as the electron approaches the core states of each atom it looks more like an atomic state. For example, in Na, the 3s valence electron in the core region is orthogonal to the tightly bound 1s and 2p electrons. In particular, its wavefunction has 2 radial nodes. As a result, the effect of the strong Coulomb potential of the ions is substantially cancelled by the oscillations of the 3s wavefunction. The real wavefunction has lots of structure, but it acts, as far as $\epsilon(\mathbf{k})$ is concerned, as if it was a smooth wavefunction in a much weaker potential, the *pseudopotential*.

There are a number of ways in which this insight is implemented in real calculations. Herring (1940) introduced one way: he expanded the wavefunction in a basis set which is automatically orthogonal to the core states treated in tight-binding, $\phi_{c,\mathbf{k}}(\mathbf{r})$. Thus, instead of plane waves we expand in:

$$e^{i\mathbf{k}\cdot\mathbf{r}} - \sum_c \phi_{c,\mathbf{k}}(\mathbf{r})\langle\phi_{c,\mathbf{k}}(\mathbf{r})|e^{i\mathbf{k}\cdot\mathbf{r}}\rangle. \tag{6.55}$$

It should be possible to use far fewer of these *orthogonalized plane waves* to get a good result than would be necessary in the plane wave basis above.

Phillips & Kleinman (1959) reformulated this scheme in a very useful way. They wrote the true wavefunction, ψ, as a smooth part, ϕ plus a strongly oscillating part which gives the orthogonality to the core states:

$$\psi = \phi - \sum_c \phi_c\langle\phi_c|\phi\rangle. \tag{6.56}$$

Now we seek an effective equation for the smooth part alone. Recall that ϕ_c is assumed to be an eigenfunction of $\hat{\mathcal{H}}$. We call the core eigenvalue ϵ_c. Then:

$$\hat{\mathcal{H}}\phi - \sum_c [\epsilon_c - \epsilon]\phi_c\langle\phi_c|\phi\rangle = \epsilon\phi$$

$$\left(\hat{\mathcal{H}} - \sum_c [\epsilon_c - \epsilon]|\phi_c\rangle\langle\phi_c|\right)\phi = \epsilon\phi. \tag{6.57}$$

This equation involves the *projection operator* onto the core states, $|\phi_c\rangle\langle\phi_c|$. Another way to put this is to say that ϕ sees an energy-dependent, non-local effective pseudopotential:

$$\hat{w} = v + \sum_c [\epsilon - \epsilon_c]|\phi_c\rangle\langle\phi_c|. \tag{6.58}$$

The second term tends to be repulsive and cancels a large part of the orginal potential, v. This is an indication of why potentials in solids look weak.

Though the pseudopotential is, in general, non-local and energy dependent, in many cases these effects are not too important, and we assume that the weak pseudopotential is an ordinary function, a *model potential*. For example, if we want to treat Na we try a simple form for the potential for the 3s states of the Na atom. Ashcroft (1968) suggested a potential of the form:

$$v = 0 \quad r < R_c$$

$$= -\frac{e^2}{r} \quad r > R_c. \tag{6.59}$$

The core radius, R_c is adjusted to fit the atomic energy. Then to transfer the potential to Na metal the same pseudopotential is used except that the effect of the other electrons is to screen the potential so that the behavior outside the core is more like $e^2 e^{-\lambda r}/r$. (Screening is treated in Chapter 8.)

Another method is based on the plane wave method above: if the pseudopotential is weak and smooth, it will have only a few large Fourier components. The *empirical pseudopotential method* takes these as adjustable parameters, and uses some constraints to fit the band structure. An example is the work of Cohen & Bergstresser (1966). They were interested in the semiconductors Si, Ge, and the similar III-V compounds such as GaAs. For example, Si has the diamond structure discussed in Chapter 3. There are two atoms per unit cell so the potential terms in Eq. (6.43) are replaced by $\mathcal{S}(\mathbf{G})v(\mathbf{G})$ where v is the potential for a single Si atom; see Problem 10. The authors restricted consideration to the first five shells of $\mathbf{G}$'s. It turns out (see problems) that only three independent coefficients are needed. These we determined by fitting to known parts of the band structure such as band gaps. These are known experimentally because, as we will see in detail below, light absorption can only take place if there is enough energy to excite an electron across a gap. So the onset of absorption gives the bandgap. The result is shown in Figure 6.6.

The state of the art in this sort of calculation is to build some of the electron-electron interactions into the pseudopotential. This is done using density-functional theory, which we will discuss later.

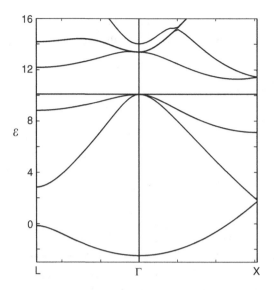

Fig. 6.6   **The band structure of Si computed using the empirical pseudopotential method. The pseudopotential form factors, from Cohen & Bergstresser (1966) are**
$w(\frac{2\pi}{a}[\pm1,\pm1,\pm1]) = -0.21$ **Ry,** $w(\frac{2\pi}{a}[\pm2,\pm2,0]) = 0.04$ **Ry,** $w(\frac{2\pi}{a}[\pm3,\pm1,\pm1]) = 0.08$ **Ry. In the last two cases the vectors gotten by permuting** x,y,z **have the same** w**. All the bands below the horizontal line are filled (valence bands) and those above are empty. Note the energy gap between empty and filled states.**

6.3.5 Metals and insulators

With some notion of band structure in hand we ask what happens, in the independent electron approximation, if we have N electrons. As in the Sommerfeld model we fill up the states in order of increasing energy to find the occupancies for a Slater determinant in the basis of the band states. There are as many states in a band as **k**'s, that is $2N$, counting spin. So each band has $2N$ states. For a material with one valence electron per atom (e.g. Na) the first band is half-filled.

Since the bands are no longer simply parabolas in k-space, the surfaces of constant energy will be non-spherical, and therefore the Fermi surface can have a complicated shape. In the nearly free electron model the surfaces will be pretty much spherical far from the zone edges, but with large distortions at the edge. Thus monovalent metals often have spherical Fermi surfaces.

Each band has a different surface associated with it. For small gaps, as the number of electrons grows, there will be some in the first band, and some in the next since the energies are very close to $\hbar^2 k^2/2m$. So, for Ca with 2 valence electrons, the first band will be partly filled, and the next will have some electrons too if the gap is small, which it is. This is called band overlap, see Figure 6.7.

We can now answer a qualitative question: why are some materials metals, and others insulators or semiconductors? In a metal electrons are free to move under the influence of external fields. It is reasonable intuitively that in order to accelerate an electron its quantum numbers must change. For weak fields they change by small amounts (we will make this

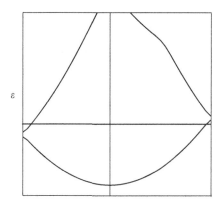

Fig. 6.7   **Band overlap for small gaps. On the horizontal axis k is plotted for two different directions as in Figure 6.3. Two bands are partly filled.**

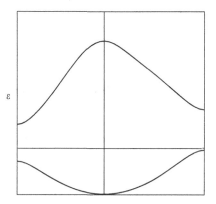

Fig. 6.8   **For large gaps, strong potentials, each band fills before another. On the horizontal axis k is plotted for two different directions as in Figure 6.3.**

quantitative later). Thus the most probable process for a band electron is to increase its **k**. However, if the next **k** is occupied, this is forbidden by the Pauli principle. The only electrons which participate in transport are those near the Fermi surface. Sometimes band gaps are large so that electrons fill up whole bands. Then a divalent material will fill a band completely, and there will be a gap between the filled and empty states. This is an *insulator*: it takes substantial energy to excite electrons from the filled to the empty states. The gap between filled and empty states can be measured optically – optical transitions between states are certainly possible. A semiconductor is an insulator with a small gap.

There is a rule (the Wilson rule) that is almost always obeyed: materials with an even number of valence electrons per unit cell can be metals or insulators. Materials with an odd number of valence electrons per unit cell must be metals. There are some exceptions to this, with NiO and some vanadium oxides being the most famous examples (Mott 1990). In these cases electron-electron interactions cause a *metal-insulator transition*. Almost always this is associated with the onset of magnetic order, i.e., the spins be ordered with a different period than the lattice. This introduces new gaps, and can make metals into insulators, as we will see later.

6.3.6 Measuring band structures and Fermi surfaces

The band picture is a very useful approximate way to think about electrons in solids. It is important to consider direct ways to measure the bands to confirm, on one hand, that they are well-defined, and that the approximation is a reasonable one, and also to check the validity of the computations using various schemes such as the pseudopotential method.

The most widely used and oldest methods rely on indirect effects of band structure on transport and optical properties. At the simplest level we can see that optical properties are important. When a photon is absorbed it excites an electron from an initial state in a filled band to a final state in an empty one. The photon gives up energy $\hbar\omega$. However, the wavelength of a photon in the visible is much larger than the lattice constant of a solid. Thus the momentum transfer of a photon to a solid is negligible on the scale of the Brillouin zone. The photon induces a "vertical" transition: the states connected are at the same place in the Brillouin zone. If a band structure has a vertical gap, Δ, that is, a minimum in $\epsilon_f(\mathbf{k}) - \epsilon_i(\mathbf{k})$, for any filled band i and empty band f, then the crystal will be *transparent* for $\hbar\omega < \Delta$. This is partial information about the band structure. We will see how to get more information from optical properties in the next chapter.

Transport measurements also give partial information, for example about *effective masses* which, as we will see below, are curvatures of bands. Once more, we are sampling parts of the band structure.

A very elegant direct method to measure bands is angle resolved photoemission spectroscopy (ARPES) and angle resolved inverse photoemission (ARIPES); see Martin (2004). The first method uses the photoelectric effect. If a photon is incident on a crystal at some angle, it can cause the emission of an electron. This requires photons whose energies are at least in the ultra-violet. The emission can be visualized as a three-step process: first the photon excites an electron (in a vertical transition), then the excited electron travels to the surface, and finally it exits the surface by giving up energy (the work function). The momentum of the emitted electron can be measured, and differs from the crystal momentum inside by one of the reciprocal lattice vectors of the surface. The energy of the excited electron inside and outside differ by the work function.

The perpendicular component of the momentum is not conserved because of the force keeping the electrons confined. However, we can also measure the energy of the emitted electron. It is reasonable to assume that the final state (of high energy) is free electron-like. Then using $\hbar\omega = \epsilon_f(\mathbf{k}) - \epsilon_i(\mathbf{k})$ we have enough information to reconstruct ϵ_i and $\mathbf{k}$. We also get information about the Fermi surface this way since the highest lying filled states can be found.

ARIPES uses inverse photoemission. That is, an electron is added to the crystal, it falls to an empty state from the high-energy band it starts in, and emits a photon. Both ARPES and ARIPES are of great interest beyond measuring band structures: they give information about interactions in cases where the independent particle model is not a good approximation.

Fermi surfaces can be experimentally measured in many other ways. For example, a neat (but seldom used) method is positron annihilation. The idea is to put positrons from a radioactive source into a metal where they slow down to be nearly at rest. The positron annihilates with an electron and usually emits two photons. If the electron is at rest as well,

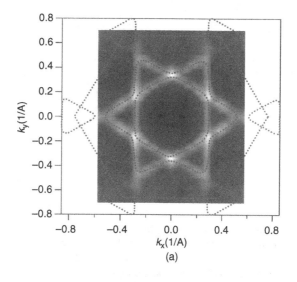

(a)

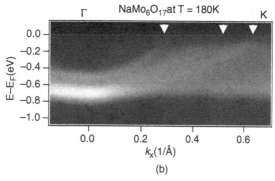

(b)

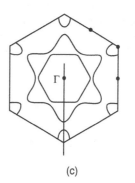

(c)

Fig. 6.9   (a) Fermi surface determined by ARPES for a quasi-two dimensional material $NaMo_6O_{17}$. The dotted lines indicate the calculated Fermi surface. (b) Three bands cross the Fermi surface at the arrows. (c) The $\mathbf{k}$ vector is along the line from the center of the surface to one of the points of the star. Courtesy of J. W. Allen.

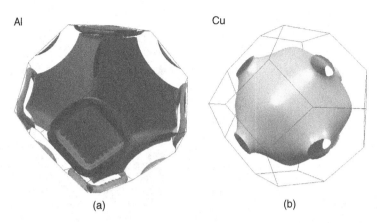

(a)   (b)

Fig. 6.10 **(a). The Fermi surface of Al which has valence 3. The first band is entirely filled. The second band gives rise to the parts of the surface in the center of the zone. These can be thought of as parts of a sphere which mapped into the Brillouin zone by translation by a G. The ring-like structures near the zone boundary are the third band. (b) The Fermi surface of Cu. From www.phys.ufl.edu/fermisurface, the Fermi surface database at the University of Florida.**

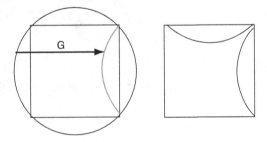

Fig. 6.11 **A two-dimensional Fermi sphere containing three electrons per atom in a square lattice. The periodic potential introduces gaps at the places where the circle cuts the zone edge. The mapping of the surface by various G's produces the surface on the right. Note that the states outside of the lines are filled; the states in the center of the zone are empty for this band.**

the photons will be "back to back," i.e., have zero net momentum. But if the electron is moving, the net momentum of the photons is not zero, and can be measured by the angle between their directions (no longer 180°). The distribution of angles gives the distribution of momenta inside the Fermi surface. Many Fermi surface measurements use other methods involving the orbits of electrons in magnetic fields, see Chapter 7.

We can get a very good account of the Fermi surfaces of many simple metals by going back to the nearly free electron model, above. If the bands are free electron-like then the Fermi surface is a sphere. However, when bands are mapped into the first Brillouin zone (cf. Figure 6.3) they consist of pieces which are parts of the sphere that are 'sliced' at the edge of the Brillouin zone. Oddly enough, that is what many Fermi surfaces do look like, for example, in Al. This was the first clue that the effective potential in many materials is weak. The Fermi surface of Al is shown in Figure 6.10(a). An illustration of the procedure of slicing and mapping in two dimensions is given in Figure 6.11. Note how the convex parts of the circle turn into concave regions when mapped, and compare to Figure 6.10(a).

The band structure of Cu is interesting. The 4s electron might be expected to give a spherical surface like that of Na. However, the 3d electron band lies in the middle of the 4s band. The resulting band mixing (or hybridization) moves the top of s band so that there are regions which touch the boundary of the zone. The resulting shape, Figure 6.10(b) is very famous. It was deduced from measurements of the anomalous skin effect (see Problem 6 in Chapter 7) in the 1950s.

Suggested reading

Second quantization is nicely treated in:
   Landau & Lifshitz (1977)
   Taylor & Heinonen (2002)
An excellent reference for modern methods in band theory with experimental and theoretical examples is:
   Martin (2004).

Problems

1. Verify that the expression for V_2 in Eq. (6.19) is correct by applying it to a trivial operator, namely a constant: set $v(\mathbf{r}_1, \mathbf{r}_2) = 1$. Then the contribution to the Hamiltonian is $N(N - 1)/2$ (say why) for a system with N electrons. Now show that the second quantized version gives $\hat{V}_2 = (1/2)(\hat{N}^2 - \hat{N})$ where $\hat{N} = \sum_j \hat{c}_j^+ \hat{c}_j$ is the operator for the total number. Then show that

$$\hat{V}_2 |\phi\rangle = [N(N - 1)/2]|\phi\rangle$$

for any state with a definite number of particles.

2. (a) The operator $\hat{\rho}(\mathbf{R}) = \sum_i \delta(\mathbf{R} - \mathbf{r}_i)$ measures the number density of electrons. Show that in second-quantized notation using plane wave orbitals:

$$\hat{\rho}(\mathbf{R}) = \frac{1}{\Omega} \sum_{\mathbf{k},\mathbf{q},s} e^{i\mathbf{q}\cdot\mathbf{R}} \hat{c}_{\mathbf{k}+\mathbf{q},s}^+ \hat{c}_{\mathbf{k},s}.$$

(b) The pair correlation function, $g(R)$ measures the likelihood of finding an electron at $\mathbf{R}$ given that one is at 0; see Eq. (3.52) and Figure 3.16. For the free electron gas show that:

$$g(R) \equiv \frac{\Omega^2}{N^2} \langle G|\hat{\rho}(\mathbf{R})\hat{\rho}(0)|G\rangle = 1 - \frac{9}{2} \left[\frac{\sin(k_F R) - k_F r \cos(k_F R)}{k_F^3 R^3} \right]^2$$

This is called the exchange hole: it is due to the statistical avoidance of electrons due to the Pauli principle. Plot the result.

3. Work out the density of states for free electrons in one, two, three, and four dimensions. In the course of this you will verify Eq. (6.27).

4. One implication of Eq. (6.29) is that there will be large contributions to $\mathcal{D}$ from *critical points* where $\mathbf{v_g} = 0$. Near such a point the energy can generically be written (changing the origin in $\mathbf{k}$ space and rescaling if necessary):

$$\epsilon(\mathbf{k}) = \epsilon_c + ak_x^2 + bk_y^2 + ck_z^2,$$

where a, b, c are ± 1. For example if $a = b = c = 1$ we are at a local band minimum, if they are all negative we are at a maximum. If some are negative and some positive the point is called a saddle point. Since the band energy is periodic in the Brillouin zone there are always critical points. (In fact, it can be shown that there are a least one minimum, one maximum, and six saddle points.)
(a) Consider the band near a minimum. Show that $\mathcal{D} \propto \sqrt{E - \epsilon_c}$ for $E > \epsilon_c$. (b) Find the result near the band maximum. c.) Consider a saddle point with $a = b = 1$; $c = -1$. Show that $\mathcal{D}$ is not singular for $E > \epsilon_c$ and has an infinite derivative as E approaches ϵ_c from below. These effects are called van Hove singularities.

5. The model of Kronig & Penney (1931) is a simple, mostly soluble model for band structure. This is a simplified version of it. Suppose we consider a one-dimensional system with repulsive δ-function potentials located at points spaced by a;

$$v(x) = \sum_n \eta \delta(x - na); \quad \eta > 0.$$

Work in a system of units where $\hbar = 2m = 1$. (a) Show that the relationship between k and ϵ is implicitly given by the equation:

$$\cos(ka) = \cos(\alpha a) - \frac{\eta}{2\alpha} \sin(\alpha a); \quad \alpha = \sqrt{\epsilon}.$$

(b) This equation only has solutions if the right-hand side is between -1 and 1. Plot the right-hand side as a function of ϵ, for some fixed value of η, and show that the allowed solutions fall in bands separated by gaps.
Hints: For x not equal to an integer the electron is free. For the region $x \in (0, a)$. Show that you can write $\psi = \sin(\alpha x) + e^{-ika} \sin(\alpha[a - x])$, using Eq. (6.38). Then show that $\psi'(x + u) - \psi'(x - u) = -\eta \psi(0)$ where $u \ll a$. Use this and the Bloch condition to prove:

$$\psi'(0) = e^{-ika} \psi(a) - \eta \psi(0).$$

This gives the equation above.

6. (a) For electrons in Na estimate E_F and k_F. You will have to look something up. (b) Write down formulas for E_F and k_F for highly relativistic electrons for which $\epsilon = pc$.

7. Derivation of the Sommerfeld expansion (a) Suppose that μ is the chemical potential, f the Fermi distribution function, and g is some smooth function of energy. Then:

$$\int_0^\infty g(\epsilon)f(\epsilon)d\epsilon = \int_0^\mu g(\epsilon)d\epsilon + \frac{\pi^2}{6}(k_B T)^2 g'(\mu) + \cdots$$

Prove this by integrating by parts. You may need the integral

$$\int_0^\infty x^2 e^x (e^x + 1)^{-2}dx = \pi^2/3.$$

(b) Work out the coefficient of T in the heat capacity of the Sommerfeld gas.

8. Work out the chemical potential for a two-dimensional electron gas. You should be able to do the integrals exactly. Exhibit the classical and quantum limits.

9. Work out the spin susceptibility of the electron gas to order $(k_B T)^2$. The first, temperature independent, term is due to W. Pauli.

Hint: in a magnetic field each electron has kinetic energy and Zeeman energy:

$$E = \epsilon(k) \pm g\mu_B B,$$

where g is the g-factor (near to 2), μ_B is the Bohr magneton, and B the magnetic field. You need a separate integral over the spin-up electrons and spin-down ones. But the chemical potential is the same for both.

10. (a) For a Bloch function, $\psi_{n,\mathbf{k}}(\mathbf{r})$, show that the charge density is periodic. (b) Define the Wannier function for the nth band in a crystal of N cells:

$$w_n(\mathbf{r}, \mathbf{R}) = \frac{1}{\sqrt{N}} \sum_{\mathbf{k}} e^{-i\mathbf{k}\cdot\mathbf{R}} \psi_{n,\mathbf{k}}(\mathbf{r}).$$

The sum is over the Brillouin zone, $\mathbf{R}$ is a lattice vector. Show that w depends only on $\mathbf{r} - \mathbf{R}$. (c) Show that

$$\langle w_n(\mathbf{r}, \mathbf{R})|w_m(\mathbf{r}, \mathbf{S})\rangle = \delta_{n,m}\delta_{\mathbf{R},\mathbf{S}}.$$

(d) Suppose that $\psi_{n,\mathbf{k}}(\mathbf{r})$ is a tight-binding wavefunction made from a single atomic wavefunction $\phi(\mathbf{r})$. What is the relationship between w and ϕ? (e) Consider a one-dimensional crystal of lattice constant a. Suppose the Bloch function is of the form $u_0(x)e^{ikx}$, where u_0 is independent of k. Find w and show that the charge density associated with w is peaked near one lattice site.

11. (a) Suppose there are several identical atoms in a unit cell each having an identical potential of interaction $v(\mathbf{r})$ with the electrons. Put

$$V_{\text{tot}} = \sum_{l,m} v(\mathbf{r} - \mathbf{R}_l - \rho_m)$$

where $\mathbf{R}_l$ are the points of the Bravais lattice, and ρ_m runs over the basis. Show that the quantity that determines the gaps in the nearly free electron model is

$$V_{\mathbf{G}} = \mathcal{S}(\mathbf{G})v_{\mathbf{G}}$$

where $\mathcal{S}(\mathbf{G}) = \sum_m e^{i\mathbf{G}\cdot\rho_m}$, the structure factor, and $v_{\mathbf{G}}$ is proportional to the Fourier transform of $v(\mathbf{r})$. (Compare Eq. (3.25).) (b) Prove that a monovalent element with the hexagonal close packed structure (hence with two electrons per unit cell) must be a metal. (c) Verify the statement made in Section 6.3.4 that only three numbers are needed to compute the Si band structure if we use the first five shells of $\mathbf{G}$'s.

12. Consider a divalent simple cubic metal. Sketch the energy bands up to the Fermi energy for $\mathbf{k}$ along the (100) direction and along the (111) direction in the free electron model. Locate the Fermi energy and put a scale on the picture. Be sure that you write each band in the form: $\epsilon(k) = (\hbar^2/2m)[\mathbf{k} + \mathbf{G}]^2$ and find the $\mathbf{G}$'s and put them on the picture.

13. Consider a two-dimensional metal with a triangular crystal structure with lattice constant a. There are two electrons per unit cell. (a) Find the reciprocal lattice and sketch the Brillouin zone. (b) Figure out the area of the Fermi disc in terms of a. Sketch it on top of the Brillouin zone, as well as you can. (c) Estimate the gap necessary to turn this material into an insulator. Express in terms of $\hbar$, m and a.

14. Suppose the E vs. $\mathbf{k}$ relation for a band of a simple cubic solid is given by

$$E(\mathbf{k}) = D(3 - \cos(ak_x) - \cos(ak_y) - \cos(ak_z))$$

where a is the lattice constant. Suppose this band is the valence band of a material with 2 electrons/unit cell. Suppose the energy gap is constant and equal to D everywhere on the surface of the zone. Is the material a metal or an insulator? Why?

15. (a) Find the tight-binding band structure for a bcc lattice with nearest neighbor overlaps. (b) Find the effective mass near the bottom of the band.

16. Try your hand at applying tight binding to NaCl and AgCl both of which have the same structure with nearest neighbor separation about 2.8 Å. Use atomic states for the ions, not the atoms. The electron affinity for Cl is 3.63 eV and the first and second ionization energies for Na and Ag are:

	E_1	E_2
Na	5.14	47.3
Ag	7.57	21.5

(a) Where, with respect to vacuum and each other do the highest occupied states of the ions lie? (b) A large correction must be made to the levels above to account for the electrostatic shift felt by an electron on an ion due to the charges of the neighbors. Estimate this effect (the Madelung energy) by finding the potential due to nearest neighbors only. (The estimate is grossly too big – use the correct answer, about ± 9eV in what follows. See Kittel (2005) for the correct way to do the problem taking all the neighbors into account.) How does the ± 9eV correction shift the energy levels of the previous part? (c) Make a guess, and say how you made it for the overlaps in the tight binding model, and roughly indicate the band widths. For which of the two compounds is the tight-binding model likely to be inadequate so that different atomic states will have to be mixed in an LCAO approach?

7 Dynamics of non-interacting electrons

The non-interacting electron model is remarkably – almost unreasonably – successful in describing many properties of condensed matter. In fact, in almost all common applications the non-interacting model is sufficient. Of course, if band-structure effects play a role, then we need to make sure that we have the right potential of interaction with the ions. As we will see later the correct potential must take electron-electron interactions into account.

In this chapter we discuss transport, that is, the way electrons move when perturbed by external fields. We will find the the non-interacting model works very well. In order to discuss this subject we need to set some background.

7.1 Drude model

After the discovery of the electron and its identification as the carrier of current in metals, a number of people treated the transport of electricity as a problem in classical gas dynamics. This is now known as the Drude model after P. Drude. It works remarkably well, and is still used for electrons in semiconductors, which are effectively classical.

Drude posed an equation of motion for classical electrons in a solid:

$$md\mathbf{v}/dt = e(\mathbf{E} + (1/c)(\mathbf{v} \times \mathbf{B})) - m\mathbf{v}/\tau. \tag{7.1}$$

Here e is the charge of the electron (a negative number), $\mathbf{E}, \mathbf{B}$ are the applied electric and magnetic fields, and τ is the *relaxation time*, namely the inverse of the rate of scattering. We now know that electrons in band states do not scatter; we think of τ as arising from scattering from inevitable impurities or imperfections in the crystal, or, at finite temperature, from "out of place" ions, that is, from lattice vibrations. The relaxation time is related to the *mean free path*, $\mathcal{L}$, by $\mathcal{L} = v\tau$. Long mean free paths in solids were a puzzle until quantum mechanics clarified that the interaction with the ions is a diffraction problem – the only scattering is Umklapp scattering.

If you are very well versed in statistical physics, you may object that the right thing to do is write a Boltzmann transport equation at this point and derive Eq. (7.1) as an average. This is a useful approach for detailed studies, and will be treated below.

7.1.1 Conductivity

We can get some quick results. For example, we can find the electrical conductivity by supposing that we are in a steady state with $\mathbf{E} \neq 0, \mathbf{B} = 0$. Then:

$$\mathbf{v} = e\tau\mathbf{E}/m$$

$$\mathbf{j} = ne\mathbf{v} = \frac{ne^2\tau}{m}\mathbf{E}. \tag{7.2}$$

Here n is the electron density, and $\mathbf{j}$ the current density. We identify the electrical conductivity, defined by $\mathbf{j} = \sigma\mathbf{E}$ as:

$$\sigma = \frac{ne^2\tau}{m}. \tag{7.3}$$

7.1.2 Hall effect

If we have both an electric and magnetic field applied to a conductor we can observe a phenomenon called the Hall effect. Suppose there is a steady state for conduction in a bar-shaped wire with $\mathbf{E}$ in the x-direction, along the bar, and $\mathbf{B}$ in the z-direction, perpendicular to the bar. The second term in Eq. (7.1) leads to a Lorentz force which makes the electrons turn in the y-direction. This leads to a pile-up of charge on the edges of the bar, and a *Hall voltage* develops in the y-direction. That is, we end up with another component of $\mathbf{E}$, namely E_y. Now write the equations of motion setting time derivatives to zero and $v_y = 0$ (since no flow is allowed across the bar):

$$mv_x/\tau = e(E_x)$$

$$0 = e(E_y - B_z v_x/c). \tag{7.4}$$

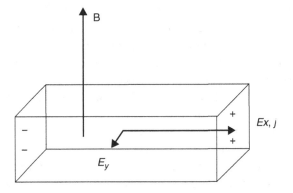

Fig. 7.1   **The Hall effect. Positive charge builds up on the back of the rod so that there is a Hall field, E_y as shown.**

Thus:

$$j_x = ne^2\tau E_x/m$$

$$E_y = -\frac{eB_z\tau}{mc}E_x. \tag{7.5}$$

Now note that j_x and E_y are both linear in τ. We can form a measurable quantity independent of the scattering time and of **B**:

$$R_H = \frac{E_y}{j_x B_z} = -\frac{1}{n|e|c}. \tag{7.6}$$

This quantity is called the *Hall coefficient*; note that in this model it is negative. This equation works well for simple metals. However, for some materials $R_H > 0$, indicating a positive charge carrier. We will see where this comes from.

There is another way to express the results above. We define the conductivity tensor by $\mathbf{j} = \Sigma\mathbf{E}$. It is quite easy to show that:

$$\Sigma = \frac{\sigma}{1+(\omega_c\tau)^2}\begin{pmatrix} 1 & -\omega_c\tau & 0 \\ \omega_c\tau & 1 & 0 \\ 0 & 0 & 1+(\omega_c\tau)^2 \end{pmatrix}. \tag{7.7}$$

In this equation the frequency $\omega_c = |e|B/mc$ is called the cyclotron frequency.

7.2 Transport in Sommerfeld theory

With quantum mechanics our notions of transport change. Electrons live in a Fermi sphere, and we have to take the Pauli principle into account. However, many of the Drude results

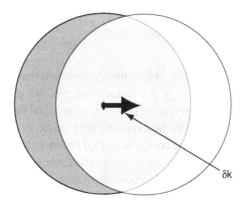

Fig. 7.2   **Displacement of the Fermi sphere corresponds to a steady current. The regions where the two spheres do not overlap gives the net current.**

persist. We will look at this in an intuitive way in this section, and be more precise below. Intuitively we assume that the effect of a small external force, $\mathbf{F} = e(\mathbf{E} + \mathbf{v} \times \mathbf{B}/c)$ is to change the momentum, $\hbar\mathbf{k}$ of the electrons by a small amount. The first thing we must note is that acceleration is forbidden for electrons in most states, because the nearby states are occupied. Only states near the Fermi surface can change. However, we can get the same result by assuming that all states change by a small amount, which amounts to shifting the Fermi sphere. Suppose a small field, $\mathbf{E}$ is applied, and scattering leads to a steady state velocity. Then we have, in time t an increase in momentum by:

$$\delta\mathbf{k} = e\mathbf{E}t/\hbar. \tag{7.8}$$

With scattering t should be set to τ since scattering limits acceleration. Now the center of the Fermi sphere is shifted by $e\mathbf{E}\tau/\hbar$. Thus, all the electrons have this amount added to their momentum, $\mathbf{k} = \mathbf{k}_o + \delta\mathbf{k}$. Now we sum the current over the shifted Fermi sphere. The net momentum of the unshifted Fermi sphere vanishes, i.e., $\sum \mathbf{k}_o = 0$, we get:

$$\mathbf{j} = \frac{e\hbar}{m\Omega} \sum \mathbf{k} = ne^2\tau/m, \tag{7.9}$$

as in Drude theory.

7.2.1 Wiedemann–Franz law

There is a striking success of the Sommerfeld theory which is related to thermal conductivity. It was observed that metals that were good conductors were good thermal conductors. That is why copper-bottom pots are used in cooking. In fact, the thermal conductivity and σ are proportional, and the proportionality constant is linear in T. Drude theory gets this almost right – off by a factor of 2, but Sommerfeld theory is amazingly accurate. Qualitatively the point is that both heat and electricity are carried by the electrons.

We recall Fourier's law and the definition of thermal conductivity, κ_T:

$$\mathbf{j}_Q = -\kappa_T \nabla T. \tag{7.10}$$

There is a standard kinetic theory argument that gives an expression for κ_T. Suppose a particle carries energy in a temperature gradient dT/dx. The mean flux of energy in the positive x-direction is $\langle (1/2)n|v_x|\rangle$, with an equal flux in the opposite direction. The energy carried is given by the heat capacity per particle, c, and the net transfer of energy as $\pm c\Delta T$. The distance that the energy is carried before being deposited is the mean free path in the x-direction: $\mathcal{L}_x = v_x\tau$. Now, $\Delta T = \mathcal{L}_x dT/dx = v_x\tau dT/dx$. Putting this together, we have the net flux of energy from both directions:

$$j_{Q,x} = -\langle (nv_x)(v_x\tau)\rangle c\frac{dT}{dx} = -n\langle v_x^2\rangle c\tau\frac{dT}{dx} = -\frac{1}{3}n\langle v^2\rangle c\tau\frac{dT}{dx}. \tag{7.11}$$

This gives the classic result:

$$\kappa_T = \frac{1}{3}C\langle v^2\rangle\tau = \frac{1}{3}Cv\mathcal{L},\tag{7.12}$$

where the last result holds for constant velocity, and C is the heat capacity per unit volume. This expression holds in insulators, where phonons carry the heat, if you put in the lattice heat capacity and the velocity of sound.

For the Sommerfeld model of metals we have calculated the heat capacity and the relevant speed is v_F. Inserting these gives:

$$\kappa_T = \frac{\pi^2 n k_B^2 T\tau}{3m}.\tag{7.13}$$

Comparing to the expression for σ we find the Wiedemann–Franz law:

$$\frac{\kappa_T}{\sigma T} = \frac{\pi^2 k_B^2}{3e^2}.\tag{7.14}$$

This says that the number on the right, the Lorenz number, is independent of the metal in which the measurements are made. The agreement with experiment is quite good. There are deviations at low temperatures because inelastic scattering affects heat and electrical transport differently.

7.3 Semiclassical theory of transport

For electrons in band states we need to rethink transport. Clearly, we cannot use Bloch states for a description since Bloch states have the same density in each unit cell, $|u|^2$. A way around this is to form wavepackets by combining electron states with different, but adjacent $\mathbf{k}$'s. There is another method that uses Wannier functions which we will not discuss.

7.3.1 Semiclassical equations of motion

Once we have a wavepacket that is more-or-less localized in both position and $\mathbf{k}$ (consistent with the uncertainty principle) we can ask how the center of the wavepacket responds to external fields. The answer is embedded in two equations which are called the *semiclassical transport equations*. For given band, n with band energy $\epsilon_n(\mathbf{k})$:

$$\dot{\mathbf{r}} = \mathbf{v} = \frac{1}{\hbar}\nabla_{\mathbf{k}}\epsilon_n(\mathbf{k})\tag{7.15}$$

$$\hbar\dot{\mathbf{k}} = e(\mathbf{E} + (1/c)(\mathbf{v}\times\mathbf{B})).\tag{7.16}$$

The first equation just says that the wavepacket moves with the group velocity. This is generally true. The second equation looks as if it says that $dp/dt = F$, i.e., Newton's second law. But things are not quite so simple in a solid where electrons are subjected to forces from the ions and $\hbar\mathbf{k}$ is not exactly the momentum.

The simplest way to understand the second equation is to make an argument about energy conservation. Suppose there is an $\mathbf{E}$ field with potential Φ. Then as the packet moves we expect that $\epsilon_l(\mathbf{k}) + e\Phi(\mathbf{r})$ should be constant. Taking the time derivative and using Eq. (7.15):

$$0 = \nabla_{\mathbf{k}}\epsilon_n(\mathbf{k}) \cdot \dot{\mathbf{k}} + e\nabla\Phi \cdot \dot{\mathbf{r}} = \nabla_{\mathbf{k}}\epsilon_n(\mathbf{k}) \cdot (\dot{\mathbf{k}} - e\mathbf{E}/\hbar). \tag{7.17}$$

This says nothing about the magnetic field part, for which a more complicated argument is necessary (Ashcroft & Mermin 1976, Marder 2000).

7.3.2 Bloch oscillations

So far we have neglected scattering. In a perfect crystal the mean free path is infinite. However, this is not the same as saying that an electron in a constant field will accelerate indefinitely. If $\mathbf{E}$ is constant $\mathbf{k}$ will increase. When a band edge is reached there will be an Umklapp. To see what this means, consider the 1d case. When $k \approx \pi/a$ the wavefunction will be a mixture of k and $k - G = k - 2\pi/a$. As k increases further, we will end up on the other side of the zone. For this purpose is it useful to use the periodic zone scheme where each band is repeated. Now $v = d\epsilon/dk$ will oscillate, and so will the center of the wavepacket. In principle the electron will accelerate periodically and emit radiation.

No solid has ever been made pure enough to observe this effect despite serious efforts. In fact, much of the early research on semiconductor superlattices (see below) was motived by the search for this effect. Bloch oscillations have been observed in an analog system, atom traps with periodic potentials provided by standing waves of laser light.

7.3.3 Effective mass

In ordinary mechanics we usually combine the equations $\dot{p} = F, \dot{r} = p/m$ to get $ma = F$. If we do that with Eq. (7.16) we get a strange result. Look at the *ith* component of the vectors:

$$a_i = \ddot{r}_i = \frac{1}{\hbar}\sum_j \frac{\partial^2\epsilon(\mathbf{k})}{\partial k_i \partial k_j}\dot{k}_j$$

$$= \sum_j \left(\frac{1}{m^*}\right)_{i,j} F_j$$

$$\mathbf{a} = \left(\frac{1}{m^*}\right) \cdot \mathbf{F}. \tag{7.18}$$

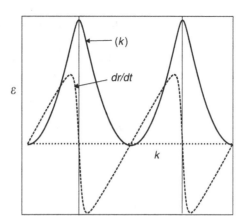

Fig. 7.3   **Band structure in the periodic zone scheme for the lowest band and group velocity in one dimension. If k increases under the influence of a constant field the electron velocity oscillates. Note that as the band edge is approached dr/dt decreases, even though $\hbar k$ increases.**

This equation defines the *inverse effective mass tensor*:

$$\left(\frac{1}{\mathsf{m}^*}\right)_{i,j} = \frac{1}{\hbar^2}\frac{\partial^2 \epsilon(\mathbf{k})}{\partial k_i \partial k_j}. \tag{7.19}$$

Note that the tensor is not diagonal so that transport is anisotropic: $\mathbf{F}$ and $\mathbf{a}$ need not be parallel. Even if the tensor is diagonal the diagonal terms need not be equal. They can be *negative* so that force and acceleration are in opposite directions. However, $(\mathsf{m}^*)^{-1}$ is symmetric, and can always be diagonalized. The inverses of the eigenvalues are called the effective masses, m^*. Flat bands have large masses because the curvature of ϵ is small. Near a band maximum the m^* will be negative.

As an example, the first unfilled band of Si (the conduction band) is important because electrons can be injected into it, often by impurities called *donors*. It has six equivalent minima along the [100] directions inside the zone. There are two masses because the constant energy surfaces near the minimum are ellipsoids of rotation. The mass corresponding to transport along [100] is called m_l and across, m_t. The values are $m_\mathrm{l} = 0.92 m_e, m_\mathrm{t} = 0.19 m_e$, in both cases less than the mass of the free electron, m_e.

7.3.4 Holes

We have seen that in cases of band overlap some bands can be nearly filled. For semiconductors electrons can be thermally excited across the band gap leaving unfilled states in the highest occupied band (called the valence band). Or electrons can be removed from the valence band by *acceptor* impurities (see below). These minority unoccupied states are called *holes*. The lowest energy state with a few holes will have the holes at the top of the band. In Figure 6.11 we saw another example: the states in the center of the zone are empty, or, alternatively, are filled with holes. In a metal with part of a band filled, we can think of holes instead of electrons.

When we apply a field the electrons in the band change their $\mathbf{k}$'s, and thus the holes move in k-space. The current can be associated with the hole alone because the electrical current for a filled band is 0 – there are always as many negative velocities as positive. Thus:

$$
\begin{aligned}
\Omega \mathbf{j} &= \sum_{\text{onehole}} (-|e|\mathbf{v}) \\
&= \sum_{\text{filled}} (-|e|\mathbf{v}) + |e|\mathbf{v}_{\text{hole}}.
\end{aligned}
\tag{7.20}
$$

Thus the current looks like that of a positive charge. Further, since the electron states at the top of the band generically have negative mass, the electron states will accelerate opposite to $-|e|\mathbf{E}$.

We can simplify our view by using the *particle-hole transformation*. This allows us to describe nearly filled bands by the following steps:

- We consider holes to be *positively charged* particles. This is why some Hall coefficients are positive.
- The energy of a hole is measured *downwards* from the top of the valence band. Accordingly, the mass of the hole changes sign to $-m^*$, i.e. positive mass near the band maximum.
- If there is an electron missing from a state with crystal momentum $\hbar\mathbf{k}$ we attribute a crystal momentum $-\hbar\mathbf{k}$ to the hole. This is because a filled band has zero net momentum so that the hole gets the missing momentum.
- The group velocity of the hole is equal to what an electron in the empty state would have.

Thus, when we excite a semiconductor we can say that we have created an electron-hole pair. Light of energy 0.8 eV causes pair production in Si.

7.4 Scattering and the Boltzmann equation

In the Drude theory we introduced the relaxation time, τ, the inverse of the scattering rate. However, in the present discussion we have neglected scattering altogether. In order to take this important effect into account we need to see how to deal with scattering of the the wavepackets that we have introduced. We follow the classic work of L. Boltzmann on transport in dilute gases; see (Huang 1987). Boltzmann introduced a distribution function that gives the density of finding a carrier at $\mathbf{r}$ with momentum $\hbar\mathbf{k}$, $f(\mathbf{r}, \mathbf{k}, t)$. To fix the normalization we put:

$$
N = \frac{1}{\Omega} \sum_{\mathbf{k}\in K} \int_{\mathbf{r}\in R} d\mathbf{r}\, f(\mathbf{r}, \mathbf{k}, t),
\tag{7.21}
$$

for the total number of electrons in a region R of space with momenta in a region K of k-space. We can do this with wavepackets as long as we do not violate the uncertainty principle.

In the absence of scattering the distribution function changes with time merely because carriers move (if there is a velocity), and change momenta (if there is a force):

$$f\left(\mathbf{r} + \frac{d\mathbf{r}}{dt}dt, \mathbf{p} + \frac{d\mathbf{k}}{dt}dt, t + dt\right) = f(\mathbf{r}, \mathbf{k}, t),$$

$$\frac{\partial f}{\partial t} + \mathbf{v} \cdot \nabla_{\mathbf{r}} f + \frac{1}{\hbar}\mathbf{F} \cdot \nabla_{\mathbf{k}} f = 0. \tag{7.22}$$

The values of $\mathbf{v}$ and $\mathbf{F} = \hbar d\mathbf{k}/dt$ are to be taken from Eq. (7.16). Now if there is scattering the positions and momenta change for other reasons. We write:

$$\frac{\partial f}{\partial t} + \mathbf{v} \cdot \nabla_{\mathbf{r}} f + \frac{1}{\hbar}\mathbf{F} \cdot \nabla_{\mathbf{k}} f = \left(\frac{\partial f}{\partial t}\right)_{\text{coll}}. \tag{7.23}$$

The collision term needs to be found.

We note first that this theory is actually easier than the Boltzmann theory in gases. In dilute gases atoms scatter from each other. In the approximation of this chapter we neglect this altogether so that all the scattering is from defects of some kind, i.e., deviations from periodicity. Recall that the Bloch states from which we make our wavepackets already fully account for the periodic potential. A common deviation from periodicity is impurities and other static defects (grain boundaries, vacancies, etc.) These scatter the electrons because there is a scattering cross-section from each such defect. In a metal this gives rise to temperature-independent scattering.

In addition thermal effects move the ions in the lattice from their average positions. This gives temperature-dependent scattering. We can think of the two kinds of scattering as being resistors in series. Thus the resistivity should be given by:

$$\rho = \rho_{\text{i}} + \rho_{\text{L}}(T). \tag{7.24}$$

Here the first term (the residual resistivity) is due to static defects such as impurities, and the second term arises from temperature-dependent electron-lattice scattering. This is called Matthiessen's rule.

In a semiconductor or an insulator there is another source of temperature dependence of the resistivity, namely that the number of carriers is thermally activated: see Problem 4. In a metal ρ typically increases with temperature, but for a pure semiconductor ρ decreases.

7.4.1 Relaxation time approximation

A simple approximation to the collision term can be written down phenomenologically. We note that in equilibrium there are no currents, and the distribution is the Fermi distribution, $f_o(\epsilon)$, Eq. (6.31). If we perturb the distribution away from equilibrium and let the system develop, the effect of scattering is to return the distribution to f_0. Now if we write $f(\mathbf{r}, \mathbf{k}, t) = f_0 + g$, it is reasonable to put:

$$\left(\frac{\partial f}{\partial t}\right)_{\text{coll}} = -\frac{f - f_0}{\tau} = -\frac{g}{\tau}. \tag{7.25}$$

We interpret τ as a measure of the scattering rate from impurities and phonons.

Electrical conductivity

From this point of view we can look again at the conductivity. Suppose that the system is in a steady current carrying state so that $\partial f / \partial t = 0 = \nabla_{\mathbf{r}} f$. We take a constant electric field to be present and set $\mathbf{B} = 0$. Now the Boltzmann equation reads:

$$\frac{e\mathbf{E}}{\hbar} \cdot \nabla_{\mathbf{k}} f = -\frac{g}{\tau}. \tag{7.26}$$

We suppose the field to be small and of the same order as g. Therefore, on the right we can put $f = f_0$. That is:

$$g = -\frac{\tau e \mathbf{E}}{\hbar} \frac{\partial f_0}{\partial \epsilon} \nabla_{\mathbf{k}} \epsilon = e\tau \left(-\frac{\partial f_0}{\partial \epsilon}\right) \mathbf{E} \cdot \mathbf{v}. \tag{7.27}$$

We have used Eq. (7.16). Recall that for low temperatures in a metal $\partial f_0 / \partial \epsilon \approx \delta(\epsilon - \mu)$. This deviation from the equilibrium distribution is precisely what is pictured in Figure 7.2.

Now we can find the electric current density in direction i by averaging over the distribution function.

$$j_i = \frac{1}{\Omega} \sum_{\mathbf{k}} e v_i f = \frac{1}{4\pi^3} \int d\mathbf{k} \, v_i g = \frac{e^2 \tau}{4\pi^3} \int d\mathbf{k} \left(-\frac{\partial f_0}{\partial \epsilon}\right) v_i v_k E_k, \tag{7.28}$$

where a sum on $k = x, y, z$ is implied. This equation defines the conductivity tensor, σ_{ik}. For an isotropic material, or if the material has cubic symmetry the tensor is proportional to the unit tensor so that $\mathbf{j}$ is parallel to $\mathbf{E}$. Further, in a metal, the integral is over the Fermi surface. Suppose $\mathbf{E}$ is in the x-direction. We have an average of v_x^2 over the Fermi surface. For cubic symmetry we get:

$$\sigma = \frac{e^2 \tau}{12\pi^3} \int d\mathbf{k} \, v^2 \delta(\epsilon_{\mathbf{k}} - \mu). \tag{7.29}$$

It is easy to show that this reduces to the Drude expression for a free electron gas.

7.4.2 Origin of the relaxation time

The relaxation time can be expressed in terms of scattering. We will continue to consider the simple case of a metal where f is a function of $\mathbf{k}$ alone. Suppose we define $P(\mathbf{k}, \mathbf{k}')d\mathbf{k}$

to be the scattering rate from state $\mathbf{k}$ to $\mathbf{k}'$. We can write:

$$
\begin{aligned}
\left(\frac{\partial f}{\partial t}\right)_{\text{coll}} &= \int d\mathbf{k}'\, P(\mathbf{k}, \mathbf{k}')[f(\mathbf{k}')(1 - f(\mathbf{k})) - f(\mathbf{k})(1 - f(\mathbf{k}))]d\mathbf{k}', \\
-\frac{g_{\mathbf{k}}}{\tau} &= \int d\mathbf{k}'\, P(\mathbf{k}, \mathbf{k}')[g(\mathbf{k}') - g(\mathbf{k})].
\end{aligned}
\tag{7.30}
$$

The first term represents scattering into state $\mathbf{k}$ from $\mathbf{k}'$ and the second is the reverse.

Now we suppose that we are dealing with elastic scattering in an isotropic material. Instead of integrating over $\mathbf{k}'$, we can use the energy, $\epsilon_{\mathbf{k}'}$ and scattering angle θ (the angle between $\mathbf{k}$ and $\mathbf{k}'$) as variables. Then:

$$
P(\mathbf{k}, \mathbf{k}')d\mathbf{k}' = \delta(\epsilon_{\mathbf{k}} - \epsilon_{\mathbf{k}'})\, d\epsilon_{\mathbf{k}'}\, \bar{p}(\theta)\, 2\pi \sin(\theta)d\theta.
\tag{7.31}
$$

The new quantity $\bar{p}$ is proportional to the cross-section for scattering. Consider the case of impurity scattering. Then the rate will be proportional to the density of impurities, n_i, and the velocity of the electrons, v_F:

$$
\bar{p}(\theta) = n_i v_F \sigma'(\theta).
\tag{7.32}
$$

Here $\sigma' = d\sigma/d\Theta$ is the differential scattering cross-section.

Now we insert Eq. (7.27), Eq. (7.31), and Eq. (7.32) into Eq. (7.30). After a bit of algebra we find:

$$
\frac{\mathbf{v}_{\mathbf{k}} \cdot \mathbf{E}}{\tau} = n_i v_F \int 2\pi \sin(\theta)d\theta\, \sigma'(\theta)(\mathbf{v}_{\mathbf{k}} - \mathbf{v}_{\mathbf{k}'}) \cdot \mathbf{E}.
\tag{7.33}
$$

We can take $\mathbf{v}_{\mathbf{k}} = v_F \hat{\mathbf{z}}$. Using isotropy we see that the component of $\mathbf{v}_{\mathbf{k}'}$ that survives the integration is along z. That is:

$$
\frac{1}{\tau} = n_i v_F \int 2\pi \sin(\theta)d\theta\, (1 - \cos(\theta))\sigma'(\theta).
\tag{7.34}
$$

Note that small angle scattering is ineffective in reducing the current.

There is also scattering in a pure material at finite temperature. We have seen in Chapter 3 that the scattering of an external incident wave from all the ions is proportional to the structure factor. We can carry this over to the electrons inside the material. The object that goes on the right-hand side of Eq. (7.32) in this case is, from Eq. (3.37):

$$
n|f_1|^2 S(\mathbf{q}).
\tag{7.35}
$$

Now suppose we are in a liquid metal. $S(\mathbf{q})$ can be measured by X-ray scattering as we have seen in Chapter 3 . We can use a screened pseudopotential (e.g. Eq. (6.59)) in the Born approximation to get the cross-section:

$$
f_1 = \frac{m}{2\pi\hbar^2} \bar{v}(\mathbf{q}).
\tag{7.36}
$$

For elastic scattering $q = 2k_F \sin(\theta/2)$. Substituting into Eq. (7.35) gives a formula due to J. Ziman for the resistivity of a liquid metal:

$$\rho = \frac{m^2}{12Ze^2 n_e \hbar^3 \pi^3} \int_0^{2k_F} dq \, q^3 S(q) |\bar{v}(q)|^2. \tag{7.37}$$

We have written the formula for valence Z and n_e is the density of the electrons.

For a crystal at finite temperature we have to consider inelastic scattering with the emission and absorption of phonons. An electron in state $\mathbf{k}$ scatters to state $\mathbf{k}'$ with emission of a phonon in $\mathbf{q} = \mathbf{k} - \mathbf{k}'$, cf. Section 5.3.2. The structure factor is the absolute square of:

$$\sum_{\mathbf{R}} e^{i\mathbf{q}\cdot\mathbf{R}} e^{i\mathbf{q}\cdot\mathbf{u}(\mathbf{R})} = \sum_{\mathbf{R}} e^{i\mathbf{q}\cdot\mathbf{R}}(1 + i\mathbf{q} \cdot \mathbf{u}(\mathbf{R}) + \cdots). \tag{7.38}$$

The first term is zero because we are scattering from state $\mathbf{k}$ to state $\mathbf{k}'$ which are both in the Brillouin zone. In other words, we have already taken the static structure factor into account in the band structure. The next term can be analyzed in normal modes using Eq. (5.78). The result is:

$$\sum_{\mathbf{R},\mathbf{k},\beta} \sqrt{\frac{1}{N}} q_{\mathbf{k},\beta}\mathbf{q} \cdot \mathbf{U}_\beta(\mathbf{k}) e^{i(\mathbf{k}+\mathbf{q})\cdot\mathbf{R}} \quad =$$

$$\sqrt{N} \sum_{\mathbf{k},\mathbf{G},\beta} \delta_{\mathbf{k}+\mathbf{q},\mathbf{G}} q_{\mathbf{k},\beta}\mathbf{q} \cdot \mathbf{U}_\beta(\mathbf{k}) e^{i\mathbf{G}\cdot\mathbf{R}}. \tag{7.39}$$

Recall that $q_{\mathbf{k}}$ is the normal mode amplitude. There are normal scattering terms (for $\mathbf{G} = 0$) and Umklapps. For normal scattering only the longitudinal mode is involved. Now:

$$S(\mathbf{q}) = \langle |q_{\mathbf{q}}\mathbf{q} \cdot \mathbf{U}(\mathbf{q})|^2 \rangle = |\mathbf{q} \cdot \mathbf{U}(\mathbf{q})|^2 \langle |q_{\mathbf{q}}|^2 \rangle. \tag{7.40}$$

We will treat the normal modes as classical, which is valid for $T > \Theta_D$. The equipartition theorem gives $\langle |q_{\mathbf{q}}|^2 \rangle = k_B T / M\omega_{\mathbf{q}}^2$. Thus, for a monovalent metal:

$$\rho = \frac{m}{e^2} \frac{k_B T}{Mc^2} \int 2\pi d\theta \, (1 - \cos(\theta)) |f_1|^2, \tag{7.41}$$

where c is the velocity of sound. The resistivity of metals for high temperatures is proportional to T. For low temperatures we must analyze q in phonon modes as in Eq. (5.86). This gives a normal process resistance that is proportional to T^5 (Ashcroft & Mermin 1976).

These results can be gotten another way by noting that since electrons scatter from lattice vibrations we should write a Hamiltonian containing terms that couple electron operators to phonon operators. This is called the electron-phonon interaction. It will be treated in Chapter 9.

7.5 Donors and acceptors in semiconductors

A remarkable, and remarkably important (from the point of view of technology) application of the ideas above is to semiconductors. A semiconductor is an insulator with a relatively small band-gap so that there is thermal generation of carriers at ordinary temperatures. The bands and band-gaps are shown in Figure 7.4 for Si. In Si the top of the valence band is in the center of the zone (the Γ point), and the bottom of the conduction band occurs at six equivalent points along the line from Γ to X. This is called an indirect gap. In GaAs the gap is direct: the bottom of the conduction band is at Γ.

7.5.1 Donors

Consider the semiconductor Si. If we add a few impurities (this is called *doping*), the electrical properties change considerably in a controllable way. This is what makes Si and other semiconductors the materials of choice for electronic devices.

The idea is as follows. For Si, valence 4, suppose we put in a few P, valence 5, a donor impurity. The chemistry of the substitution is such that four of the valence electrons of the P are used up in covalent bonds, and the other one is liberated into the conduction band making Si doped in this way an *n-type* semiconductor – the n is for the negative charge of the conduction electron. The fact that this happens is what makes Si so useful: if the bonding were different we would have a less interesting situation, and we would not be able to control the conduction electron density.

However, this is not the whole story. The P impurity has positive charge left over, and the conduction electron will be attracted to it. Thus we have a single electron in the band,

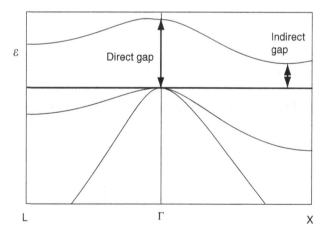

Fig. 7.4   **Direct and indirect gaps in Si. The band structure a portion of the structure shown in Figure 6.6. The direct gap is 0.8 eV.**

and a single fixed charge. It is clear that we can have a bound state which is just like the ground state of the H atom with two differences:

- The mass will be the effective mass, m^*. As we have seen, for the Si conduction band $m^* < m_e$ so the state will be less bound. For this anisotropic case we need to take an average of m_l and m_t.
- The Coulomb potential of the donor is screened, as are all potentials inside a dielectric, with dielectric constant ϵ. This weakens the potential and makes the state less bound.

The upshot of this is that the donor state is bound by an amount that we can get by using the usual formula for the ground state of H. This energy is measured from the band edge, and is given by:

$$E_0 = -\frac{m^* e^4}{2\epsilon^2 \hbar^2}. \tag{7.42}$$

For Si $\epsilon \approx 12$. The binding energy turns out to be about 20 meV. This means that at room temperature the donor will be thermally ionized. The binding energy is much less than the band gap, hence the name *shallow* impurity. This H atom is very big since it is weakly bound. The size of the 1s wavefunction is about 80 Å, 160 times as big as in hydrogen.

We have explained the state of the donor by using the ideas of semiclassical transport theory. If we were to formulate a quantum-mechanical theory of the donor, we would look for a solution for the band-structure problem with a slowly varying potential (due to the charge of the donor atom screened by the dielectric constant of the semiconductor) superimposed. Such a theory has been worked out and goes by the name of *effective mass theory*. The result is that the wavefunction for the donor state is composed of an envelope function which is the solution of the hydrogen problem with $m \rightarrow m^*$ and $e^2 \rightarrow e^2/\kappa$ multiplied by the Bloch wavefunction for the conduction band edge. For a full description see Kohn (1957).

7.5.2 Acceptors

We can also dope Si with a valence 3 impurity like Al. This makes an *acceptor* impurity. Once more it is energetically favorable for the impurity to make four covalent bonds. To do this it robs the valence band of an electron, making a hole. The whole story of donors is repeated, standing on its head, with holes bound to the negative impurity. At room temperature the acceptors in Si are ionized, so that there are holes in the valence band. This is a *p-type* material.

7.6 Excitons

There is an excited state of semiconductors or insulators which is the result of the pair production mentioned above. If we shine light on a semiconductor which has $\hbar\omega$ greater than a gap, we get an electron-hole pair. As we noted in the previous chapter, the momentum of the photon, $\hbar\mathbf{k}$ which is transferred to the crystal is very small on the scale of the Brillouin zone. The photon causes a "vertical transition" in the band structure. Thus optical absorption sets in at the *direct gap*.

However, there is actually absorption below the gap energy because there is a final-state interaction between the electron and the hole. That is, we can have a bound state of the electron and the hole because they are oppositely charged. In a direct-gap semiconductor like GaAs, excitons decay rapidly by radiative transitions. However, in an indirect-gap material the lowest exciton is much longer lived (on the order of microseconds) because crystal momentum cannot be conserved in a purely radiative decay – remember that the photon has negligible $\mathbf{k}$. It is necessary to also emit a phonon to take up the momentum.

Excitons are important in many optical properties. They are created by the coupling between the electric field of incident light and the dipole moment of the electron-hole pair. Thus they represent a wave of polarization just as in the case of the ionic dipoles of Section 5.1.7. Excitons can mix with incident light to make another kind of polariton.

Excitons are used to make quantum-well lasers. An exotic phenomenon occurs in Si at low temperatures: excitons are long lived and can be made in great numbers. By pumping the crystal continuously there can be a steady-state population of excitons which, in certain conditions, condense into a liquid called the *electron-hole liquid*; see Problem 1 of Chapter 9.

7.7 Semiconductor devices

Almost all of modern microelectronics is based on the fact that we can tailor the conduction of Si or GaAs by suitable doping with donors and acceptors. For example, we can make p-n junctions by putting two materials together. This turns out to be a diode rectifier. Transistors, solar cells, light-emitting diodes (LED's) semiconductor lasers, and many other devices are all variations on this theme. This is a vast field and whole textbooks and courses are easily available. To give a flavor of the subject we will describe some aspects of two devices, the p-n junction, and the Metal-Oxide-Semiconductor Field Effect Transistor (MOSFET).

There are some elementary considerations from statistical mechanics that we will need. In equilibrium there will be a few thermally generated electrons and holes. We can consider their generation to be a chemical reaction $n + p \leftrightarrow 0$. The law of mass action of elementary chemistry says that the product of the concentrations for a given semiconductor depends on temperature alone:

$$np = K(T). \tag{7.43}$$

If we add lots of electrons by doping with donors we suppress the number of holes. In an *n*-type material electrons are called majority carriers and the few holes present according to Eq. (7.43) are called minority carriers.

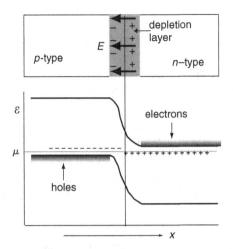

Fig. 7.5 **Top: the p-n junction with the space charge region giving rise to an internal electric field. Bottom: the bands edges bend so that the electro-chemical potential is the same on both sides of the junction.**

Consider an *intrinsic* material, that is, one with very few impurities. We ask for the location of the chemical potential. Recall the Fermi distribution function, Eq. (6.31) and its temperature dependence, Figure 6.2. The mean occupancy at the chemical potential is 1/2. Now apply this to an intrinsic semiconductor at finite temperature. The number of electrons and holes is equal and the electrons have energies above the edge of the conduction band, and the holes below the top of the valence band. It is not hard to believe (and easy to show – see problems) that the chemical potential will be around the middle of the gap.

Now consider an *n*-type material. Once more we have thermal generation of carriers, but now the donor states are partially filled. In this case the chemical potential will be pinned near the donor level, namely near the bottom of the conduction band. Similarly, in a *p*-type material the chemical potential will be near the top of the valence band.

Now suppose that we grow a material so that *n* regions and *p* regions are in contact. This is a p-n junction. When electron transfer is allowed between the two parts of the junction the chemical potentials must line up so that the conduction band edge in the *n*-type is aligned with the valence band edge in the *p*-type. The bands must "bend" at the junction: see Figure 7.5. The way this occurs is that electrons spill into the *p* region and fill the holes on some of the acceptors. This exposes the negative acceptor impurity. Similarly, holes move into the *n* side and expose donors. Thus there is a region with few carriers (the depletion zone) and an electric field. In equilibrium there will be thermally generated electrons and holes on the right. The electrons will be accelerated by the field of the diode to move into the *n* region. This current, I_g, must be balanced by an "uphill" current in the opposite direction, I_r which is thermally activated to climb the barrier and recombine with holes in the *p* region. The same is true of the holes in the opposite direction.

This device can serve as a light emitting diode (LED). Suppose we apply a bias so that electrons move to the left (forward bias). This lowers the barrier so that I_r is increased.

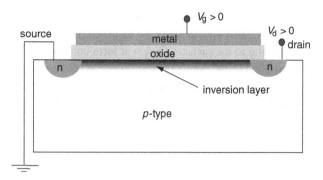

Fig. 7.6

A sketch of a MOSFET.

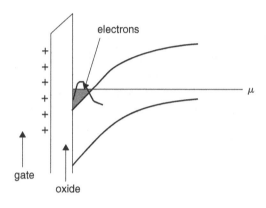

Fig. 7.7

Formation of an inversion layer. Electron energy is plotted as a function of distance from the surface. When the gate voltage is large enough the conduction band bends below the chemical potential so that electrons can be injected. The bold line is the electron wave function in the direction perpendicular to the layer.

Now there is a net current across the junction since I_g is unaffected. If the semiconductor is direct gap many of the electrons will fall into holes and emit light.

In a solar cell, light falling on the same device at zero bias increases the number of electron hole pairs. Now I_g is increased so that excess electrons end up in the n region and holes in the p region. Now a forward bias is created, and current runs around the external circuit. Solar cells can be made from both direct and indirect gap semiconductors.

A MOSFET is based on a similar kind of band bending. In a common version of this device a p-type piece of Si is coated with an oxide and topped with a metal gate. If the gate voltage is large enough the bands will bend so that there is a channel where electrons are stable. That is, we have channel of negative carriers in a p-type material. This is called an *inversion layer*. Contacts at each end of the layer provide a source and drain for electrons. The point of the device is that the gate voltage controls that width of the inversion layer and the current through the device. Small changes in gate voltage produce large changes in current. See Figures 7.6 and 7.7.

The most interesting situation for fundamental physics occurs when the inversion layer is very thin. Then the electrons are confined in a V-shaped potential well in the direction perpendicular to the surface. They will form bound states in this case; see Figure 7.7. If the splitting between the perpendicular energy levels is greater than $k_B T$ the electron degree of freedom perpendicular to the surface is frozen out, whereas they are still free to propagate along the surface. Thus we have a *two-dimensional* electron gas.

Two-dimensional electron gases can also be formed in heterostructures, namely layers of different materials grown on top of one another. We will not go into detail about this, but it is not hard to produce a similar inversion layer at the surface between pure GaAs and n-type AlGaAs.

7.8 Large magnetic fields

Some of the most interesting results in electron transport involve orbits in large magnetic fields. We recall some results of classical mechanics: A particle with charge e in a magnetic field, $\mathbf{B}$, will move in a circular orbit with frequency $\omega_c = |e|B/mc$ (the cyclotron frequency) in the plane perpendicular to $\mathbf{B}$. The particle freely translates along the direction of the field so that the general motion is a helix. What we mean by a large $\mathbf{B}$ is a field such that $\omega_c \tau \ll 1$ so that the particle can execute many rotations before scattering.

7.8.1 Cyclotron orbits in solids

For free electron-like metals like Na the behavior in a magnetic field is exactly like that of free electrons up to the use of an effective mass. However, when the Fermi surface intersects zone edges Umklapps become important. In this case electrons in a static field can move in *open orbits*, that is, they are transported in some direction rather than move in circles. In this case electrical conductivity depends on $\mathbf{B}$; this is called magnetoresistance. This is an easily observable analog of Bloch oscillations.

Cyclotron orbits in solids reflect the shape of the Fermi surface. In a magnetic field alone and without scattering electrons move on surfaces of constant energy and constant momentum perpendicular to $\mathbf{B}$. Furthermore, the equation of motion in a magnetic field (cf. Eq. (7.16)):

$$\hbar \dot{\mathbf{k}} = \frac{e}{c}(\mathbf{v} \times \mathbf{B}), \tag{7.44}$$

means that the real-space orbit is a rotated image of of the k-space orbit. Thus up to the scale factor

$$dk = (|e|B/\hbar c)dr, \tag{7.45}$$

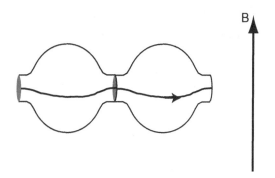

Fig. 7.8   A Fermi surface that intersects the zone boundaries (gray circles) can give rise to open orbit rather than the closed orbits of free electrons. The process may be viewed as an Umklapp — the momentum of the electron is transferred to the other side of the zone, or by repeating zones, as shown. In any case, the real-space orbit corresponds to a velocity in a more-or-less constant direction.

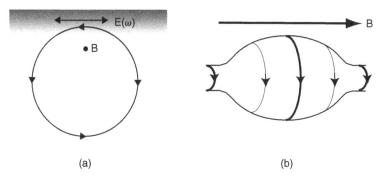

(a)   (b)

Fig. 7.9   (a)The Azbel–Kaner geometry for cyclotron resonance in metals. The high-frequency electric field cannot penetrate beyond the skin depth (shown in gray). The magnetic field is out of the paper. Every time an electron in a cyclotron orbit passes through the skin depth it is accelerated by $\mathbf{E}$. Successive accelerations will be in phase and lead to large absorption of power if the period of the orbit is a multiple of the period of $\mathbf{E}$, that is if $\omega = n\omega_c$. (b) A set of orbits. The extremal orbits are the heavy lines.

orbits are the same shape in real space and k-space. The shapes can be visualized as intersections of the Fermi surface with planes of constant $k_{||}$, where $k_{||}$ is the crystal momentum in the $\mathbf{B}$ direction.

A classic experiment to observe Fermi surfaces is cyclotron resonance. A magnetic field and a radio frequency electric field with frequency ω are applied parallel to the surface of a metal. The electric field penetrates up to a skin depth in the metal. Then it can pump energy into electrons whose periods are multiples of the period of the field because in this case the field will accelerate the electrons every time they pass into the skin depth. See Figure 7.9.

In general there will be orbits with many different frequencies coming from the different values of $k_{||}$. The ones that dominate the resonance are the ones whose frequency is a maximum or a minimum as a function of $k_{||}$. The reason for this is clear: if a set of electrons move into the skin depth and their frequencies vary linearly with $k_{||}$, then some will be ahead of the mean and some behind so that there is phase cancellation. However, around the extrema there is no cancellation so that there is substantial absorption. Finding the dominant resonances for different orientations of **B** gives a good deal of indirect information about the Fermi surface, and has been used to sort out some relatively complex shapes.

7.8.2 Quantization of orbits

At very high fields another effect comes into play, the fact that the areas of the orbits are quantized. This effect is important only at low temperatures and high fields so that the spacing between the energy levels is more than the thermal smearing of the Fermi distribution function. We will see below that the energy spacing is $\hbar\omega_c$ so that the criterion is that $k_B T \ll \hbar\omega_c$. If this is the case the sharp energy level leads to periodic variations in induced magnetic moments (the de Haas–van Alphen effect) or electrical resistance (Shubnikov–de Haas effect). Both of these are used to measure Fermi surfaces. We will briefly review these methods. For a more complete treatment see (Ziman 1962, Ziman 1972).

An elegant way to approach the subject was introduced by L. Onsager and I. M. Lifshitz. They used the Bohr phase integral formula from the "old" quantum mechanics (which is valid for large quantum numbers). It is the quantization of the action integrated around a closed orbit:

$$\oint \mathbf{p} \cdot d\mathbf{r} = \oint \left(\hbar \mathbf{k} + \frac{e}{c} \mathbf{A} \right) \cdot d\mathbf{r} = h\left(l + \gamma\right). \tag{7.46}$$

In this equation we have used the fact that the canonical momentum, **p**, is the kinetic momentum, $\hbar \mathbf{k}$, plus the contribution from the electromagnetic field, $e\mathbf{A}/c$ where **A** is the vector potential. That is, $\nabla \times \mathbf{A} = \mathbf{B}$. In the equation $l = 0, 1, 2, \ldots$ and γ is a constant phase factor which is $1/2$ for free electrons.

The two terms in Eq. (7.46) are easy to work out. The first comes from the time integral of Eq. (7.44):

$$\oint \hbar \mathbf{k} \cdot d\mathbf{r} = \frac{e}{c} \oint \mathbf{r} \times \mathbf{B} \cdot d\mathbf{r} = -\frac{e}{c} \mathbf{B} \cdot \oint \mathbf{r} \times d\mathbf{r}. \tag{7.47}$$

However, $\oint \mathbf{r} \times d\mathbf{r}$ is twice the area, $\mathcal{A}$, of the orbit since $|\mathbf{r} \times d\mathbf{r}|/2$ is the area of a triangle swept out as the orbit is traversed. (The direction of the vector $\mathcal{A}$ is normal to the plane of the orbit.) Thus:

$$\oint \hbar \mathbf{k} \cdot d\mathbf{r} = -2\,\mathbf{B} \cdot \mathcal{A} = -2\Phi. \tag{7.48}$$

Here Φ is the flux of the magnetic field through the orbit. The other also involves Φ by Stokes' theorem:

$$\oint \mathbf{A} \cdot d\mathbf{r} = \int \nabla \times \mathbf{A} \cdot d\mathbf{S} = \mathbf{B} \cdot \mathcal{A} = \Phi. \tag{7.49}$$

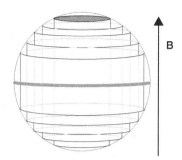

Fig. 7.10   Visualization of the quantized orbits in k-space as a series of tubes whose areas are quantized. When B increases the area of each tube increases. When a tube passes through the extremal area $\mathcal{S}$ (heavy line) there is a change in the derivative of the energy with respect to B.

Thus:

$$\Phi = B\mathcal{A}_n = (l + \gamma)\Phi_0; \quad \Phi_0 = hc/|e|. \tag{7.50}$$

The quantity Φ_0 is known as the flux quantum.

Since orbits in real space are quantized, orbits in k-space are as well. Using the scale factor of Eq. (7.45) we find:

$$\mathcal{S}_n = \frac{2\pi|e|B}{\hbar c}(l + \gamma). \tag{7.51}$$

This equation has an important interpretation. We can think of the quantized orbits in k-space as being a set of "tubes" that hold the electrons. See Figure 7.10. Once more, the intersection of the tubes with the Fermi surface gives the orbits. As B is increased there will come a point when a tube will pass through an extremal area of the Fermi surface and empty. The electrons will rearrange into tubes of smaller l. This gives an abrupt change in the field derivative of the energy of occupied states, and thus a change in the magnetic moment. Therefore the magnetic moment becomes oscillatory as B changes. The period of the oscillations is found by finding the magnetic field such that $\mathcal{S}_n(B_n) = \mathcal{S}$ where $\mathcal{S}$ is the extremal area. Thus:

$$\frac{1}{B_{l+1}} - \frac{1}{B_l} = \frac{2\pi|e|}{\hbar c\mathcal{S}}. \tag{7.52}$$

This oscillation in $1/B$ is observed in de Haas–van Alphen experiments so that the extremal area $\mathcal{S}$ can be measured. This experiment played an important historical role in sorting out Fermi surfaces.

7.8.3 Landau levels

For the case of free electrons we can go further and solve for the energy levels and wavefunctions for electrons in a **B** field. This was first done by L. Landau. Landau levels are treated in most texts on quantum theory. Note that electrons also have spin, so that the Zeeman interaction with the external field shifts all the energy levels by $\pm g\mu_B H$. We are solving for the energy levels for each spin.

We need to solve:

$$\hat{\mathcal{H}}\psi = \frac{1}{2m}(\mathbf{p} - e\mathbf{A}/c)^2\psi = \epsilon\psi. \tag{7.53}$$

The kinetic energy term is consistent with the discussion above about the distinction between the kinetic momentum and the canonical momentum. Suppose that $\mathbf{B}$ is along z. Then we can use a gauge (the Landau gauge) where $\mathbf{A} = (-By, 0, 0)$.

The Schrödinger equation now reads:

$$-\frac{\hbar^2}{2m}\left[-\left(\frac{1}{i}\frac{\partial}{\partial x} - \frac{eB}{\hbar c}y\right)^2 + \frac{\partial^2}{\partial y^2} + \frac{\partial^2}{\partial z^2}\right]\psi = \epsilon\psi. \tag{7.54}$$

The solution may be written $\psi = \exp{(ikx)}\exp{(iqz)}\phi(y)$. The new function $\phi(y)$ satisfies:

$$-\frac{\hbar^2}{2m}\left[-\left(k - \frac{eB}{\hbar c}y\right)^2 + \frac{\partial^2}{\partial y^2} - q^2\right]\phi = \epsilon\phi. \tag{7.55}$$

Rewriting this gives;

$$-\frac{\hbar^2}{2m}\phi'' + \frac{m\omega_c^2}{2}(y - y_0)^2\phi = \left(\epsilon - \frac{\hbar^2 q^2}{2m}\right)\phi; \quad y_0 = \frac{\hbar c}{|e|B}k. \tag{7.56}$$

This equation is that of a harmonic oscillator with frequency ω_c and origin shifted to y_0. The shift is proportional to the quantum number k. It is useful to write $y_0 = l_m^2 k$ where $l_m = (\hbar c/|e|B)^{1/2}$ is called the magnetic length.

Using well-known results for the harmonic oscillator we have:

$$\epsilon(l, q) = \left(l + \frac{1}{2}\right)\hbar\omega_c + \frac{\hbar^2 q^2}{2m}, \tag{7.57}$$

where $l = 0, 1, 2, \ldots$ Note that E is independent of k. We can interpret this as the energy of a quantized oscillator plus the kinetic energy along $\mathbf{B}$. For a strictly two-dimensional system (which will come up below) we set $q = 0$. These states are Landau levels. We can think of them as representing states with free motion in the x-and z-directions and confinement in the y direction around a mean position y_0.

For each pair l, q the state is highly degenerate because there are many values of k. This degeneracy will be of considerable interest. We can count the states by assuming there is a finite strip in the y direction of width W. Since $0 < y < W$ we need $0 < k < W/l_m^2$. If we use periodic boundary conditions in the x-direction we have, as usual, $k = 2\pi m/L$ where L is the length of the strip. Thus we can count the states by setting $m < LW/2\pi l_m^2$. That is the degeneracy of each l state is:

$$LWB|e|/hc = \Phi/\Phi_0. \tag{7.58}$$

That is, the number of states for each q is the flux through the strip, Φ, in units of the flux quantum; cf. Eq. (7.50). The degeneracy increases with B.

This may all seem rather disconcerting since the classical picture is of circular orbits with no distinction between the x and y directions. Further, we could have used other gauges which give the same $\mathbf{B}$ but treat x and y differently. For example, we might have started with:

$$\mathbf{A} = (0, Bx, 0) \quad \text{or} \quad \frac{1}{2}(By, -Bx, 0). \tag{7.59}$$

The second choice is called the symmetric gauge. The answer to the question lies in the degeneracy. We can find a linear combination of the wavefunctions confined in the y direction which are confined in both directions (and look more like classical orbits) or even interchange x and y.

It is useful to solve the problem explicitly in the symmetric gauge. We work in two dimensions. Suppose we scale both coordinates in terms of l_m, so that $u = x/l_m, v = y/l_m$. Then the wave equation becomes:

$$-\frac{\hbar\omega_c}{2}\left[(\partial_u + iv/2)^2 + (\partial v - iu/2)^2\right]\phi(u, v) = \epsilon\phi(u, v). \tag{7.60}$$

Note that there is a term $-(u^2 + v^2)/4$ in the operator on the right. This motivates the substitution $\phi = \exp[-(u^2 + v^2)/4]\psi$. A bit of algebra shows that:

$$-\frac{\hbar\omega_c}{2}\left[\partial_u^2 + \partial_v^2 - (u - iv)(\partial_u + i\partial_v) - 1\right]\psi = \epsilon\psi. \tag{7.61}$$

Now it is natural to replace u, v with the independent complex variables $z = u + iv, z^* = u - iv$. Then we have:

$$\frac{\hbar\omega_c}{2}\psi - [4\partial_z\partial_{z*} - 2z^*\partial_{z*}]\psi = \epsilon\psi. \tag{7.62}$$

If ψ is a function of z alone, then the operator on the left gives zero. Recall that analytic functions of a complex variable can be taken to be a function of z (and not z^*). Thus any analytic function of z gives zero when the operator $4\partial_z\partial_{z*} - 2z^*\partial_{z*}$ is applied, so that it is a solution of Eq. (7.62) with eigenvalue $\hbar\omega_c/2$. Thus the eigenfunctions for the lowest Landau level are of the form (restoring ordinary units):

$$\phi = e^{-|z|^2/4l_m^2}f(z), \tag{7.63}$$

where $f(z)$ is any analytic function. A useful basis set is the powers of z from which f can be made by a power series. Thus, for the lowest Landau level, $z^m e^{-|z|^2/4l_m^2}$ is a set of eigenfunctions. We will use these below. Higher Landau levels can be constructed from Eq. (7.62), but we will not do this.

For the two-dimensional case we can define a parameter which gives the number of filled Landau levels:

$$\nu = \frac{n}{n_B}; \quad n_B = \frac{|e|B}{hc}. \tag{7.64}$$

Here n is the number of electrons per unit area. When ν is an integer we say that there are ν filled Landau levels in the ground state. Otherwise the ground state will have a partially filled level.

The statements that we have made above about the dominance of extremal orbits in de Haas–van Alphen experiments can be verified explicitly for the case of free electrons. This is a nice exercise in applied mathematics; the student can consult the literature (e.g. Ziman (1972)) for details. Other explicit calculations can be done using these results.

7.8.4 The integer quantum Hall effect

K. von Klitzing, and collaborators (Von Klitzing, Dorda & Pepper 1980) did an experiment that revealed fascinating new features of motion in magnetic fields. This is now known as the *integer quantum Hall effect*. The experiment used the methods described above to create a two-dimensional electron gas in a p-type Si MOSFET. They connected voltage and current probes as shown in Figure 7.11. The quantities measured are the total current, I, the longitudinal resistance, $R_{xx} = V_L/I$, and the Hall resistance, $R_{xy} = V_H/I$. The temperature was very low, of order 1 Kelvin, and the magnetic field very large, of order 10 Tesla. In these conditions the spacing between Landau levels is large compared to $k_B T$. A typical measurement of R_{xx} and R_{xy} is shown in Figure 7.12.

In order to get some perspective on the results we return to Eq. (7.7) in two dimensions. In the limit of large scattering time the tensor reduces to:

$$\left(\begin{array}{cc} 0 & \frac{-n|e|c}{B} \\ \frac{n|e|c}{B} & 0 \end{array} \right) = \left(\begin{array}{cc} 0 & \frac{-\nu e^2}{h} \\ \frac{\nu e^2}{h} & 0 \end{array} \right). \tag{7.65}$$

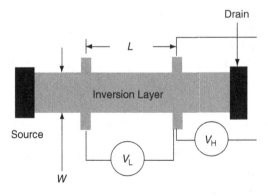

Fig. 7.11 **A schematic view of a four terminal method of measuring the quantum Hall effect. In the original von Klitzing experiment the number of electrons in the inversion layer was tuned by changing the gate voltage on a MOSFET. In more recent experiments using heterostructures (such as Figure 7.12) the magnetic field is tuned.**

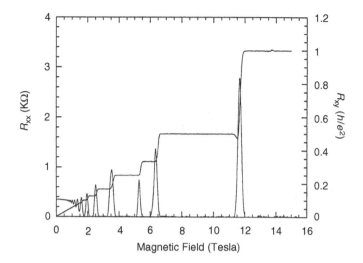

Fig. 7.12 **The integer quantum hall effect in a two-dimensional electron gas formed in a GaAs-AlGaAs heterostructure at 50 mK. The longitudinal resitance R_{xx} and the Hall resistance, R_{xy}, are plotted against magnetic field. Courtesy of C. Kurdak.**

The inverse of this tensor gives the resistivities, $\rho_{xx} = (W/L)R_{xx}, \rho_{xy} = R_{xy}$ where W is the width of the sample and L is its length. Inverting Eq. (7.65) we find:

$$\rho_{xx} = 0; \quad \rho_{xy} = \frac{h}{ve^2}. \tag{7.66}$$

Note the curious fact that σ_{xx} and ρ_{xx} are both zero in this system.

This expression both agrees and disagrees with experiments such as Figure 7.12 in startling ways. In the case that v is an integer – filled Landau levels – there is perfect agreement for ρ_{xy}. The plateaus in the Figure are *exactly* at $h/e^2 j, j = 1, 2, \ldots$ In fact, the accuracy is unexpected: the plateaus give a very accurate and reproducible value for the *quantum resistance $h/e^2 = 25812.807$ ohms*. In fact, the value measured in the quantum Hall effect is now used as a resistance standard! On the other hand, the experiment does not agree at all with Eq. (7.66) for partially filled levels. The free electron theory would give the linear behavior that is seen in the experiment only for small field.

The key to understanding the experiment is to realize that the environment of the electrons in a semiconductor inversion layer is far from ideal. Paradoxically, the disorder in the layer makes the accurate measurement of h/e^2 possible. In the presence of disorder we must think of the electrons in the sharp Landau levels as being subject to a random potential so that energies between the Landau energies will have allowed states. Near the ideal energies there will be many states. However, large, rare, fluctuations of the potential can spread out in energy of order $\hbar\omega_c/2$. In these deep wells the electrons are likely to be bound or *localized*. The density of states of free-electron Landau levels is a series of δ-functions at $(l + 1/2)\hbar\omega_c$. For the disordered material we may imagine that the real density of states is as shown in Figure 7.13.

If the chemical potential is in the gray regions there is no longitudinal current carried by those states because they are localized. The current remains constant until the next band of

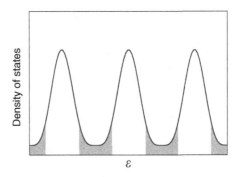

Fig. 7.13 Assumed density of states for the quantum Hall effect. For no disorder the density of states is a series of δ functions at $(l + 1/2)\hbar\omega_c$. Disorder spreads the peaks and introduces localized states (shown shaded) between the bands of extended states.

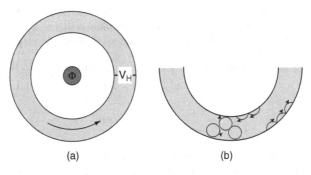

Fig. 7.14 (a) A version of the Laughlin geometry. There is a fixed magnetic field, B, perpendicular to the plane. (b) Currents due to cyclotron orbits in the interior of the sample cancel, but the edge orbits (that reflect from the walls) carry the current. The edge currents on the inside and the outside of the ring do not cancel since the chemical potential differs by eV_H between the edges.

delocalized states is encountered. This is the region of the plateaus. However, it is far from clear that the steps in R_{xy} should be what they were for free electrons.

A clue to what is happening was given by R. Prange who introduced a bound state from a localized potential into a Hall layer; see Prange & Girvin (1990). Oddly enough, the current was not affected by the bound state: there were fewer electrons to carry the current, but the ones that were left moved faster so that the drift current was conserved.

Laughlin argument

Laughlin (1981) gave a very clever argument based on gauge invariance which explains the extreme accuracy of the quantum Hall steps. Consider the idealized geometry shown in Figure 7.14. Suppose that the radius of the ring is large. Then we can consider the x-direction to be around the ring and the y-direction to be radial. We need to remember that we now have periodic boundary conditions on x with period L. In what follows we will neglect ordinary electrical resistance as in Eq. (7.65). We can control the current around the ring by changing the flux, Φ, through a solenoid that threads the ring since a change in flux gives rise to an emf, $-(1/c)\partial\Phi/\partial t$. This emf can do work and change the internal energy, E, of the system:

$$\frac{\partial E}{\partial t} = -V_L I = \frac{I}{c}\frac{\partial\Phi}{\partial t} \quad\Rightarrow\quad I = c\frac{\partial E}{\partial\Phi}. \tag{7.67}$$

Now suppose that we change the flux by $\Delta\Phi$ and the vector potential by $\Delta\mathbf{A}$. We can take $\Delta\mathbf{A}$ to be in the azimuthal direction and fix its value by finding the flux by Stokes theorem: $\oint \Delta\mathbf{A} \cdot d\mathbf{l} = \Delta\Phi$. Thus:

$$\Delta A = \Delta\Phi/L. \tag{7.68}$$

Now we ask for the effect on the current. To this end we examine the effect on the quantum states. Note that the extra vector potential appears only in the kinetic energy term in the Schrödinger equation:

$$\frac{1}{2m}\left(\mathbf{p} - \frac{e}{c}[\mathbf{A} + \Delta\mathbf{A}]\right)^2. \tag{7.69}$$

The transformation:

$$\psi \to \Psi = e^{\frac{ie}{\hbar c}\int \Delta A\,dx}\psi = e^{\frac{ie}{\hbar c}\Delta A x}\psi, \tag{7.70}$$

removes the extra term from the equation. However, we must check that the new wavefunction satisfies the boundary conditions. For the localized states this is not a problem: they simply change their phase and still contribute nothing to I. However the extended wavefunctions are a problem: Ψ is not single valued unless we can go around the ring (a distance L) and return to the same value. Thus the phase must change by a multiple of 2π. By using Eq. (7.68) and the definition of the flux quantum we see:

$$\frac{e}{\hbar c}\Delta AL = 2n\pi; \quad \Delta\Phi = n\Phi_0. \tag{7.71}$$

This means that when an integer number of flux quanta are injected all of the wavefunctions are unchanged. If we start in the localized regime – on a plateau – we must return to it.

If the flux change is less than Φ_0 the wavefunctions do change. We can get some insight into what happens by noting two things. First, consider any two-dimensional sheet where electrons move in cyclotron orbits. Then in the interior of the sheet the current cancels. All of the net flow is along the edge, as in Figure 7.14 (b). The edge currents should be expected to continue to carry all the current even if there is disorder. Next, consider what happens for free electrons when we add ΔA. If we look at Eq. (7.55) we see that we get the same result if we shift k by $-e\Delta A/\hbar c$. Recall that the allowed values of k are $2\pi m/L$ so that adding a flux quantum means that A shifts by Φ_0/L and k shifts by $2\pi/L$. That is each state shifts to the next. Likewise there is a shift of y_0 in Eq. (7.56) by $\Delta A/B$. The flux acts a pump and moves the electrons across the strip, and adding one flux quantum moves an electron in at one edge and another out at the other edge for each occupied level.

An electron is moved by the pump through potential V_H for each occupied level, and this does work $\Delta E = j|e|V_H$, where j is the number of occupied levels. We can evaluate the current in Eq. (7.67) by replacing the derivative by a finite difference:

$$I = c\frac{\Delta E}{\Delta \Phi} = \frac{je^2}{\hbar}V_H. \tag{7.72}$$

This gives the Hall resistance:

$$R_{xy} = \frac{h}{e^2 j}, \tag{7.73}$$

which is the result of von Klitzing et al (1980).

7.8.5 Fractional quantum Hall effect

Subsequent to the work described above Tsui, Stormer & Gossard (1982) pushed the experiment to still higher magnetic fields and lower temperatures. By using GaAs-AlGaAs heterostructures rather than a MOSFET they were able to work with samples with extraordinarily small scattering. The way they did this is called *modulation doping*. The electrons come from donors and they put the donors into the AlGaAs part of the heterostructure only. The inversion layer which carries the Hall current is in the GaAs region. Thus the carriers are distant from the donors and do not scatter from them very much. They observed Hall plateaus at fractional values of ν, for example, at $\nu = 1/3$. More than 30 such states have been observed.

This observation led to considerable theoretical and experimental activity. It became evident that the new observations resulted from interactions between the electrons. Suppose all the electrons are in the ground Landau level. The term in the energy that we have not considered so far is the Coulomb repulsion between the electrons, $\sum e^2/r_{ij}$. Laughlin (1983), and susequently, J. Jain (see (Jain 2000)), constructed wavefunctions that minimize the repulsion and that seem to be remarkably accurate representations of the real situation. The two approaches agree for 1/3 filling.

We start by constructing the wavefunction for free electrons in the lowest Landau level. Our orbitals will be those developed above, $\phi_m = z^m e^{-|z|^2/4l_m^2}$. This is a useful set of functions for two reasons: first, they are eigenfunctions of the angular momentum in the z direction (as may be easily shown). Also, for quantum number m the maximum of the probability density is at $r^2 = 2ml_m^2$, so that we can fill states up to a maximum m in a disc-shaped sample. Suppose we do this. Then the wavefunction of the ground state of N

electrons is the Slater determinant:

$$\Psi_1(\mathbf{r}_1, \mathbf{r}_2, \ldots, \mathbf{r}_N) = \begin{vmatrix} 1 & 1 & \ldots & 1 \\ z_1 & z_2 & \ldots & z_N \\ \ldots & \ldots & \ldots\ldots & \\ z_1^{N-1} & z_2^{N-1} & \ldots & z_N^{N-1} \end{vmatrix} e^{-\sum_j |z_j|^2/4l_m^2}$$

$$= \prod_{j<k}(z_j - z_k)e^{-\sum_j |z_j|^2/4l_m^2}. \tag{7.74}$$

The last line of the equation follows from the observation that the expression is of a classic form called the *Vandermonde determinant*. Since the filled ground Landau level has a gap of $\hbar\omega_c$ to the next excited state, we can take this as a good approximation of the state with electron-electron interactions present. We can view the wavefunction as "rigid" against perturbations. Note that the factor $z_j - z_k$ means that Ψ_1 vanishes if two electrons are at identical points, as it must for fermions.

For the case $\nu = 1/3$ the number of electrons is $1/3$ of what is necessary to fill the level. What Laughlin did for this case was to replace each factor $z_i - z_j$ by $(z_i - z_j)^3$. Because of the odd power this is still an acceptable fermion wavefunction. Clearly, the electrons stay out of each others way even more effectively than in the filled Landau level. Numerical calculations on small systems show that this wavefunction is very close to the exact result.

Jain rewrote Laughlin's wavefunction in an interesting way:

$$\Psi_3 = \prod(z_j - z_k)^2 \Psi_1, \tag{7.75}$$

where Ψ_1 is the wavefunction which would occur for a filled Landau level with N electrons. He interpreted this formula by thinking of an electron in a partially filled level as capturing parts of the magnetic flux and making *composite fermions*. In our example, each electron is bound to thin solenoids that have two flux quanta. Recall from above that carrying an electron around a loop enclosing a solenoid containing a flux quantum generates a phase factor of 2π. In this wavefunction, carrying any other electron around the jth one generates a factor of 4π, hence a flux of $2\Phi_0$. When electrons are bound in this way they automatically stay out of each others way, as we have already seen. Now we can think of the composite fermions as being weakly interacting particles.

Further, the new particles see a smaller field than the electrons – the part that they have not captured. We can estimate this by smearing out the captured flux and subtracting it from B. The effective captured field is the density of electrons multiplied by the number of captured quanta, in our case $2\Phi_0 n$. Thus there is an effective field:

$$B^* = B - 2\Phi_0 n. \tag{7.76}$$

Now the composite fermions have filling factor $\nu^* = n\Phi_0/|B^*|$ from Eq. (7.64). (Note that B^* can be negative, i.e. antiparallel to B.) Thus we have:

$$\nu = \frac{\nu^*}{2\nu^* \pm 1}. \tag{7.77}$$

Thus we can have $\nu^* = 1$ when $\nu = 1/3$. Since the composite fermions are in a filled Landau level, we expect a gap to the next state, and thus a plateau when disorder is added. Further, we expect the whole pattern of Figure 7.12 to repeat inside the lowest Landau level. This remarkable behavior is observed.

An astonishing property of the fractional quantum Hall effect is that excitations around the states of the fractional effect have fractional charge. Laughlin's argument is as follows: suppose we increase the field a bit above what is necessary to have $\nu = 1/3$ by adding one flux quantum. We can think of the flux quantum as being confined in a thin solenoid at z_0 so that we have created a "quasi-hole" at that point. A candidate for the new wavefunction would have a factor of $\prod_j(z_j - z_0)$ by the same argument as above.

This quasi-hole acts as if it has charge $|e|/3$. Roughly speaking, if we were to simply remove an electron at z_0 we would increase the charge by $|e|$. If we leave the factor of $\prod_j(z_j - z_0)^3$ in the wavefunction we would have three flux quanta. However, these need not be bound together, and could just as well be $\prod_j(z_j - z_1)(z_j - z_2)(z_j - z_3)$ so we have three quasi-holes at z_1, z_2, z_3 which share the charge. The fractionization of charge is also indicated by the arguments above for the integer effect: since the quantization of the plateaus can be traced to the fact that electrons have discrete charge e, if we have a fractional plateau we should conclude that we are moving fractional charge, e.g. $e/3$ across the sample.

In fact, direct experiments on current fluctuations have shown that the carriers in the fractional case do act exactly as if they are fractionally charged. For small currents there are fluctuations because of the *shot noise* in the arrival of charge, namely the usual $\sqrt{M}$ fluctuations of any random signal composed of the sum of M random events. That is, the fluctuations of the square of the current are proportional to the current itself. By measuring the current fluctuations we get a measure of the charge of the carriers. In the experiment (Saminadayar, Glattli, Jin & Etienne 1997), a charge of $e/3$ is clearly seen.

Suggested reading

A good reference for electron transport theory is:
    Ziman (1962).
For semiconductors see:
    Yu & Cardona (2001).
The engineering literature on semiconductor devices is huge and growing. Here is a classic reference:
    Sze (1981).
Quantum Hall effect:
    Prange & Girvin (1990)
    Jain (2000).

Problems

1. (a) Use the Drude model to calculate the ac conductivity, $\sigma(\omega)$ by applying an electric field $\mathbf{E} = \mathbf{E}_\circ e^{-i\omega t}$. You should find a complex quantity whose real part gives the in-phase current, and whose imaginary part gives the out-of-phase current. Display the real and imaginary parts. (b) Similarly, calculate the polarization, $\mathbf{P}$, i.e., the dipole moment per unit volume. (c) Construct the frequency dependent dielectric constant: $\epsilon(\omega) = 1 + 4\pi P_\circ / E_\circ$. Once more you should get a complex quantity. Write the real and imaginary parts. (d) Show that $\epsilon = 1 + 4\pi i\sigma(\omega)/\omega$. (e) Show that for $\tau \to \infty$ $\epsilon \to 1 - \omega_p^2/\omega^2$, where $\omega_p = (4\pi n e^2/m)^{1/2}$. In the next chapter we will see these formulas again in a different context.

2. We discussed isolated donors in semiconductors. When a semiconductor is heavily doped the donor wavefunctions can overlap, and form an *impurity band*. This turns the semiconductor into a metallic conductor, and is a form of the Mott metal-insulator transition. (a) Work out a formula for the effective Bohr radius, a^*, of a donor state. (b) GaN has $\kappa = 10$ and $m_e^* = 0.2\ m$ where m is the mass of the electron. InAs has $\kappa = 15, m_e^* = 0.024\ m$. Find a^* for each case. (c) Argue that the transition occurs when $n_d^{1/3} a^* = C$ where C is a constant of order unity and n_d is the donor density. (Empirically, $C \approx 0.26$). (d) Suppose GaN and InAs are both doped so that $n_d = 2 \times 10^{17}$ cm^{-3}. Which is a metal and which an insulator?

   In fact, there is an enormous amount of physics in the innocent constant C. The Mott transition depends on electron-electron interactions in a complicated way. We will allude to some of the issues in a later chapter.

3. Model a semiconductor as having a parabolic conduction band with mass m_e and a valence band with mass m_h. Measure energies from the top of the valence band, and take the gap to be E_G. (a) Show that the density of states is $\mathcal{D} \propto m_e^{3/2}\sqrt{E - E_G}$ for $E > E_G$ and $m_h^{3/2}\sqrt{-E}$; for $E < 0$. Find the proportionality constant. (b) In an intrinsic semiconductor all of the N electrons are in the valence band at $T = 0$ and there are no donors or acceptors. Introduce a cutoff energy $-E_c$ so that $N = \int_{-E_c}^{0} dE\ \mathcal{D}(E)$. Show that

$$N_h = \int_{-E_c}^{0} dE\ \mathcal{D}(E)(1 - f(E)) = \int_{E_G}^{\infty} dE\ \mathcal{D}(E)f(E) = N_e.$$

The left-hand side is the number of holes in the valence band, and the right is the number of electrons in the conduction band. (Explain why). (c) Now suppose that $\beta\mu \gg 1$, $\beta(E_G - \mu) \gg 1$ and $\beta E_c \gg 1$. Prove that the chemical potential is given by:

$$\mu = \frac{1}{2}E_G + \frac{3}{2}k_BT \ln(m_h/m_e).$$

The chemical potential is near the middle of the gap, as claimed in the text. This also shows that the first two assumptions above were valid if the temperature is not too high. (e) Prove Eq. (7.43) and find $K(T)$.

4. Repeat the previous problem for the case of N_d donors. Now the density of states has another piece, $N_d\delta(E - E_d)$. The total number of electrons is $N + N_d$. Assume that

$\beta(\mu - E_d) \gg 1$. Prove that the chemical potential is pinned near the donor level and that the density of electrons in the conduction band is given by:

$$n \propto \sqrt{N_d}e^{-\beta E_B/2},$$

where $E_B = E_G - E_d$ is the binding energy of the donor.

5. In the text we discussed excitons as bound electron-hole pairs in semiconductors (Wannier excitons). There is another kind, the Frenkel exciton, which occurs in molecular crystals. You may think of it as an excited state of the molecules which hops from site to site – note that the excitation hops. (The coupling which leads to excitation transfer is usually electromagnetic.) Consider a wavefunction for a linear chain of molecules:

$$\Psi_n(r_1, r_2, ...r_N) = \phi_1(r_1)\phi_2(r_2)...\chi_n(r_n)\phi_{n+1}(r_{n+1})...\phi_N(r_N).$$

In this expression, $\phi_i(r_i)$ is the ground state wavefunction for the electron which is on molecule i and $\chi_n(r_n)$ is the excited state at site n. The hopping matrix element (which you can assume to be non-zero only for nearest neighbors) is called Δ:

$$\Delta = \langle \Psi_n|\hat{\mathcal{H}}|\Psi_{n\pm1}\rangle = \langle \chi_n(r_n)\phi_{n\pm1}(r_{n\pm1})|\hat{\mathcal{H}}|\phi_n(r_n)\chi_{n\pm1}(r_{n\pm1})\rangle.$$

The only other matrix element of H which you need consider is $\epsilon = \langle \Psi_n|\hat{\mathcal{H}}|\Psi_n\rangle$, the molecular excited state energy. Write down energy eigenfunctions assuming $\hat{\mathcal{H}}$ acts only in the space of the Ψ_n. Sketch the energy band of the excitons. Use periodic boundary conditions. You can assume that $\langle \Psi_n|\Psi_m\rangle = \delta_{m,n}$.

   Hint: the eigenfunctions must obey Bloch's theorem. You might compare tight-binding for electrons.

6. Consider an anisotropic Fermi gas for which the relation between energy and momentum is given by:

$$E = (\hbar^2/2)(k_x^2/m_1 + k_y^2/m_1 + k_z^2/m_2).$$

(a) Find the Fermi energy as a function of density by figuring out the volume in k-space of the Fermi surface, and dividing by the volume per k-vector $(2\pi)^3/(2\Omega)$. (Where does the 2 in the denominator come from?) (b) Find the density of states and identify the quantity that plays the role of m^* in it. (c) Sketch a real space cyclotron orbit for B along x or y.

   Hint: It is well but not widely known that the volume of an ellipsoid defined by $1 = x^2/a^2 + y^2/b^2 + z^2/c^2$ is $4\pi abc/3$.

7. Azbel–Kaner resonance depended on the mean free path of the electrons being larger than the skin depth, which can happen at low temperatures in pure metals. The skin depth itself is interesting in this regime. (a) Derive the classical formula for the skin depth. Take a metal whose surface is in the $x - y$ plane. Assume an external ac electric field is $E_x e^{-i\omega t}$. Take the z-axis to point into the metal. Find E_x just inside the metal, and also **B**. Then use Maxwell's equations, neglecting the displacement current. You should find that $E_x(z) \propto e^{-(1+i)z/\delta}$, where $\delta = c/\sqrt{2\pi\sigma\omega}$. (b) The derivation above assumed that the mean free path of the electrons, $\mathcal{L} = v_F\tau$, is smaller than δ. However, if $\mathcal{L}$ is large, we must suppose that conduction is *non-local* because the electron "remembers" the

electric field over distances of order $\mathcal{L}$. A. B. Pippard showed how to estimate δ in this case by pointing out that the only electrons which can be significantly accelerated by **E** are those that remain inside the skin depth for a distance of order $\mathcal{L}$, i.e. those whose trajectories are almost parallel to the surface. Estimate the density of such electrons, n_e, i.e. those whose velocity makes an angle of less than $\delta/\mathcal{L}$ with the x-axis. (You should get $n_e/n = \gamma \delta/\mathcal{L}$, where γ is a number of order unity.) (c) Argue that σ in the formula in (a) should be replaced by $\sigma_e = \sigma(n_e/n)$. Then solve for δ. You should find $\delta \propto (\mathcal{L}c^2/\sigma\omega)^{1/3}$. (d) Show that this *anomalous* skin depth does not depend on τ, but only on Fermi surface properties.

The proper way to work out this effect is to solve the Boltzmann equation together with Maxwell's equations. See Kittel (1963).

8. (a) Find the density of states for Landau levels in three dimensions. (b) Find the same quantity in two dimensions.

9. Prove the Vandermonde relation:

$$\begin{vmatrix} 1 & 1 & \ldots & 1 \\ z_1 & z_2 & \ldots & z_N \\ \ldots & \ldots & \ldots\ldots & \\ z_1^{N-1} & z_2^{N-1} & \ldots & z_N^{N-1} \end{vmatrix} = \prod_{1 \leq j < k \leq N} (z_j - z_k).$$

Hint: you could try mathematical induction. Begin the induction step by subtracting columns until there is a 1 in the upper left corner, and expand by minors. Then factor out $\prod_{i=1,N}(z_N - z_1)$. Then start scaling and subtracting rows starting from the bottom. You should be able to produce the next smaller Vandermonde determinant.

8 Dielectric and optical properties

The transport theory of the previous chapter dealt mostly with the response to static fields. Here we allow the fields to vary with time. In order to be consistent, we describe the electrons with Maxwell's equations. Since electrons are charged we are taking some of the electron-electron interaction into account. As we will see later, only the long-wavelength properties can be got at this way. First we will deal with the so-called *longitudinal* response, namely the response to the introduction of an external charge density which may be time and space-dependent. Later we will turn to the *transverse* response to external fields such as those in a light wave.

8.1 Dielectric functions

We start with the simple observation that a fixed positive charge in a system of mobile electrons *polarizes* the environment. It attracts the electrons so that there is an induced negative charge. The net field produced by the positive charge plus the induced charge is smaller than the positive charge would produce alone: this is called *screening*. This is familiar in electrostatics; thus we try to think about these effects as a generalization of electrostatics.

We start by being quite general. Suppose we have a medium and we introduce an external charge density, ρ_e. We will suppose that the external charge is small, and that we can assume the medium is linear. What we are doing here is an example of a general method called *linear response theory*.

The simplest form of perturbation will be a single Fourier component in space and time, with wavevector $\mathbf{q}$ and frequency ω. Since the medium is linear, we can find the response to a general perturbation by adding up Fourier components:

$$\rho_e(\mathbf{r}, t) = \int \frac{d\mathbf{q}}{(2\pi)^3} e^{i\mathbf{q}\cdot\mathbf{r}} \int \frac{d\omega}{2\pi} e^{-i\omega t} \rho_e(\mathbf{q}, \omega). \tag{8.1}$$

Correspondingly, the electric and displacement fields, $\mathbf{E}, \mathbf{D}$ and the induced charge density, ρ_i will be represented by their Fourier components.

The subset of Maxwell's equations that we need are the following:

$$\nabla \cdot \mathbf{D} = 4\pi\rho_e$$

$$\mathbf{D} = \mathbf{E} + 4\pi\mathbf{P} \equiv \epsilon\mathbf{E}$$

$$\rho_i = -\nabla \cdot \mathbf{P} \tag{8.2}$$

The induced charge density, ρ_i is the result of the polarization. The dielectric constant in this equation is the conventional one. We will introduce a more general quantity in Fourier space.

Using Eq. (8.1) we can write:

$$i\mathbf{q} \cdot \mathbf{D} = 4\pi\rho_e$$

$$i\mathbf{q} \cdot (\mathbf{D} - 4\pi\mathbf{P}) = i\mathbf{q} \cdot \mathbf{E}. \tag{8.3}$$

Therefore:

$$i\mathbf{q} \cdot \mathbf{E} = 4\pi(\rho_e + \rho_i). \tag{8.4}$$

Generalizing ordinary electrostatics we define the dielectric function by:

$$\mathbf{D}(\mathbf{q}, \omega) = \epsilon(\mathbf{q}, \omega)\mathbf{E}(\mathbf{q}, \omega). \tag{8.5}$$

Therefore:

$$\epsilon(\mathbf{q}, \omega) = \frac{\rho_e}{\rho_e + \rho_i}. \tag{8.6}$$

From Eq. (8.6) we have several useful expressions:

$$\frac{1}{\epsilon} = 1 + \frac{\rho_i}{\rho_e}$$

$$\epsilon = 1 + \frac{4\pi i \rho_i}{\mathbf{q} \cdot \mathbf{E}}. \tag{8.7}$$

There are several points that should be made about this definition.

- If $\epsilon(\mathbf{q}, \omega)$ is not a constant, but depends on $\mathbf{q}, \omega$, then the response of the medium is non-local in space, and delayed in time. These are real physical effects.
- If we use the convention that the charge densities and fields are complex quantities, ϵ will be complex. The meaning of this is as follows: the real part of ϵ corresponds to polarization in phase with the electric field, as is usual. The imaginary part means that part of the polarization is out of phase. That is, we have a *loss* term.

To see how this arises recall the continuity equation for the electric current:

$$\frac{\partial \rho_i}{\partial t} + \nabla \cdot \mathbf{J} = 0$$

$$\mathbf{q} \cdot \mathbf{J}(\mathbf{q}, \omega) = \omega\rho_i(\mathbf{q}, \omega). \tag{8.8}$$

Ohm's law defines the conductivity, $\sigma(\mathbf{q}, \omega)$:

$$\mathbf{J} = \sigma(\mathbf{q}, \omega)\mathbf{E}(\mathbf{q}, \omega). \tag{8.9}$$

If we take the dot product of this expression with $\mathbf{q}$, and use Eq. (8.7) we find:

$$\epsilon(\mathbf{q}, \omega) = 1 + \frac{4\pi i \sigma(\mathbf{q}, \omega)}{\omega}. \tag{8.10}$$

That is, the imaginary part of ϵ corresponds to the real part of σ, or to ohmic loss.

We note that Poisson's equation, $\nabla^2 \phi = -4\pi\rho$, where ϕ is the electrostatic potential, becomes, in Fourier space:

$$\phi(\mathbf{q}) = \frac{4\pi\rho}{q^2}. \tag{8.11}$$

We will distinguish between ϕ_e, which has as its source the external charge density alone, ρ_e, and the total ϕ which arises from $\rho_e + \rho_i$.

We now need to introduce the physics of the medium. We do this via the *susceptibility* or response function, χ. This specifies the linear response of the system to ρ_e. We define it in the following way:

$$\rho_i(\mathbf{q}, \omega) = \chi(\mathbf{q}, \omega)\phi_e = \frac{4\pi}{q^2}\chi\rho_e. \tag{8.12}$$

From Eq. (8.7):

$$\frac{1}{\epsilon} = 1 + \frac{4\pi\chi}{q^2}$$

$$\mathrm{Im}\frac{1}{\epsilon} = \frac{4\pi}{q^2}\mathrm{Im}\chi. \tag{8.13}$$

8.2 The fluctuation-dissipation theorem

The previous section contained a series of definitions. In order to deal with the physics of a real system, we need to compute something concrete. The most convenient function to consider is χ.

Suppose we apply a weak external perturbation to the electron gas. For example we can insert a time and space dependent external charge density, $\rho_e(\mathbf{r}, t)$. We will consider each Fourier component separately. The external charge will induce a charge in the system. The operator for this quantity is given in Problem 2 of Chapter 6:

$$\hat{\rho}(\mathbf{r}) = \sum_q \hat{\rho}(q)e^{i\mathbf{q}\cdot\mathbf{r}}; \quad \hat{\rho}(q) = \frac{1}{\Omega}\sum_{\mathbf{k},\sigma} \hat{c}^+_{\mathbf{k}+\mathbf{q},\sigma}\hat{c}_{\mathbf{k},\sigma}. \tag{8.14}$$

Note that we follow the somewhat confusing convention that $\hat{\rho}$ is a *number* density operator. The charge density is $e\langle\hat{\rho}(\mathbf{r})\rangle$.

If we proceed, we will have technical trouble because the model that we are considering has no explicit damping (e.g. from impurities). If we drive such a system with a sinusoidal external field, transients will never die out. In order to avoid spurious effects we use the trick of *adiabatic turning on*. That is, we assume that the system was in equilibrium in the distant past, and introduce the external charge slowly:

$$\rho_e(\mathbf{r}, t) = \bar{\rho}_e e^{i\mathbf{q}\cdot\mathbf{r}} e^{-i\omega t} e^{\eta t} + \text{c.c.} \tag{8.15}$$

Here η is a small positive number and c.c. means complex conjugate.

The coupling to the system is the Coulomb interaction of the external potential and the induced charge density:

$$\delta\hat{\mathcal{H}} = e \int d\mathbf{r}\phi_e(\mathbf{r}, t)\hat{\rho}_i(\mathbf{r}) = e\frac{4\pi\Omega}{q^2}\bar{\rho}_e\hat{\rho}_i(-\mathbf{q})e^{-i\omega t}e^{\eta t} + \text{c.c.} \tag{8.16}$$

We now use time-dependent perturbation theory. If $|G\rangle$ is the ground state, then the wavefunction can be written as a mixture of the excited states of the unperturbed system, labeled by m:

$$|\Psi\rangle = e^{-iE_o t/\hbar}|G\rangle + \sum_m a_m(t)e^{-iE_m t/\hbar}|m\rangle. \tag{8.17}$$

The Schrödinger equation implies that:

$$a_m(t) = \int_{-\infty}^{t} \frac{d\tau}{i\hbar}\langle m|\delta\hat{\mathcal{H}}|G\rangle e^{-i\omega_m \tau}, \tag{8.18}$$

where $\omega_m = (E_m - E_o)/\hbar$.

We are interested in the expectation value of $\hat{\rho}_i(\mathbf{q}, t) = e^{i\hat{\mathcal{H}}t}\hat{\rho}(\mathbf{q})e^{-i\hat{\mathcal{H}}t}$. Since there is no induced charge density in the ground state, we find, from Eq. (8.17) to the lowest non-vanishing order:

$$\langle\Psi|\hat{\rho}_i(\mathbf{q}, t)|\Psi\rangle = \sum_m [a_m^*\langle m|\hat{\rho}_i(\mathbf{q}, t)|G\rangle + \langle G|\hat{\rho}_i(\mathbf{q}, t)|m\rangle a_m]. \tag{8.19}$$

Combining what we have so far we get for the induced charge density:

$$\rho_i = e\langle\Psi|\hat{\rho}_i(\mathbf{q}, t)|\Psi\rangle = e^2\frac{4\pi\Omega\rho_e}{\hbar q^2}\sum_m |\langle m|\hat{\rho}_i(\mathbf{q})|G\rangle|^2$$

$$\times \left[\frac{1}{\omega + \omega_m + i\eta} - \frac{1}{\omega - \omega_m + i\eta}\right]e^{-i\omega t} + \text{c.c.} \tag{8.20}$$

The term that multiplies $e^{-i\omega t}$ gives us $\chi(\mathbf{q}, \omega)$.

What we have done so far is exact for a small external potential. It is expressed in terms of the (unknown) ground and excited states of the unperturbed system. For any integral whose integrand is smooth enough (see problems):

$$\frac{1}{x - i\eta} = \text{P}\frac{1}{x} + i\pi\delta(x). $$

This is called the Dirac relation. Using this in Eq. (8.20) we get:

$$\mathrm{Im} \frac{\rho_i(\mathbf{q}, \omega)}{4\pi\rho_e/q^2} \equiv \mathrm{Im}\chi(\mathbf{q}, \omega) = -\Omega e^2 \sum_m \frac{\pi}{\hbar}|\langle m|\hat{\rho}_i(\mathbf{q})|G\rangle|^2 \delta(\omega - \omega_m). \qquad (8.21)$$

We have used the fact that $\omega > 0, \omega_m > 0$.

The right-hand side of Eq. (8.21) involves the dynamic structure factor, cf. Eq. (5.141), which gives the scattering cross-section:

$$S(\mathbf{q}, \omega) = \Omega^2 \sum_m |\langle m|\hat{\rho}_i(\mathbf{q})|G\rangle|^2 \delta(\omega - \omega_m). \qquad (8.22)$$

The relationship in Eq. (8.21) is between a response function, χ and a scattering function, S.

$$\mathrm{Im}\chi(\mathbf{q}, \omega) = -\frac{\pi e^2}{\hbar\Omega} S(\mathbf{q}, \omega). \qquad (8.23)$$

This is an example of a more general theorem called the *fluctuation-dissipation theorem*. In this case it relates the out-of-phase response, which gives dissipation, to scattering, which is a measure of density fluctuations. Response to an external field is related to fluctuations in equilibrium, in the absence of an external field.

8.2.1 Response function for free electrons

We can evaluate the expression in Eq. (8.20) if we know the matrix elements of the density operator of Eq. (8.14). For the free Fermi gas this operator has matrix elements only if the excited state has a hole in $\mathbf{k}$ and a particle in $\mathbf{k} + \mathbf{q}$. Thus:

$$\chi^\circ = -\frac{e^2}{\hbar\Omega}\sum_{\mathbf{k}} f(\mathbf{k})(1 - f(\mathbf{k} + \mathbf{q}))\left[\frac{1}{\omega + (\epsilon_\circ(\mathbf{k} + \mathbf{q}) - \epsilon_\circ(\mathbf{k}))/\hbar + i\eta}\right.$$
$$\left. - \frac{1}{\omega - (\epsilon_\circ(\mathbf{k} + \mathbf{q}) - \epsilon_\circ(\mathbf{k}))/\hbar + i\eta}\right]. \qquad (8.24)$$

In this expression $\epsilon_\circ$ is the kinetic energy of a free electron. By changing variables in the first term, $\mathbf{K} = -(\mathbf{k} + \mathbf{q})$ and using the fact that $\epsilon_\circ(-\mathbf{k}) = \epsilon(\mathbf{k})$ it is easy to see that this expression is equivalent to:

$$\chi^\circ = \frac{e^2}{\Omega}\sum_{\mathbf{k}} \frac{f(\mathbf{k}) - f(\mathbf{k} + \mathbf{q})}{\hbar\omega - (\epsilon_\circ(\mathbf{k} + \mathbf{q}) - \epsilon_\circ(\mathbf{k})) + i\hbar\eta}. \qquad (8.25)$$

We will discuss the properties of this expression below. For the moment, we note a simple fact which we will need immediately, $\lim_{\mathbf{q} \to 0} \chi^\circ(\mathbf{q}, 0) \neq 0$.

8.2.2 Response function for Bloch electrons

For electrons in a periodic potential the charge density operator has a different form:

$$\hat{\rho}(\mathbf{q}) = \frac{1}{\Omega} \sum_{\mathbf{k},n,m} \hat{c}^+_{\mathbf{k}+\mathbf{q},n} \hat{c}_{\mathbf{k},m} \langle \phi_{\mathbf{k},n} | e^{i\mathbf{q}\cdot\mathbf{s}} | \phi_{\mathbf{k},m} \rangle, \tag{8.26}$$

where $\phi_{\mathbf{k},m}$ is the Bloch function of the mth band. Thus with a periodic potential we have:

$$\chi^\circ = \frac{e^2}{\Omega} \sum_{\mathbf{k},n,m} \frac{[f(\mathbf{k},n) - f(\mathbf{k}+\mathbf{q},m)] |\langle \phi_{\mathbf{k},n} | e^{i\mathbf{q}\cdot\mathbf{s}} | \phi_{\mathbf{k},m} \rangle|^2}{\hbar\omega - (\epsilon_n(\mathbf{k}+\mathbf{q}) - \epsilon_m(\mathbf{k})) + i\hbar\eta}. \tag{8.27}$$

The response function has intra-band transitions, $m = n$ and inter-band transitions $m \neq n$. In an insulator there are only the latter because the density operator has matrix elements only if the excited state has a hole in a filled band at $\mathbf{k}, m$ and a particle in an empty band at $\mathbf{k} + \mathbf{q}, n$.

8.2.3 Screening

The physical effect we are looking for is *screening*. This should weaken the potential of an external point charge at long ranges. That is, the singularity in the Coulomb potential as $\mathbf{q} \to 0$ should be reduced. From ordinary electrostatics this should occur by making the replacement (for the static case):

$$\phi(\mathbf{q}) \to \frac{\phi(\mathbf{q})}{\epsilon(\mathbf{q},0)}. \tag{8.28}$$

However, from Eq. (8.7), we see that the singularity gets worse:

$$\frac{\phi(\mathbf{q})}{\epsilon(\mathbf{q},0)} = \frac{4\pi}{q^2} \Big[1 + \frac{4\pi\chi^\circ(\mathbf{q},0)}{q^2} \Big]. \tag{8.29}$$

We need to do better to get the physics right.

8.3 Self-consistent response

As a guide to improving our treatment of response we look at long-wavelength effects that we can treat from a macroscopic viewpoint.

8.3.1 Self-consistent static screening

Wave-vector dependent screening as a physical phenomenon has been discussed for many systems for nearly a century. The oldest treatment was that of P. Debye and E. Hückel who were interested in the behavior of ions in an electrolyte. They thought of the ions as a dilute classical gas of charges. Their method was carried over to atoms by L. Thomas and E. Fermi. We discuss this method for the uniform electron gas here. In Chapter 9 we will apply this method to atoms.

Suppose we have a static external charge density, ρ_e inserted into the electron gas. Poisson's equation for the total potential reads:

$$\nabla^2 \phi(\mathbf{r}) = -4\pi[\rho_e(\mathbf{r}) + \rho(\mathbf{r}) - en_\circ]. \tag{8.30}$$

The second term is the total charge density of the electrons and the last is the positive background, $n_\circ = N/\Omega$. Recall that in our convention, e is a negative number.

We need to find $\rho(\mathbf{r}) \equiv en(\mathbf{r})$. We can combine the last two terms as $\rho_i = en_i = e(n - n_\circ)$, i.e., the induced density. Then Eq. (8.30) reads:

$$\nabla^2 \phi(\mathbf{r}) = -4\pi[\rho_e(\mathbf{r}) + \rho_i(\mathbf{r})] = -4\pi\rho_e(\mathbf{r}) - 4\pi en_i(\mathbf{r}). \tag{8.31}$$

In order to keep things straight, suppose that the external charge is positive, so that we expect ρ_i to be negative. Thus $n_i(\mathbf{r})$ and ϕ are positive. Suppose that the external charge is slowly varying, so that each part of the electron gas is uniform over distances large compared to k_F^{-1}. Then each part of the electron gas is effectively uniform.

Now the total energy of the electrons near the external charge contains two terms, the negative electrostatic energy, $e\phi$, plus the kinetic energy. As usual, we fill up states of total energy until the last filled state is at the chemical potential, μ; see Figure 8.1. Thus:

$$\frac{\hbar^2}{2m}[3\pi^2 n(\mathbf{r})]^{2/3} + e\phi(\mathbf{r}) = \mu. \tag{8.32}$$

Now suppose that ϕ is small. Also note that the chemical potential must be the Fermi energy of the unpolarized gas far from the external charge, $\mu \to \hbar^2[3\pi^2 n_\circ]^{2/3}/2m$. Thus:

$$\frac{\hbar^2}{2m}[3\pi^2 n_\circ]^{2/3}(1 + n_i/n_\circ)^{2/3} \approx \mu + \frac{\pi^2\hbar^2}{mk_F}n_i = \mu - e\phi$$

$$n_i \approx \frac{mk_F}{\pi^2 m}|e|\phi = |e|\frac{\mathcal{D}(E_F)}{\Omega}\phi. \tag{8.33}$$

Now we can insert this into Eq. (8.31):

$$\nabla^2 \phi(\mathbf{r}) = -4\pi\rho_e(\mathbf{r}) + 4\pi e^2 \frac{\mathcal{D}(E_F)}{\Omega}\phi(\mathbf{r})$$

$$\nabla^2 \phi(\mathbf{r}) - \lambda^2\phi(\mathbf{r}) = -4\pi\rho_e(\mathbf{r}). \tag{8.34}$$

This defines the *Thomas–Fermi screening parameter*:

$$\lambda^2 = 4\pi e^2 \frac{\mathcal{D}(E_F)}{\Omega}. \tag{8.35}$$

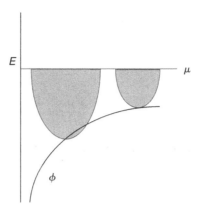

Fig. 8.1 **A schematic picture illustrating the Thomas–Fermi approximation. The electron states are filled to the same maximum total energy at different *r*, which corresponds to different densities, *n(r)*. The density difference is, for small changes in ϕ, given by $\Delta n = e\Delta\phi\mathcal{D}(E_F)/\Omega$.**

The solution to Eq. (8.34) is quite simple in Fourier space:

$$\phi(\mathbf{q}) = \frac{4\pi\rho_e(\mathbf{q})}{q^2 + \lambda^2}. \tag{8.36}$$

In particular, for an external point charge Q, at the origin, we have a screened Coulomb potential, as above:

$$\phi(\mathbf{r}) = \frac{Qe^{-\lambda r}}{r}.$$

If we now compare to Eq. (8.28) above, we find the dielectric function in the Thomas–Fermi limit:

$$\epsilon(\mathbf{q}, 0) = 1 + \lambda^2/q^2. \tag{8.37}$$

Note the steps that we have followed: the induced charge density was determined by a simple approximation for the response, but as a response to the *total* potential. Then the total potential was determined self-consistently.

8.3.2 Plasma oscillations

We can apply the same method of self-consistently determining the potential in another context, that of long-wavelength dynamic oscillations. Suppose we apply a positive external charge again, but now oscillating in time. Once more, we require that the variation be slow compared to the natural scales of the electron gas. The external charge causes the electrons to move. They obey Newton's equation of motion:

$$m\dot{\mathbf{v}} = e\mathbf{E} = -e\nabla\phi. \tag{8.38}$$

The current, $\mathbf{J}$, is $ne\mathbf{v} \approx n_\circ e\mathbf{v}$. Further, the current obeys the continuity equation, $\dot{\rho}_i = -\nabla \cdot \mathbf{J}$. Thus:

$$\frac{d\mathbf{J}}{dt} = -\frac{n_\circ e^2}{m}\nabla\phi$$

$$-\frac{d^2\rho_i}{dt^2} = \frac{4\pi n_\circ e^2}{m}(\rho_e + \rho_i). \tag{8.39}$$

We have used Eq. (8.31). Now solve for ρ_i:

$$\frac{d^2\rho_i}{dt^2} + \omega_p^2\rho_i = -\omega_p^2\rho_e. \tag{8.40}$$

The characteristic frequency, $\omega_p = \sqrt{4\pi n_\circ e^2/m}$ is called the plasma frequency. (The same effect as the one we are discussing occurs in classical plasmas). Now suppose the time variation is $e^{-i\omega t}$. We get:

$$\rho_i(-\omega^2 + \omega_p^2) = -\omega_p^2\rho_e$$

$$\epsilon(0,\omega) = \rho_e/(\rho_i + \rho_e) = 1 - \frac{\omega_p^2}{\omega^2}. \tag{8.41}$$

This is a useful expression for the dielectric function of a metal. This behavior is that of an undamped resonance oscillation, as Eq. (8.40) shows. Physically, what is happening is that the accumulations of charge density act like parallel plate capacitors. The total field causes the electrons to oscillate with the natural frequency, ω_p.

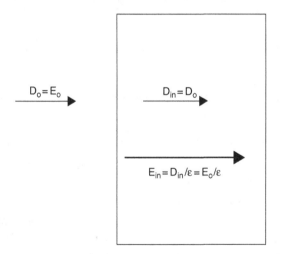

Fig. 8.2   The slab geometry. In a plasma oscillation the electrons move back and forth with respect to the positive background. The restoring force due to the polarization charges at the boundaries generates the plasma oscillation. If $q \neq 0$ then there are, in effect, slabs of charge spaced by $2\pi/q$ and the effect is the same.

The effect is simplest in a slab geometry. If the slab of metal is excited with an external field, the response is for the electrons to slosh back and forth with respect to the ions. As they build up at the boundaries, they generate an electric field that causes them to move the other way: (see Figure 8.2). An external electric field, E_o gives rise to a field inside the slab of $E_o/\epsilon(\omega)$. At $\omega = \omega_p$ there is a divergent response because of the zero of $\epsilon(\omega)$. The resonance frequency is that of a *longitudinal normal mode* of the system. We have already seen an example of this in Eq. (5.64).

8.4 The RPA dielectric function

Armed with the insights of the previous section we are now in position to improve the approximation of the dielectric function and give a correct description of screening. To do this we return to Eq. (8.31) in Fourier space:

$$q^2\phi(\mathbf{q},\omega) = 4\pi\rho_e(\mathbf{q},\omega) + 4\pi\rho_i(\mathbf{q},\omega)$$
$$= 4\pi\rho_e(\mathbf{q},\omega) + 4\pi\chi(\mathbf{q},\omega)\phi(\mathbf{q},\omega). \tag{8.42}$$

Therefore:

$$\left(1 - \frac{4\pi\chi}{q^2}\right)\phi(\mathbf{q},\omega) = \frac{4\pi}{q^2}\rho_e = \phi_e$$

$$\phi(\mathbf{q},\omega) = \frac{\phi_e}{1 - 4\pi\chi/q^2} \equiv \frac{\phi_e}{\epsilon}$$

$$\epsilon(\mathbf{q},\omega) = 1 - \frac{4\pi}{q^2}\chi. \tag{8.43}$$

The self-consistent version of the dielectric function is now known once we know χ. For this we use χ° given in Eq. (8.24) or Eq. (8.25). The nature of this approximation is to consider the electron gas to be non-interacting as far as local correlations are concerned, but to take the long-range parts of the interaction into account. The result is the RPA dielectric function:

$$\epsilon(\mathbf{q},\omega) = 1 + \frac{4\pi e^2}{\Omega q^2}\sum_{\mathbf{k}}\frac{f(\mathbf{k}) - f(\mathbf{k}+\mathbf{q})}{\epsilon_\circ(\mathbf{k}+\mathbf{q}) - \epsilon_\circ(\mathbf{k}) - \hbar\omega - i\hbar\eta}. \tag{8.44}$$

(The acronym RPA means Random Phase Approximation. It refers to a method of deriving Eq. (8.44) which we do not discuss here; see (Pines 1999).)

It is sometimes useful to define a screened response function. This is the screened response to the *external* field. We can get this by comparing Eq. (8.13) to Eq. (8.43). We introduce the abbreviation $v = 4\pi/q^2$. Then:

$$1 + v\chi_{\text{scr}} = \frac{1}{1 - v\chi^\circ}; \quad \chi_{\text{scr}} = \frac{\chi^\circ}{1 - v\chi^\circ}. \tag{8.45}$$

We can recover the macroscopic results of the previous sections from Eq. (8.44). For example, putting $\omega = 0, \mathbf{q} \to 0$ we find:

$$\epsilon(\mathbf{q}) \to 1 + \frac{4\pi e^2}{\Omega q^2} \sum_{\mathbf{k}} \frac{f(\mathbf{k}) - f(\mathbf{k} + \mathbf{q})}{\hbar \mathbf{k} \cdot \mathbf{q}/m}. \tag{8.46}$$

However, $f(\mathbf{k}) - f(\mathbf{k} + \mathbf{q}) \approx -\mathbf{q} \cdot \nabla_{\mathbf{k}} f(\mathbf{k}) = (\mathbf{q} \cdot \hbar \mathbf{k}/m)\delta(\epsilon_\circ(\mathbf{k}) - E_F)$. Inserting this into Eq. (8.46) we get:

$$\epsilon(\mathbf{q}) = 1 + \frac{4\pi e^2}{\Omega q^2} \sum_{\mathbf{k}} \delta(\epsilon_\circ(\mathbf{k}) - E_F) = 1 + \frac{4\pi e^2 \mathcal{D}(E_F)}{\Omega q^2} = 1 + \frac{\lambda^2}{q^2}. \tag{8.47}$$

This is the Thomas-Fermi dielectric function. Similarly we can show that $\lim_{\mathbf{q} \to 0} \epsilon(\mathbf{q}, \omega) = 1 - \omega_p^2/\omega^2$.

The integrals involved in evaluating the full dielectric function are complicated, but pose no real challenge. The expression for the full function can be found in several of the references. For $\omega = 0$:

$$\epsilon(\mathbf{q}, \omega) = 1 + \frac{4\pi e^2}{q^2} \left[\frac{1}{2} + \frac{4k_F^2 - q^2}{8k_F q} \ln \left| \frac{2k_F + q}{2k_F - q} \right| \right]. \tag{8.48}$$

There is a lot of physics in Eq. (8.44). We now discuss some things that are direct results of this expression.

8.4.1 Zeros of ϵ

As we remarked above the zeros of the dielectric function correspond to normal modes of the system. In particular, if we plot the solutions of $\epsilon(\mathbf{q}, \omega) = 0$ in the form $\omega_p = \omega(q)$ we have the dispersion relation for plasmons.

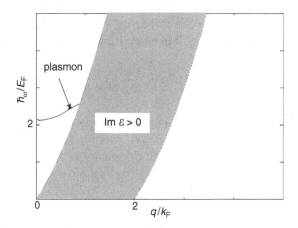

Fig. 8.3   **A sketch of the ω, q plane. The region where $\mathrm{Im}\,\epsilon(q, \omega) \neq 0$ is shown along with the plasmon dispersion relation.**

8.4.2 Damping

There are regions of the ω, q plane where the denominator of Eq. (8.44) can be 0. As we have seen above, these are regions where there is lossy response. By examining the expression, it is clear that these are regions corresponding to the frequency matching the energy of an electron-hole pair.

$$\hbar\omega = \epsilon_\circ(\mathbf{k} + \mathbf{q}) - \epsilon_\circ(\mathbf{k}). \tag{8.49}$$

In these regions real excitations of the system are possible, not merely polarization. In particular, when the plasmon dispersion relation meets this region, the plasmons have a finite lifetime; this is called *Landau damping*.

8.4.3 Friedel oscillations

If q is large enough, the factor $1 - f(\mathbf{k} + \mathbf{q})$ in Eq. (8.24) does not restrict the integration, because $\epsilon_\circ(\mathbf{k} + \mathbf{q})$ will be bigger than E_F for all $k < k_F$. This occurs when $q \geq 2k_F$. We expect that there will be a change in the nature of the function at this value of q. Note that at $q = 2k_F$ the vector $\mathbf{q}$ just spans the Fermi sphere. A look at $\epsilon(q)$ in Eq. (8.48) shows that there is a singularity in the derivative at this point.

If we return to the expression for the screening of a point charge:

$$\phi(\mathbf{r}) = \int d\mathbf{q} e^{i\mathbf{q}\cdot\mathbf{r}} \frac{4\pi Q}{q^2 \epsilon(q)}, \tag{8.50}$$

it is not hard to believe (and to show) that the singularity picks out the value $q = 2k_F$. The result of this is that the potential and the screening charge have an unexpected behavior at large distances:

$$\phi(\mathbf{r}) \to \frac{\cos(2k_F r)}{r^3}. \tag{8.51}$$

There are halos of charge of both signs surrounding the impurity. These *Friedel oscillations* are observed in NMR experiments. They are a general phenomenon, not an artifact of the RPA. In fact, not only point charges but also surfaces generate oscillations. They are an example of the Gibbs phenomenon: whenever we have the Fourier transform of a sharp structure (in this case, the Fermi surface) we find oscillations.

J. Friedel first derived this result by using scattering theory for the conduction electrons interacting with the impurity. The argument is as follows: suppose we change the geometry and put a scattering potential (it could be a Coulomb potential, as above, or something else) at the center of a spherical sample of radius R. Far from the origin each electron will have a wavefunction which reflects the scattering through *phase shifts* η_l. From standard quantum mechanics we can write the wavefunction for each electron as:

$$\psi_{k,l,m}(r) \to \sqrt{\frac{2}{R}} \frac{\sin(kr - l\pi/2 + \eta_l)}{r} Y_{lm}. \tag{8.52}$$

The spherical harmonic is evaluated at the angle between $\mathbf{k}$ and $\mathbf{r}$. The boundary condition at the edge of the sphere can be taken to be $\psi(r, \theta, \phi) = 0$. Thus:

$$kR - l\pi/2 + \eta_l = M\pi. \tag{8.53}$$

Now the number of states per unit interval of k, for each spin and each angular momentum is R/π.

Let us assume that the phase shifts are independent of k and write the additional electron density at some large radius due to the scattering center. We do this by comparing with the case with no scattering, i.e., $\eta_l = 0$.

$$\Delta n = 2 \sum_{l,m} \frac{R}{\pi} \frac{2}{R} \int_0^{k_F} dk \, \frac{1}{r^2} [\sin^2(kr - l\pi/2 + \eta_l) - \sin^2(kr - l\pi/2)]$$

$$= \frac{4}{\pi r^2} \sum_l (2l+1) \int_0^{k_F} dk \, [\sin^2(kr - l\pi/2 + \eta_l) - \sin^2(kr - l\pi/2)]$$

$$\propto \sum_l (2l+1)(-1)^l \frac{\cos(2k_F r + \eta_l) \sin(\eta_l)}{r^3}. \tag{8.54}$$

Friedel sum rule

The description of a scattering center by phase shifts gives rise to a useful relation which we will digress to derive. If we examine the boundary condition of Eq. (8.53) we note that the phase shifts can tell us if electrons are localized near the scattering center. For example, for an impurity of charge Z, we know that there will be Z electrons near the center to screen it.

We see that the boundary condition means that the allowed k's near k_F are shifted by $-\eta_l(k_F)/R$. Thus the total number of new states drawn into the sphere is the number of states per unit k which we have found to be 2 (for spin) times $(2l + 1)R/\pi$ times this quantity. Summing over all the angular momentum channels gives:

$$\Delta N = \frac{2}{\pi} \sum_l (2l+1)\eta_l(k_F). \tag{8.55}$$

Thus, for example, for s-wave scattering there are η_0/π states near the center for each spin.

8.4.4 RPA dielectric function for Bloch electrons

We can easily generalize the RPA dielectric function to electrons in a band structure by using Eq. (8.27).

$$\epsilon(\mathbf{q}, \omega) = 1 + \frac{4\pi e^2}{\Omega q^2} \sum_{\mathbf{k},n,m} \frac{[f(\mathbf{k}, n) - f(\mathbf{k} + \mathbf{q}, m)]|\langle \phi_{\mathbf{k}+\mathbf{q},n}|e^{i\mathbf{q}\cdot\mathbf{s}}|\phi_{\mathbf{k},m}\rangle|^2}{\epsilon_n(\mathbf{k} + \mathbf{q}) - \epsilon_m(\mathbf{k}) - \hbar\omega - i\hbar\eta}. \tag{8.56}$$

The new feature of this expression is the inter-band terms. We can have energy absorption if $\hbar\omega = \epsilon_n(\mathbf{k} + \mathbf{q}) - \epsilon_m(\mathbf{k})$.

8.4.5 Nesting

We noted above that the singularity in Eq. (8.48) gives rise to Friedel oscillations. In fact, more dramatic things can happen in special systems.

To see this, we note that we can look at the singularity in $\epsilon(\mathbf{q}, 0)$ in another way. The denominator in Eq. (8.44) (setting $\omega = 0$) is close to zero if the energies $\epsilon_o(\mathbf{k} + \mathbf{q})$, $\epsilon_o(\mathbf{k})$ are nearly equal. This will occur if they are on opposite sides of the Fermi surface and if there is time-reversal invariance so that $\epsilon_o(\mathbf{k}) = \epsilon_o(-\mathbf{k})$. For example if $\mathbf{k} = -\mathbf{q}/2, q = 2k_F$ we have a singularity in the integrand of Eq. (8.44) for free electrons. This occurs at one point in the integration so that the singularity is rather mild, as in Eq. (8.48). However, if we have a general band structure, the corresponding effect in Eq. (8.56) can be much stronger, as Figure 8.4 shows. On the right we see a Fermi surface that has a flat portion so that many vectors can connect one side with another. This is known as *nesting* and it can give rise to a singularity in the dielectric function or in the response function. A singularity in response means that the system is unstable. This has remarkable consequences.

A singularity of this type means that the crystal can spontaneously distort. Suppose it does: if we apply a field (from the distortion) at the nesting wavevector $\mathbf{q}$ then the electrons will see a potential that can open up new gaps in the band structure *exactly* at the Fermi surface. The electrons will lower their kinetic energy due to the new gap, and the distortion can be stabilized by paying positive lattice energy. This occurs in many materials and is called a *charge density wave* (Grüner 1994). The material whose Fermi surface is shown in Figure 6.9 has a charge-density wave transition at about 200 K. An examination of this figure will show numerous portions of the Fermi surface that can nest.

Note that the wavelength of the distortion need not be commensurate with the lattice so that the whole density wave could slide with respect to the lattice. This gives rise to strange

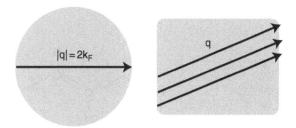

Fig. 8.4 The free electron Fermi surface has a nesting point which gives rise to an integrable singularity in χ, left side. However, band structures that give rise to perfect nesting, as on the right, can give rise to singularities that are not integrable.

features in the conductivity. A different kind of density wave, the spin density wave, is closely related, but depends on the electron-electron interaction.

A situation in which we necessarily have nesting is in one dimension. The Fermi point at $-k_F$ is exactly connected to the one at k_F by $q = 2k_F$. It is easy to see the singularity in this case. Suppose that $T = 0$ and $q \approx 2k_F$. Then:

$$\chi^{\circ} \propto \sum_{k} \frac{f_k(1 - f_{k+q})}{\epsilon_{\circ}(k + q) - \epsilon_{\circ}(k)} \propto \int_{-k_F} dk \, \frac{1}{\hbar v_F(2k + q)} \propto \ln(|q - 2k_F|). \qquad (8.57)$$

In this expression we have linearized the electron energy near the Fermi points: $\epsilon_{\circ}(k) \approx E_F + \hbar v_F(k - k_F)$ for $k \approx k_F$, and similarly for k near the other Fermi point.

This effect was discovered by R. Peierls from a different point of view. He thought of a one-dimensional metal with a half-filled band (one electron per atom) and lattice constant a. Then he considered the possibility of a commensurate distortion of the sort shown in Figure 8.5 where every other atom has a static displacement by δa. This raises the energy of the system by the strain energy which is proportional to $(\delta a)^2$. However, now there is a Fourier component of the lattice potential at a reciprocal lattice vector, $g = 2\pi/2a$ associated with the new lattice period, $2a$. It is easy to see that the magnitude of this component, $V_{\pi/a}$ will be of order δa. Thus there is a gap at the Fermi surface, $|V_{\pi/a}|^2 \propto (\delta a)^2$ from Eq. (6.49). A gap lowers the kinetic energy of all the electrons in the filled states. In the problems you will show that this energy is of order $(\delta a)^2 \ln(\delta a)$. That is, the lowering of the electronic energy always wins, and the system distorts. The system changes from a metal to an insulator.

This distortion is called a Peierls distortion. In his book, *More Surprises in Theoretical Physics* (Peierls 1991), Peierls noted that he found this effect when "tidying material" during the preparation of a textbook. He thought of it as a theoretical curiosity, and never imagined that it could be observed. This is the surprise: the Peierls distortion is routinely observed in quasi-one-dimensional materials such as blue bronzes. As shown in Figure 8.5 we can think of the linear polymer polyacetylene, which has alternating single and double bonds, as a Peierls system.

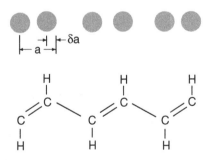

Fig. 8.5 **Top: the distortion of a one-dimensional metal that opens a gap at the Fermi surface gives rise to the Peierls effect. Bottom: the polymer polyacetylene has a half-filled band formed by the π-bonded electrons. It may be thought of as a Peierls insulator.**

8.5 Optical properties of crystals

Electromagnetic radiation interacts with condensed matter. Solids have interesting colors and can be transparent or opaque. In this section we will look at these phenomena using the tools that we have developed above.

8.5.1 Optical constants

In elementary treatments of optics a central role is played by the *refractive index* n which is given by $n = \sqrt{\epsilon}$ for dielectric materials. In the previous section we generalized ϵ to include a real and imaginary part, corresponding to polarization and loss. We will do the same here for the refractive index.

Let us consider Maxwell's equations in a material medium with no external charge and no magnetic properties so that $\mathbf{B} = \mathbf{H}$. They read:

$$\nabla \cdot \mathbf{E} = 4\pi \rho_i$$

$$\nabla \cdot \mathbf{B} = 0$$

$$\nabla \times \mathbf{E} = -(1/c)\partial \mathbf{B}/\partial t$$

$$\nabla \times \mathbf{E} = (1/c)\partial \mathbf{E}/\partial t + 4\pi \mathbf{J}/c. \tag{8.58}$$

Now suppose that there is a plane wave in the z direction, polarized in the x direction and of frequency ω. Then there is no induced charge density, and the wave equation follows:

$$\frac{d^2 E}{dz^2} + \frac{\omega^2}{c^2} E = -\frac{4\pi i \omega}{c^2} J. \tag{8.59}$$

Assume that $J = \sigma(\omega)E$. We neglect the $\mathbf{q}$ dependence here since we are interested in visible light so that the wavelength is very long on the scale of the crystal. Then:

$$\frac{d^2 E}{dz^2} + \frac{\omega^2}{c^2}\left(1 + \frac{4\pi\sigma}{\omega}\right)E = 0. \tag{8.60}$$

The factor in parentheses is $\epsilon(\omega)$.

The solution to this is a damped wave (we restore the time dependence):

$$E = E_0 \exp(i[q\, \mathsf{N}z - \omega t]); \quad q = \omega/c; \quad \mathsf{N}^2 = \epsilon(\omega). \tag{8.61}$$

This is the generalization we need. The definition $\mathsf{N} = \sqrt{\epsilon} = \sqrt{1 + 4\pi\sigma/\omega}$ implies that N has a real part, n and an imaginary part, conventionally called k. The spatial dependence of the wave is:

$$E = e^{-i\omega t} e^{inqz} e^{-kqz}. \tag{8.62}$$

We can also calculate the reflectivity of a piece of material on which a wave falls. The computation is the same as given in Jackson (1999). The result is:

$$R = \left| \frac{N-1}{N+1} \right|^2 = \frac{(n-1)^2 + k^2}{(n+1)^2 + k^2}. \tag{8.63}$$

As an example, consider a free electron metal with negligible scattering. For the longitudinal response we derived $\epsilon(\omega) = 1 - \omega_p^2/\omega^2$. However, for $\mathbf{q}$ very small the transverse response must be the same, so the result holds here too. Now for $\omega < \omega_p$, ϵ is negative. The corresponding refractive index, $N = \sqrt{\epsilon}$ is pure imaginary. By the result above, Eq. (8.63), this implies $R = 1$, total reflection, which is the case for metals. However, for $\omega > \omega_p$ the situation is reversed. Metals are transparent in the ultraviolet, for frequencies above the plasma frequency.

8.5.2 RPA transverse dielectric function

We can repeat the linear response theory we did above for the transverse case. We are interested in the response to an external electromagnetic wave which is most conveniently represented by a vector potential $\mathbf{A}$. We assume no external charge. Then we can use the transverse guage $\nabla \cdot \mathbf{A} = 0, \mathbf{E} = -(1/c)\partial \mathbf{A}/\partial t$. Suppose $\mathbf{A} = \mathbf{e}A$ where $\mathbf{e}$ is a unit vector in some direction.

The coupling Hamiltonian to the external perturbation is well known from quantum theory:

$$(eA/mc)\mathbf{e} \cdot \hat{\mathbf{p}}. \tag{8.64}$$

Now we need to calculate a response function. A useful object to look for is the induced current, $\mathbf{J}$. We can get the response function we need by setting $\mathbf{J} = \sigma \mathbf{E}$. The details of the calculation are exactly what was given above for the longitudinal case.

Once we get σ we can write down the dielectric function using $\epsilon = 1 + 4\pi\sigma/\omega$. The result is:

$$\epsilon(\mathbf{q}, \omega) = 1 + \frac{8\pi e^2 \hbar^2}{\Omega m^2} \sum_{\mathbf{k},n,m} \frac{[f(\mathbf{k}, n) - f(\mathbf{k}+\mathbf{q}, m)]}{(\epsilon_n(\mathbf{k}+\mathbf{q}) - \epsilon_m(\mathbf{k}))^2}$$

$$\times \frac{|\langle \phi_{\mathbf{k}+\mathbf{q},n}|e^{i\mathbf{q}\cdot\mathbf{s}}\mathbf{e} \cdot \hat{\mathbf{p}}|\phi_{\mathbf{k},m}\rangle|^2}{(\epsilon_n(\mathbf{k}+\mathbf{q}) - \epsilon_m(\mathbf{k}) - \hbar\omega - i\hbar\eta)}. \tag{8.65}$$

This function can be shown to be the same as the longitudinal function for $\mathbf{q} \to 0$.

8.5.3 Interband transitions

An interesting feature of Eq. (8.65) is its imaginary part which corresponds to absorption. From the Dirac relation,

$$1/(x - i\eta) = \mathrm{P}(1/x) + i\pi\delta(x),$$

we see that there is an imaginary part only if (for $\mathbf{q} = 0$):

$$\epsilon_n(\mathbf{k}) - \epsilon_m(\mathbf{k}) = \hbar\omega. \tag{8.66}$$

These are the band-to-band transitions that we have mentioned above. The formalism we have given allows us to calculate them in detail.

We choose a few simple examples. Si is a semiconductor, and is transparent if $\hbar\omega$ is less than the direct gap. However, this is in the infrared. In the visible there are many allowed transitions. Thus ϵ_R and ϵ_i are comparable, so that n and k are both large. Thus the reflectivity is large, and Si looks metallic even though it is a semiconductor. The red color of Cu is due to d bands below the Fermi surface. There can be transitions in the red from filled states to empty states. Materials that have large band gaps such as NaCl or diamond are transparent because there are no possible transitions.

Suggested reading

Our treatment of dielectric functions follows:
   Nozieres & Pines (1999)
For the fluctuation-dissipation theorem see:
   Chaikin & Lubensky (1995)
   Plischke & Bergersen (1994)
For a nice treatment of optical properties see:
   Yu & Cardona (2001)

Problems

1. (a) Suppose that the relationship between $\mathbf{D}$ and $\mathbf{E}$ is non-local (a field at $\mathbf{r}'$ causes a polarization at $\mathbf{r}$) and delayed (a field at time t' gives rise to a *later* response at time t). Then write:

   $$\mathbf{D}(\mathbf{r}, t) = \int d\mathbf{r}' \int_{-\infty}^{t} dt' \bar{\epsilon}(\mathbf{r} - \mathbf{r}', t - t') \mathbf{E}(\mathbf{r}', t').$$

   Find a relationship between $\bar{\epsilon}(\mathbf{r}, t)$ and $\epsilon(\mathbf{k}, \omega)$. Hint: Take the Fourier transform in space and time of the equation above using Eq. (8.1). (b) Use the fact that $\bar{\epsilon}$ is a real number to show that ϵ_R is even in ω and ϵ_I is odd.

2. In this problem we will analytically continue the response functions that we have studied into the complex ω plane so that $\omega = \omega_R + i\omega_I$. (a) Use the result of the previous problem to show that a response function such as $\epsilon(\omega)$ is analytic for $\omega_I > 0$. Here, and in what follows, we will suppress the $\mathbf{q}$ dependence. (b) Define $\delta\epsilon = \epsilon(\omega) - \epsilon(\infty)$. Consider a contour that follows the real axis from $-W$ to W and closes in the upper half plane with

a semicircle. Take W larger than the frequencies of interest. Then show that:

$$\delta\epsilon(\omega) = \oint \frac{d\omega'}{2\pi i} \frac{\delta\epsilon(\omega')}{\omega' - \omega - i\eta},$$

with $\eta > 0$. (c) Assume that $\epsilon_i(\infty) = 0$. Write the real and imaginary parts of this equation in the limit $\eta \to 0; W \to \infty$ to show:

$$\epsilon_R(\omega) - \epsilon_R(\infty) = \frac{2}{\pi} P \int_0^\infty d\omega' \frac{\omega' \epsilon_I(\omega')}{\omega'^2 - \omega^2}$$

$$\epsilon_I(\omega) = \frac{2\omega}{\pi} P \int_0^\infty d\omega' \frac{\epsilon_R(\omega') - \epsilon_R(\infty)}{\omega'^2 - \omega^2}. \tag{8.67}$$

Here P means the principal part of the integral. These are called the *Kramers–Kronig* relations and show that the real part of ϵ determines the imaginary part, and vice-versa.

3. Kubo relations: in the text we described the response of a many-electron system to an external electrostatic potential. This calculation can be generalized by considering an external force, f, which couples to an operator $\hat{\alpha}$ as follows:

$$\delta\hat{\mathcal{H}} = -\int d\mathbf{r} f(\mathbf{r}, t)\hat{\alpha}(\mathbf{r}).$$

This will induce a response in an operator $\hat{\beta}$ at time t so that:

$$\langle \hat{\beta}(\mathbf{r}) \rangle = \int_{-\infty}^t dt \int d\mathbf{r}' \, \chi(\mathbf{r}, \mathbf{r}', t - t') f(\mathbf{r}', t');$$

compare Problem 1. We can assume that $\langle \hat{\beta}(t = -\infty) \rangle = 0$, and that the system starts in state $|i\rangle$. (a) Show that:

$$\langle \hat{\beta}(t) \rangle = \frac{i}{\hbar} \int_{-\infty}^t dt \int d\mathbf{r}' \, f(\mathbf{r}', t')$$

$$\times \sum_m [e^{i\omega_{mi}\tau} \langle i|\hat{\beta}|m\rangle \langle m|\hat{\alpha}|i\rangle - e^{-i\omega_{mi}\tau} \langle i|\hat{\alpha}|m\rangle \langle m|\hat{\beta}|i\rangle].$$

Here $\omega_{im} = (E_i - E_m)/\hbar; \tau = t' - t$. (b) Express this in terms of the Heisenberg operators $\hat{\beta}(t) = e^{i\hat{\mathcal{H}}t/\hbar} \hat{\beta} e^{-i\hat{\mathcal{H}}t/\hbar}$, and similarly for $\hat{\alpha}$. Hint: figure out the matrix element, $\langle m|\hat{\beta}(t)|i\rangle$. (c) Prove the Kubo relation:

$$\chi(\mathbf{r}, \mathbf{r}', t - t') = \frac{i}{\hbar} \langle i|[\hat{\beta}(\mathbf{r}, t), \hat{\alpha}(\mathbf{r}', t')]|i\rangle e^{-\eta(t - t')},$$

if $t > t'$ and zero otherwise.

4. Continuation of previous problem. We now assume that we average over initial states with a Boltzmann weight to get a thermal average:

$$\langle \hat{\alpha} \rangle_T = \sum_i e^{-\beta E_i} \langle i|\hat{\alpha}|i\rangle.$$

(a) Take $\hat{\beta} = \hat{\alpha}$. Define $S(\mathbf{r}, \mathbf{r}', t) = \langle \hat{\alpha}(\mathbf{r}, t)\hat{\alpha}(\mathbf{r}') \rangle_T$. Show that:

$$\mathrm{Im}\,\chi(\omega) = \frac{\pi(1 - e^{-\beta\hbar\omega})}{\hbar} S(\omega.)$$

This is a more general form of the relation between response and scattering. (Hint: the key step is to show for thermal averages that if $\bar{S}(t) = \langle \hat{\alpha}(0)\hat{\alpha}(t) \rangle_T$ then $\bar{S}(\omega) = e^{-\beta\hbar\omega} S(\omega)$.)
(b) Define the noise correlator:

$$\Psi(t) = \frac{1}{2}\langle \hat{\alpha}(t)\hat{\alpha} + \hat{\alpha}\hat{\alpha}(t) \rangle_T.$$

As in the previous part, show that:

$$\Psi(\omega) = \hbar \coth(\beta\hbar\omega/2)\,\mathrm{Im}\,\chi(\omega) \to \frac{2k_BT}{\omega}\mathrm{Im}\,\chi(\omega).$$

The last expression is the classical limit, $\hbar \to 0$. This, particularly in the classical limit, is what is usually called the fluctuation-dissipation theorem. A noise correlation function is proportional to the dissipative part of the response.

5. A famous example of the fluctuation-dissipation theorem of the previous problem is Johnson-Nyquist noise. Consider a circuit that contains a resistor at temperature T. The resistor will produce a fluctuating voltage $V(t)$. Johnson observed, and Nyquist explained, that the noise power is proportional to the absolute temperature. This kind of noise is "white"; that is, the magnitude of $|V(\omega)|^2$ is a constant over a large frequency interval. Note that we need to imagine that we observe the noise over a finite time, say $\tau = 1/df$. The exact result is $S_V = 4Rk_BT$ where R is the resistance and S_V is the power spectral density of the noise voltage, that is $S_V = 2|V(\omega)|^2|df$. Derive this either from the previous problem, or by the following steps: (a) Suppose the resistor is in series with a capacitor, C, and can be thought of as a random voltage source giving voltage $V(t)$. Find the voltage *on the capacitor*, $V_C(\omega)$ in terms of $V(\omega)$ by solving the circuit. (b) Show from equipartition that $\langle (V_C(t)^2 \rangle_T = k_BT/C$. (c) Use the Weiner–Khintchine theorem,

$$\langle X^2(t) \rangle = \int_0^\infty S_X d\omega/(2\pi),$$

to find S_V.

6. This problem deals with screening in the classical case (Debye–Hückel screening). This theory is used in the chemistry of electrolytes. (a) Consider a positive charge $\delta\rho_e$ in a sea of mobile charges, $-q$ of average density n_0 and a neutralizing background qn_0. Suppose U is the electrostatic potential. Justify the following equations:

$$\nabla^2 U = -4\pi(\rho_e + \rho_i); \rho_e = qn_0 + \delta\rho_e$$

$$\rho_i = -qn_0 \exp(qU/k_BT).$$

(b) Solve by a Fourier transformation for the case $\rho_i = 0$, no screening. You should find for the bare potential

$$U_b(k) = 4\pi\delta\rho_e(k)/k^2.$$

(c) Now turn on the screening, and assume U is small. Show that you get $U = U_b/\epsilon(k)$; $\epsilon(k) = 1 + \lambda^2/k^2$. You must find λ. (d) Find $U(r)$, the screened potential in real space, for the case where $\delta\rho_e = Q\delta(r)$, a point charge.

7. Consider a metal for which we have constructed an atomic pseudopotential $v_a(\mathbf{r})$ such as in Eq. (6.59). The object that enters band structure calculations is $v(\mathbf{k}) = v_a(\mathbf{k})/\epsilon(\mathbf{k})$ which is the screened Fourier component (see Eq. (6.43)). Show that $\lim_{\mathbf{k}\to 0} v(\mathbf{k}) = -2E_F/3$.

8. Another way to get plasma oscillations: Suppose that a slab of metal is exposed to an external electric field perpendicular to the slab which varies at frequency ω. Write, as in electromagnetic theory, $E_{in} = E_{out}/\epsilon(\omega)$. Show that E_{in} becomes very large near $\omega = \omega_p = \sqrt{4\pi ne^2/m}$ where n is the number density of electrons. You should get:

$$\epsilon(\omega) = 1 - \omega_p^2/\omega^2.$$

Hint: suppose the electrons in the slab move uniformly back and forth. Then there is an induced surface charge, and an induced electric field inside the slab as in an ordinary capacitor. Write Newton's equation for the electrons with the total field, induced plus external, as a restoring force.

9. Show that $\lim_{\omega\to 0} \epsilon_{RPA}(0, \omega) = 1 - \omega_p^2/\omega^2$.

10. The Peierls argument for the distortion of a one-dimensional metal with a half-filled band: (a) Suppose that a lattice distorts as in Figure 8.5. Use the result of Problem 10 in Chapter 6 to show that the gap at the Fermi surface is proportional to δa^2 for small δa. (b) Use Eq. (6.48) to show that the change in the electron kinetic energy which results from the distortion has a term that is proportional to $\delta a^2 \log(\delta a)$ for small distortion. (c) Minimize the total energy with respect to δa to show that the metal always gains energy by distorting.

Hint: The positive elastic energy of the distortion is $\propto K\delta a^2$ where K is an elastic constant. Also, the integral for the electron energy is a bit simpler if you differentiate with respect to δa first.

11. Prove the Dirac relation:

$$\lim_{\eta\to 0} \int_{-\infty}^{\infty} dx \frac{f(x)}{x - i\eta} = P \int_{-\infty}^{\infty} dx \frac{f(x)}{x} + i\pi f(0).$$

Here P stands for principal part. You should suppose that the function f is analytic in a strip containing the real axis and that η is positive.

12. In this problem use the Drude results from Problem 1 of Chapter 7. (a) Show that if $\omega\tau \ll 1, \omega \ll \omega_p$ then $|n| = |k|$. (b) Derive the Hagen–Rubens relation: $R \approx 1 - 2(\omega/2\pi\sigma)^{1/2}$.

Electron interactions

Electrons in metals interact via the Coulomb interaction. The size of this effect is not negligible: it is of order e^2/a where a is the mean distance between electrons. For a metal this ≈ 3 eV, that is, comparable to E_F. An approach to the problem would be to treat the interaction as a perturbation. However, the perturbation is strong and long-ranged, and this gives rise to many difficulties, such as divergences in finite orders of perturbation theory. The problems can be overcome by an application of the methods of quantum field theory to the problem, and is treated in books on many-body theory. We will not pursue this approach here, but try to get some of the central conclusions with physical reasoning.

At first glance, we might be tempted to conclude that the sharp Fermi surface that we have discussed, and which is readily observed, should be wiped out by the interaction. The much smaller energy $k_B T$ does lead to smearing of the Fermi distribution; it is natural to conclude that this is true for interactions as well. This natural conclusion is not correct in three dimensions. The situation in one dimension is complicated, and the Fermi surface does disappear, leading to a state called the Luttinger liquid. In three dimensions the sharp Fermi surface survives the presence of interactions. This is an important phenomenon which demands explanation.

9.1 Fermi liquid theory

To understand the situation L. D. Landau introduced the notion of a "Fermi Liquid" in 1956; see (Nozieres & Pines 1999). He considered the simplest case, a collection of fermions in a uniform background. If the particles are charged, as electrons are, it is necessary to assume that the background is positively charged so that the system is neutral. In effect, the ions have been averaged out to produce a uniform background. This is called the "jellium model". In this model momentum $\mathbf{p}$ is conserved, and the interparticle interaction does not change the total momentum of interacting pairs.

9.1.1 Quasi-particles

Landau started with the Sommerfeld model. As we have seen there is a filled Fermi sphere in the ground state, and the excitations around the ground state are particles, i.e. electrons excited so that $\epsilon > E_F, |p| > p_F$, and holes in states with $\epsilon < E_F, |p| < p_F$. These are eigenstates of the non-interacting Hamiltonian. Low energy excited states are sometimes called elementary excitations.

Now imagine that we slowly turn on the interaction between the particles. Elementary quantum mechanics shows that eigenstates develop into eigenstates in the absence of degeneracy. It is natural to suppose that the Sommerfeld ground state will develop into the ground state of the interacting system. This defines a "normal Fermi liquid". (This can fail if there is a phase transition, such as the superconducting transition.)

As for the excited states, the excited particle states with momentum **p** will develop into an eigenstate in the presence of interactions which is called a *quasi-particle*. The holes will become *quasi-holes*. This will be correct unless a single excitation becomes many excitations. In any case, for the non-interacting system the Fermi momentum, p_F, sharply divided the particles and holes. Since momentum is conserved, this is still true in the interacting system. That means the *the volume of the Fermi sphere is independent of interactions*.

Luttinger (1960) proved this statement from microscopic theory: he showed that the volume of the Fermi surface is constant to all orders in perturbation theory. The Fermi surface is well defined because there is a discontinuity in the momentum distribution, $\langle \hat{c}_{p\sigma}^+ \hat{c}_{p,\sigma} \rangle$. In the non-interacting gas this is given by $f(\mathbf{p}) = 1, p < p_F$, $f(\mathbf{p}) = 0, p \geq p_F$, so that the discontinuity is unity. In the interacting gas it is less than one, but not zero. This set of statements is called the Luttinger theorem. We will not go this route here, but rely on Landau's qualitative argument. However, the key to the proof is given in the next section.

We can define the quasi-particle energy and their interactions in the following way. Suppose f_p is the occupancy of the exact (unknown) eigenstates of the system; in the ground state we have f_p^0. Since the quasi-particles are low-lying excitations we can put $\delta f_{\mathbf{p}} = f_{\mathbf{p}} - f_{\mathbf{p}}^0$ as the quasi-particle momentum distribution function. So for low-lying states:

$$E\{f_{\mathbf{p}}\} = E_0 + \sum_{\mathbf{p}} \epsilon(\mathbf{p})\delta f_{\mathbf{p}} + \frac{1}{2}\sum_{\mathbf{p},\mathbf{p}'} F(\mathbf{p},\mathbf{p}')\delta f_{\mathbf{p}}\delta f_{\mathbf{p}'}. \qquad (9.1)$$

The first term counts the quasi-particles, and the second quasi-particle interactions. We should note that the energy of a quasi-particle is the change in the energy when we occupy one state:

$$\epsilon(\mathbf{p}) = \frac{\delta E}{\delta f_{\mathbf{p}}}. \qquad (9.2)$$

The quasi-particle energy is measured with respect to the ground state. Thus, for the non-interacting system $\epsilon = p^2/2m - E_F$.

9.1.2 Lifetime of a quasi-particle

How can we understand that low-lying excitations of the non-interacting system will retain their sharply defined character, and not develop into a messy collection of excitations? (In one dimension this is exactly what happens: the real excitations do not resemble quasi-particles at all.) If the *lifetime* of the excitations is short compared to the time required to

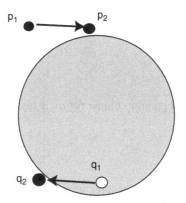

Fig. 9.1   **Estimating the lifetime of a quasi-particle.**

turn on the interactions then we can expect to see a complicated situation with excitations that have little in common with particles or holes. However, if $|p - p_F|$ is small enough, we can show that the lifetime diverges. Thus, we can still turn on the interactions slowly, and preserve the Fermi surface.

To see this, start with the non-interacting system, and treat the electron-electron interaction as a perturbation. Consider an excited state consisting of one electron at $\mathbf{p}_1$ above the Fermi surface by a small amount, $\Delta = |p_1 - p_F|$. Note that $\epsilon = p_1^2/2m - p_F^2/2m \approx \Delta p_F/m$. This state can decay by scattering to state $\mathbf{p}_2$, $p_2 > p_F$ and making a particle-hole pair starting with an electron at $\mathbf{q}_1, q < p_F$ which is promoted to state $\mathbf{q}_2, q_2 > p_F$. See Figure 9.1.

Now to estimate the lifetime we use Fermi's Golden rule.

$$\frac{1}{\tau(p_1)} = \frac{2\pi}{\hbar} \sum_{p_2, q_1, q_2} |M|^2 \delta(\mathbf{p}_1 + \mathbf{q}_1 - \mathbf{p}_2 - \mathbf{q}_2)$$
$$\delta(p_1^2 - p_2^2 + q_1^2 - q_2^2) f_{q_1}(1 - f_{q_2})(1 - f_{p_2}). \qquad (9.3)$$

Here f is the $(T = 0)$ Fermi distribution function, and M is a matrix element for the scattering. The δ functions assure conservation of energy and momentum. Suppose that M is constant near the Fermi surface. Then we have:

$$\frac{1}{\tau(p_1)} \propto \int_{q_1 < p_F} d\mathbf{q}_1 \int_{q_2 > p_F} d\mathbf{q}_2 (1 - f_{p_2}) \delta(p_1^2 - p_2^2 + q_1^2 - q_2^2). \qquad (9.4)$$

The constraints from energy conservation are quite severe. Consider the volume of p-space over which the integration takes place. We need:

$$p_1^2 + q_1^2 = p_2^2 + q_2^2$$
$$p_1^2 - p_F^2 = p_2^2 + q_2^2 - q_1^2 - p_F^2$$
$$\geq p_F^2 - q_1^2. \qquad (9.5)$$

We used $q_2 > p_F, p_2 > p_F$. Now, since all the momenta are near p_F we can put:

$$2p_F(p_1 - p_F) \geq 2p_F(p_F - q_1)$$
$$\Delta \geq p_F - q_1. \tag{9.6}$$

Thus the integration volume for q_1 is less than $4\pi p_F^2 \Delta$. In the same way we can show that:

$$p_1 - p_F \geq q_2 - p_F. \tag{9.7}$$

The volume for the other integration is also less than $4\pi p_F^2 \Delta$. Taken together this implies that:

$$\frac{1}{\tau(p_1)} \approx C\Delta^2, \tag{9.8}$$

where C is a constant. That is, the lifetime goes quickly to zero as the Fermi surface is approached. A measure of how well defined the quasi-particle is the ratio of the rate of decay to the energy:

$$\frac{1/\tau}{\epsilon} \propto \Delta. \tag{9.9}$$

9.1.3 One dimension

Fermi liquid theory does not apply to one-dimensional systems. Part of the reason is the Peierls instability of Section 8.4.5. There is a further instability associated with pairing which we will see later when we look at superconductivity. The result is that the Fermi surface is destroyed, and the electrons form a Luttinger liquid. The excitations of the liquid are exotic objects called spinons and holons; see Giamarchi (2004). There is some evidence for these effects in recent experiments on carbon nanotubes.

It is often stated that the estimate of the previous section of the quasi-particle lifetime breaks down in one dimension. The reason given is that if we linearize the electron energy around the Fermi surface conservation of momentum and energy are the same so that the restrictions on scattering are fewer. However, we never used conservation of momentum above, and the result of a naive calculation is the same in one dimension as in three.

9.1.4 Quasi-particle mass

Landau showed that many of the results of Sommerfeld theory such as the specific heat being proportional to T are preserved under interactions. For a good account of this, see Nozieres & Pines (1999). There is one result that we will prove which gives an example of how Fermi liquid theory goes. Return to the definition of $\epsilon(\mathbf{p})$, the quasi-particle energy. In

Fermi liquid theory we use another definition of effective mass (different from the one we saw in band theory):

$$\nabla_{\mathbf{p}}\epsilon = \frac{\mathbf{p}}{m^*}. \tag{9.10}$$

Of course, for $\epsilon = p^2/2m - E_F$, the free electron gas, we recover $m^* = m$.

Now we will derive a relationship between the mass, a property of ϵ, and the interaction function, $F(\mathbf{p}, \mathbf{p}')$. We are supposing that the Fermi liquid is translation invariant. Consider the total mass current in some state $|s\rangle$. Since momentum is conserved by the interactions it must be given by;

$$\mathbf{j} = \left\langle s \left| \frac{\sum_i \mathbf{p}_i}{m} \right| s \right\rangle = \frac{\mathbf{p}}{m}, \tag{9.11}$$

where $\mathbf{p}$ is the total momentum. We will consider a state with one quasi-particle excited. In that case $\mathbf{p}$ is the quasi-particle momentum.

We can get the current another way. Suppose all of the momenta in the system are increased by $\mathbf{q}$. This is equivalent to viewing the system in a moving frame of reference with speed $-\mathbf{q}/m$. The only change in $\hat{\mathcal{H}}$ is in the kinetic energy term. The interactions depend on relative positions, and are unchanged in the moving frame. Therefore the change in the energy of any state is:

$$dE = \left\langle s \left| \frac{\mathbf{q} \cdot \sum \mathbf{p}_i}{m} + \frac{q^2}{2m} \right| s \right\rangle = \mathbf{q} \cdot \mathbf{j} + \mathcal{O}(q^2). \tag{9.12}$$

Neglecting the last term we find, for each component of $\mathbf{q}$:

$$dE/dq_l = j_l. \tag{9.13}$$

Consider the change in energy for the state with one quasi-particle. There are two parts to the change. The quasi-particle itself changes its energy by:

$$\epsilon(\mathbf{p} + \mathbf{q}) - \epsilon(\mathbf{p}) \approx \mathbf{q} \cdot \nabla_{\mathbf{p}}\epsilon = \mathbf{q} \cdot \frac{\mathbf{p}}{m^*}. \tag{9.14}$$

There is another change, however. The Fermi sphere is shifted by $\mathbf{q}$ so that excitations are created, as shown in Fig. 9.2. There is a change in the energy due to the interaction of the quasi-particle with the excitations:

$$\sum_{\mathbf{p}'} F(\mathbf{p}, \mathbf{p}')\delta f_{\mathbf{p}'}. \tag{9.15}$$

Now we find $\delta f_{\mathbf{p}'}$ by referring to Fig. 9.2. In the unperturbed liquid we have the usual Fermi function, $f_{\mathbf{p}} = 1, p < p_F$. Note that $df/dp = -\delta(p - p_F)$. For the perturbed liquid we need $|\mathbf{p} - \mathbf{q}| < p_F$. Now:

$$\delta f_{\mathbf{p}} = -\mathbf{q} \cdot \nabla_{\mathbf{p}} f_{\mathbf{p}} = -\mathbf{q} \cdot \nabla_{\mathbf{p}} p \frac{df}{dp} = \mathbf{q} \cdot \frac{\mathbf{p}}{p}\delta(p - p_F) \tag{9.16}$$

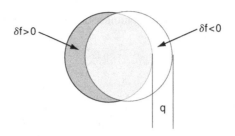

Fig. 9.2   **Excitations created by shifting the Fermi sea.**

Putting everything together we find:

$$\mathbf{j} = \frac{\mathbf{p}}{m} = \frac{\mathbf{p}}{m^*} + \sum_{\mathbf{p}'} F(\mathbf{p}, \mathbf{p}') \frac{\mathbf{p}'}{p} \delta(p' - p_F). \tag{9.17}$$

Taking the inner product with **p** we deduce:

$$\frac{1}{m^*} = \frac{1}{m} - \sum_{\mathbf{p}', s'} \cos(\theta) F(\mathbf{p}, s; \mathbf{p}', s') \delta(p' - p_F)/p_F, \tag{9.18}$$

where θ is the angle between **p** and **p**′, and we have restored the spin indices. We can think of the extra term in the mass as a kind of drag, or *backflow* due to the interaction of the liquid with the quasi-particle.

9.2 Many-electron atoms

We now turn to microscopic theory. A good place to start is not in metals but in the best-understood many-electron system, the many-electron atom.

9.2.1 Hartree and Hartree–Fock theory

The simplest many electron system is the He atom with just two electrons. Suppose we consider the ground state, and begin, as we do in solids, with states of the non-interacting system. In this case the orbitals that play the role of the plane waves of Sommerfeld theory are the Bohr orbitals of atomic physics. For He in the ground state we have one such wavefunction, the 1s function for an atom with $Z = 2$, ϕ_{1s}. If we have two electrons, they can both be in this state provided one has spin up, the other spin down. Thus we are led to a 2-electron function:

$$\Phi_0(\mathbf{r}_1, \mathbf{r}_2) = \frac{1}{\sqrt{2}} \phi_{1s}(\mathbf{r}_1) \phi_{1s}(\mathbf{r}_2)[\alpha(1)\beta(2) - \beta(1)\alpha(2)], \tag{9.19}$$

where α is a spinor for spin up, and β for spin down.

The Hamiltonian is:

$$\hat{\mathcal{H}} = \frac{1}{2m}\left(p_1^2 - \frac{Ze^2}{r_1}\right) + \frac{1}{2m}\left(p_2^2 - \frac{Ze^2}{r_2}\right) + e^2/r_{12} \equiv \hat{\mathcal{H}}_1 + \hat{\mathcal{H}}_2 + e^2/r_{12}. \quad (9.20)$$

Here $r_{12} = |\mathbf{r}_1 - \mathbf{r}_2|$. Note that for this light atom there is no dependence on spin at all in $\hat{\mathcal{H}}$. For heavier atoms there is spin-orbit coupling, but we will always neglect that effect.

The wavefunction in Eq. (9.19) is not an eigenfunction of $\hat{\mathcal{H}}$, but we can estimate its energy by simply taking an expectation value:

$$\langle 0|\hat{\mathcal{H}}|0\rangle = 2\epsilon_{1s} + \int d\mathbf{r}_1 d\mathbf{r}_2 |\phi_{1s}(\mathbf{r}_1)\phi_{1s}(\mathbf{r}_2)|^2 e^2/r_{12}. \quad (9.21)$$

This amounts to first-order perturbation theory. It is easy to understand the extra term; it is just the Coulomb repulsion of the charge densities of the two electrons. It is called the (direct) Coulomb integral.

D. Hartree improved on perturbation theory by introducing the *self-consistent field* method for atoms. He pointed out that the wavefunctions that we used are not really appropriate because each electron sees the charge of the nucleus as if the other electron were not present. He proposed to find better orbitals, ϕ in the following way: each ϕ sees the average field of the others.

$$\left[\hat{\mathcal{H}}_1 + e^2 \int d\mathbf{r}_2 \frac{e^2}{r_{12}}|\phi(\mathbf{r}_2)|^2\right]\phi(\mathbf{r}_1) = \epsilon\phi(\mathbf{r}_1). \quad (9.22)$$

The wavefunction is found by iteration: a trial ϕ is put in the integral, then the equation is solved to get a new trial function, and so on, until self-consistency is achieved. It is easy to see that this equation arises from a variational calculation: take the expectation of $\hat{\mathcal{H}}$ with a wavefunction of the form of Eq. (9.19) (with ϕ_{1s} replaced by ϕ) and vary with respect to ϕ^* until an extremum is found. The eigenvalue, ϵ, is a Lagrange multiplier which preserves the normalization of ϕ. This yields Eq. (9.22).

For an n electron atom with orbitals labeled by j, the generalization is:

$$\left[\hat{\mathcal{H}}_1 + \sum_j e^2 \int d\mathbf{r}_2 \frac{e^2}{r_{12}}|\phi_j(\mathbf{r}_2)|^2\right]\phi_i(\mathbf{r}_1) = \epsilon\phi(\mathbf{r}_1). \quad (9.23)$$

For the general case, however, the approximation is suspect since it does not arise from the variation of a properly antisymmetric wavefunction, but rather from the product $\prod_j \phi_j$. (For the two electron case, it is the spinors that give the antisymmetry.) The general case was investigated by J. Slater and V. Fock who showed that there are extra terms called *exchange energies* that Hartree missed. We have already seen this effect in Chapter 1, and, in particular, Problem 2 in that chapter.

The calculation is most easily done by using second quantization. Suppose i now labels the quantum numbers of the occupied spatial orbitals, and σ their spins so that the many-particle state is $\Psi = \prod_{i,\sigma} \hat{c}^+_{i,\sigma}|0\rangle$. The Hamiltonian is:

$$\hat{\mathcal{H}} = \sum_i \left(-\frac{\hbar^2}{2m}\nabla_i^2 - \frac{Ze^2}{r_i} \right) + \frac{1}{2}\sum_{i\neq j}\frac{e^2}{r_{ij}}. \tag{9.24}$$

We seek the expectation value of $\hat{\mathcal{H}}$. The rule in Eq. (6.16) gives the first two terms:

$$\langle\Psi|\hat{\mathcal{H}}_1|\Psi\rangle = \sum_{i,\sigma}\int d\mathbf{r}\phi^*_{i,\sigma}(\mathbf{r})\left(-\frac{\hbar^2}{2m}\nabla^2 - \frac{Ze^2}{r} \right)\phi_{i,\sigma}(\mathbf{r}). \tag{9.25}$$

In order to evaluate the interaction term we use Eq. (6.19). We need:

$$\langle\Psi|\hat{c}^+_{i\sigma}\hat{c}^+_{j\sigma'}\hat{c}_{l,\sigma'}\hat{c}_{k\sigma}|\Psi\rangle = \delta_{i,k}\delta_{j,l} - \delta_{i,l}\delta_{j,k}\delta_{\sigma,\sigma'}. \tag{9.26}$$

This is true because the creation and destruction operators must occur in pairs, and the minus sign in the second term is from the anticommutation relations. Thus:

$$\langle\Psi|\hat{\mathcal{H}}_2|\Psi\rangle = \sum_{i,j,\sigma,\sigma'}\frac{1}{2}\int d\mathbf{r}_1 d\mathbf{r}_2 \frac{e^2}{r_{12}}|\phi_{i,\sigma}(\mathbf{r}_1)|^2|\phi_{j,\sigma'}(\mathbf{r}_2)|^2$$
$$- \sum_{i,j,\sigma}\frac{1}{2}\int d\mathbf{r}_1 d\mathbf{r}_2 \phi^*_{i\sigma}(\mathbf{r}_1)\phi^*_{j\sigma}(\mathbf{r}_2)\frac{e^2}{r_{12}}\phi_{i\sigma}(\mathbf{r}_2)\phi_{j\sigma}(\mathbf{r}_1). \tag{9.27}$$

In this equation the first term is the Hartree energy. It can be written in terms of the electron density:

$$E_{\mathrm{H}} = \frac{e^2}{2}\int d\mathbf{r}d\mathbf{r}'\frac{n(\mathbf{r})n(\mathbf{r}')}{|\mathbf{r}-\mathbf{r}'|}; \quad n(\mathbf{r}) = \sum_{i,\sigma}|\phi_{i,\sigma}(\mathbf{r})|^2. \tag{9.28}$$

The second term is the exchange energy.

We need to maintain that the orbitals for the same spin be orthonormal: $\langle i\sigma|j\sigma\rangle = \delta_{i,j}$. We do this with Lagrange multipliers $\epsilon_{i,j,\sigma}$. We are led to seek the solution of:

$$\frac{\delta}{\delta\phi^*_i}\left(\langle\Psi|\hat{\mathcal{H}}|\Psi\rangle - \sum_{k,l,\sigma}\epsilon_{k,l,\sigma}\langle i\sigma|j\sigma\rangle \right) = 0. \tag{9.29}$$

The result is:

$$\hat{\mathcal{H}}_1\phi_{i,\sigma}(\mathbf{r}_1) + V_{\mathrm{H}}\phi_{i,\sigma}(\mathbf{r}_1) + V_X\phi_{i,\sigma} = \sum_j\epsilon_{i,j,\sigma}\phi_{j,\sigma}(\mathbf{r}_1). \tag{9.30}$$

Here:

$$V_H \, \phi_{i,\sigma}(\mathbf{r}_1) = \left[\sum_{j,\sigma'} \int d\mathbf{r}_2 \frac{e^2}{r_{12}} |\phi_{j,\sigma'}(\mathbf{r}_2)|^2 \right] \phi_{i,\sigma}(\mathbf{r}_1),$$

$$V_X \, \phi_{i,\sigma}(\mathbf{r}_1) = \left[- \sum_{j} \int d\mathbf{r}_2 \phi_{j\sigma}^*(\mathbf{r}_2) \frac{e^2}{r_{12}} \phi_{i\sigma}(\mathbf{r}_2) \right] \phi_{j\sigma}(\mathbf{r}_1). \tag{9.31}$$

The first term is the Hartree potential, just as in Eq. (9.23). The second term, the exchange potential, is a non-local operator. The sums are over the occupied orbitals. Note that we need not exclude the term $j = i$ in either sum since the "self-interaction" term exactly cancels.

The right-hand side of Eq. (9.30) can be simplified by noting that any linear combination of the $\phi_{j,\sigma}$ will do just as well since we are dealing with a Slater determinant. We use this freedom to diagonalize the matrix $\epsilon_{i,j,\sigma}$. The standard form for the Hartree–Fock equation follows:

$$\hat{\mathcal{H}}_1 \phi_{i,\sigma} + V_H \phi_{i,\sigma} + V_X \phi_{i,\sigma} = \epsilon_{i,\sigma} \phi_{i,\sigma}. \tag{9.32}$$

This equation looks like a Schrödinger equation, but it must be interpreted carefully. It is a self-consistent equation which is solved by iteration. The eigenvalues ϵ_i are discussed in the problems.

The solution of the Hartree–Fock equation is quite difficult because of the self-consistency step and because V_X is non-local. Slater suggested replacing V_X by an effective local operator: in fact, he used the expression we will derive below for a uniform electron gas, Eq. (9.42) with the density in that expression replaced by $\sum_{\text{occ}} |\phi_k(\mathbf{r})|^2$. In effect, he was replacing the atom by a uniform electron gas at the local density. This is an interesting point of view which will also be used in the next section, and later, from another angle, in density functional theory, Section 9.5.

The Hartree–Fock (HF) approximation is not exact. The rest of the energy is referred to as correlation energy. The name comes from the observation that in HF the electron in an orbital sees the average of the charge density of the other electrons in the Hartree term. Further, the exchange term is negative only because electrons (of the same spin) avoid each other automatically in a Slater determinant. However, in a real system electrons of both spin avoid each other; that is, their motions are correlated.

In atomic physics there is a natural way to compute the correlation energy: a given configuration is a single Slater determinant, but other, higher energy configurations, can be mixed in by perturbation theory. This is called the configuration interaction approach. The situation is more difficult in metals since there is no gap between configurations.

9.2.2 The Thomas–Fermi atom

There is another way to look at atoms which is most appropriate for large Z. This is to continue the point of view that we introduced in Chapter 8 by viewing the (spherically symmetric) density of electrons in an atom as defining a local Fermi wavevector, $k_F(r) = 3\pi^2(n(r))^{1/3}$, and a local Fermi energy $E_F(r) = \hbar^2 k_F(r)^2/2m$. In an atom there is a negative electrostatic potential energy, $e\phi(r)$ so the chemical potential is $\mu = E_F + e\phi(r)$. For a neutral atom we take $\phi(\infty) = 0$. This means that the electron density is given by Eq. (8.32), $3\pi^2 n(r) = (2m(\mu - e\phi)/\hbar^2)^{3/2}$. In order to have $n(\infty) = 0$ we need $\mu = 0$. We also have the Poisson equation:

$$\nabla^2 \phi = \frac{1}{r}(r\phi(r))'' = -4\pi en(r) = -4\pi \left(\frac{1}{3\pi^2} \left[\frac{2m|e|}{\hbar^2} \right]^{3/2} \right) \phi(r)^{3/2}, \qquad (9.33)$$

with the boundary condition $\lim_{r \to 0} r\phi(r) = Z|e|$ so that the nuclear potential is correct.

Now use atomic units $\hbar = m = |e| = 1$ and change variables so that $r = xbZ^{-1/3}$; $b = (3\pi/4)^{2/3}/2$ and put $\phi = Z\chi/r$. The result is the Thomas–Fermi differential equation:

$$\frac{d^2\chi}{dx^2} = \frac{\chi^{3/2}}{x^{1/2}}. \qquad (9.34)$$

This equation must be solved numerically. It is a reasonable approximation for large atoms except very near the nucleus and very far out in the tail of the distribution; in both cases the density is changing too fast to allow the local density approximation to make sense. For more details and comparison with experiment see Landau & Lifshitz (1977).

E. Leib and B. Simon (see (Lieb 1981)) proved that in the limit of large Z Thomas–Fermi is exact. As we will see later, the Thomas–Fermi approximation is the leading term for density-functional theory. There have been extensions of the theory which bear notice. P. Dirac added local exchange to the theory, and C. von Weiszäcker noted that the local-density approximation could be improved by adding a term involving the gradient of the density. This sort of gradient correction is also used in density-functional theory.

9.3 Metals in the Hartree–Fock approximation

Now we return to metals. The simplest model for interacting charged electrons in metals would ignore the ions altogether. However, this won't do, since the negative charge of the electrons would not be balanced, and the whole system would be unstable. The simplest useful model is to imagine that the positive charge of the ions is smeared out into a uniform "jelly." The number density of the background is $n_+ = N/\Omega$ for N electrons in volume Ω.

We need to use a mathematical trick to avoid problems with the long-range of the Coulomb interaction. We suppose that the interaction energy between two electrons is not Coulombic but screened: $v(r) = e^2 e^{-\lambda r}/r$. At the end of the calculation we put $\lambda = 0$. This trick allows us to go to large Ω and keep Coulomb energies finite, rather than having to deal with a finite system throughout.

9.3.1 Hamiltonian

The Hamiltonian has four terms, kinetic energy, electron-electron interaction, electron-background interaction, and the interaction of the background with itself.

$$\hat{\mathcal{H}} = \sum_i \frac{p_i^2}{2m} + \frac{1}{2} \sum_{i \neq j} v(r_{ij})$$

$$- \sum_i \int d\mathbf{r}\, n_+(\mathbf{r}) v(|\mathbf{r} - \mathbf{r}_i|)$$

$$+ \frac{1}{2} \int d\mathbf{r}\, d\mathbf{r}'\, n_+(\mathbf{r}) v(|\mathbf{r} - \mathbf{r}'|) n_+(\mathbf{r}'). \tag{9.35}$$

It is quite simple to work out the last two terms, using the fact that n_+ is, in fact, a constant. In the course of this we need an integral, $\int d\mathbf{r}\, v(r)$. This is a special case of an integral we will need below:

$$\tilde{v}(q) = \int d\mathbf{r}\, e^{i\mathbf{q} \cdot \mathbf{r}} v(r) = \frac{4\pi e^2}{q^2 + \lambda^2}. \tag{9.36}$$

Using this integral, it is easy to see that the sum of the last two terms is:

$$-\frac{1}{2} \frac{4\pi e^2}{\lambda^2} \frac{N^2}{\Omega}. \tag{9.37}$$

Now we can write the Hamiltonian in second-quantized form in the basis of plane wave states, $\phi_{\mathbf{k},s} = \frac{1}{\Omega} e^{i\mathbf{k} \cdot \mathbf{r}} \chi$, where χ is a spinor:

$$\hat{\mathcal{H}} = \sum_{k,s} \epsilon_\circ(k) + \frac{1}{2\Omega} \sum_{k,K,q,s,s'} \tilde{v}(q) \hat{c}_{k+q,s}^+ \hat{c}_{K-q,s'}^+ \hat{c}_{K,s'} \hat{c}_{k,s} - \frac{1}{2} \frac{4\pi e^2}{\lambda^2} \frac{N^2}{\Omega}. \tag{9.38}$$

Here, $\epsilon_\circ(k) = \hbar^2 k^2 / 2m$.

Now examine the $\mathbf{q} = 0$ part of the second term using the anticommutation relations, and neglecting a term of order $1/N$:

$$\frac{1}{2\Omega} \sum_{k,K,s,s'} \tilde{v}(0) \hat{c}_{k,s}^+ \hat{c}_{K,s'}^+ \hat{c}_{K,s'} \hat{c}_{k,s} = \frac{1}{2\Omega} \tilde{v}(0) \sum_{k,s} \hat{c}_{k,s}^+ \hat{c}_{k,s} \sum_{K,s'} \hat{c}_{K,s'}^+ \hat{c}_{K,s'}$$

$$= \frac{1}{2} \frac{4\pi e^2}{\lambda^2} \frac{N^2}{\Omega}. \tag{9.39}$$

This exactly cancels the term from Eq. (9.37) (the Hartree term). From now on we take $\mathbf{q} \neq 0$ in the sum in Eq. (9.38), and drop the last term.

9.3.2 Ground-state energy in Hartree–Fock

To get the Hartree–Fock expression for the energy in this case, we need only take the expectation value of $\hat{\mathcal{H}}$. This is unlike the case in atoms. In the uniform jellium system plane waves are still good orbitals (they are the only ones that are translation invariant), so the extra step of minimizing the Hamiltonian with respect to the orbitals isn't needed.

Thus, the HF expression for the ground-state energy is:

$$E_\circ = \frac{3}{5}NE_F + \frac{1}{2\Omega} \sum_{k,K,q,,s,s'} \tilde{v}(q) \langle G|\hat{c}_{k+q,s}^+ \hat{c}_{K-q,s'}^+ \hat{c}_{K,s'} \hat{c}_{k,s}|G\rangle. \tag{9.40}$$

The first term is the kinetic energy of the free-electron gas, which we worked out above, and $|G\rangle = \prod_{k<k_F,s} \hat{c}_{k,s}^+|0\rangle$.

Now, exactly as in the atomic case, we can work out the Hartree and exchange energies. The operator destroys at $\mathbf{k}, s$ and $\mathbf{K}, s'$, and creates at $\mathbf{k} + \mathbf{q}, s$; $\mathbf{K} - \mathbf{q}, s'$. In order to have a non-zero expectation value of the operator we need to start and end with the filled Fermi sphere. This can happen two ways:

(i) $\mathbf{k} + \mathbf{q} = \mathbf{k}, \mathbf{K} - \mathbf{q} = \mathbf{K}$. That means $\mathbf{q} = 0$, but that term has been cancelled.
(ii) $\mathbf{k} + \mathbf{q} = \mathbf{K}, \mathbf{K} - \mathbf{q} = \mathbf{k}, s = s'$. That is, the operator exchanges two electrons. This is the exchange term.

Now we can rearrange the operators using anticommutation relations:

$$\langle G|\hat{c}_{K,s}^+ \hat{c}_{k,s}^+ \hat{c}_{K,s} \hat{c}_{k,s}|G\rangle = -\langle G|\hat{c}_{K,s}^+ \hat{c}_{K,s} \hat{c}_{k,s}^+ \hat{c}_{k,s}|G\rangle = -f_{k,s}f_{K,s}, \tag{9.41}$$

where, as usual, f is the Fermi distribution function at $T = 0$.

We now can take $\lambda = 0$, and write the exchange energy for the real Coulomb interaction:

$$E_x = -\frac{2\pi e^2}{\Omega} \sum_{k,K,s} \frac{f_{K,s}f_{k,s}}{|\mathbf{k} - \mathbf{K}|^2} = -\frac{3}{4\pi}Ne^2k_F. \tag{9.42}$$

We leave the details of the integration as an exercise. The exchange energy, as in the atomic case, is the reduction of the Coulomb repulsion due to the "statistical repulsion" of electrons of like spin. The pair correlation function of two electrons is reduced by 1/2 at the origin by the antisymmetry of the wavefunction. This is sometimes called the "exchange hole"; see Problem 2 of Chapter 6.

The energy in the HF approximation is:

$$E_\circ = \frac{3}{5}NE_F - \frac{3}{4\pi}Ne^2k_F. \tag{9.43}$$

9.3.3 quasi-particle properties in Hartree–Fock

We now can find the quasi-particle properties. The ground-state energy can be written:

$$E_\circ = \sum_{k,s} \epsilon_\circ(k)f_{k,s} - \frac{2\pi e^2}{\Omega} \sum_{k,K,s} \frac{f_{K,s}f_{k,s}}{|\mathbf{k} - \mathbf{K}|^2}. \tag{9.44}$$

Thus:

$$\epsilon(k) = \frac{\delta E_\circ}{\delta f_{k,s}} = \epsilon_\circ(k) - \frac{2\pi e^2}{\Omega} \sum_{K,s} \frac{f_{K,s}}{|\mathbf{k} - \mathbf{K}|^2}$$

$$F(k, s; K, s') = \frac{\delta^2 E_\circ}{\delta f_{k,s} \delta f_{K,s'}} = -\delta_{,s,s'} \frac{4\pi e^2}{\Omega |\mathbf{k} - \mathbf{K}|^2}. \tag{9.45}$$

These expressions signal a serious difficulty with HF. The first expression can be integrated:

$$\epsilon(k) = \epsilon_\circ(k) - \frac{e^2}{2\pi} \left[\frac{k_F^2 - k^2}{k} \ln \left| \frac{k_F + k}{k_F - k} \right| + 2k_F \right]. \tag{9.46}$$

This expression has a singularity in the derivative at $k = k_F$, i.e., $1/m^*$ diverges at the Fermi surface. This is an unphysical result. It would lead to a specific heat that is proportional to $T/|\ln(T)|$, which is not observed.

The singularity in the mass is even easier to see from Eq. (9.18). In that equation we showed that the correction to m^* was proportional to the integral over the Fermi surface of $F(\mathbf{p}, \mathbf{p}') \cos(\theta)$, where θ is the angle between $\mathbf{p}, \mathbf{p}'$. The HF expression for F involves $|\mathbf{k} - \mathbf{K}|^2 = 2k_F^2[1 - \cos(\theta)]$. Then we have:

$$m/m^* = 1 + C \int_{-1}^1 dx \frac{x}{1 - x}, \tag{9.47}$$

where $x = \cos(\theta)$. The integral diverges. Note for future reference that if the potential were screened ($\lambda \neq 0$) we would have a convergent result.

We will need to go beyond HF to get an acceptable theory of jellium. The key to understanding what is wrong is involved precisely with screening. In addition to statistical correlations, there are real correlations between electrons: even electrons of parallel spins avoid one another. The effective potential seen far from an electron in a metal is weaker than Coulomb because the electron is surrounded by an exchange-correlation hole: that means there is effectively positive charge near the electron, and this screens the field. The reduction in energy beyond exchange is the correlation energy.

9.4 Correlation energy of jellium

The Hartree–Fock theory of the previous section has many good features. It shows how the electron-electron interactions lower the energy of the metal, for instance. It is also equivalent to first-order perturbation theory. A reasonable strategy to improve HF would be to go to higher orders in perturbation theory.

This turns out to be difficult and complicated. The reason is that the Coulomb interaction is not particularly well-behaved, and the divergences that we saw in HF get worse in higher orders. In order to proceed down this route a number of people used the formalism of quantum field theory (which has similar, but not identical divergences). A key step is to organize the perturbation theory in Feynman diagrams.

M. Gell-Mann and K. Brueckner carried out this program for the jellium model; see Mahan (2000). They found that by summing up a class of divergent diagrams (one-loop diagrams) they found a convergent, physically reasonable result. This is an expansion of the theory in inverse density around the Sommerfeld limit. At very high density, the kinetic energy dominates, and the interaction is small. The conventional expansion parameter is r_s, the mean distance between electrons in units of the Bohr radius:

$$\frac{4\pi(r_s a_\circ)^3}{3} = \frac{1}{n}.$$ (9.48)

What Gell-Mann and Brueckner showed is that the ground-state energy of jellium can be written:

$$\frac{E_\circ}{N} = \frac{2.21}{r_s^2} - \frac{0.916}{r_s} + 0.062\ln(r_s) - 0.096... \quad \text{(Ryd)}$$ (9.49)

The first two terms are the kinetic energy and the exchange energy (as in HF). The next are the the first two terms in an expansion of the correlation energy in r_s. Unfortunately, real metals have $2 \leq r_s \leq 6$, so this expression is not directly useful for metals. The formalism of Feynman diagrams is complicated, and we will not develop it here.

The results above can be got by another route which is physically interesting. It is called the dielectric function approach. The point of this method is that we already know a good deal about the response properties of the electron gas – things like screening and plasma oscillations. These are, strictly speaking, interaction properties, though we introduced them by arguments about macroscopic electrostatics. As we will see, these are enough to give an account of the high-density electron gas.

9.4.1 Ground state and quasi-particles in the RPA

We now show that the microscopic physics of the electron gas can be worked out using the RPA dielectric function of Chapter 8.

Ground-state energy

The RPA dielectric function immediately gives us an expression for the interaction energy of the electron gas. Consider the second term of Eq. (9.40):

$$\langle G|V_{\text{int}}|G\rangle = \frac{1}{2\Omega} \sum_{k,K,q,,s,s'} \tilde{v}(q)\langle G|\hat{c}_{k+q,s}^+\hat{c}_{K-q,s'}^+\hat{c}_{K,s'}\hat{c}_{k,s}|G\rangle$$

$$= \frac{1}{2\Omega} \sum_{k,K,q,,s,s'} \tilde{v}(q) \langle G | (\hat{c}_{k+q,s}^+ \hat{c}_{k,s} \hat{c}_{K-q,s'}^+ \hat{c}_{K,s'} - \hat{n}_{k+q} \delta_{K+q,k} \delta_{,s,s'}) | G \rangle$$

$$= \frac{1}{2\Omega} \sum_q \tilde{v}(q) [\Omega^2 \langle G | \hat{\rho}_q^+ \hat{\rho}_q | G \rangle - N]$$

$$= \frac{1}{2\Omega} \sum_q \tilde{v}(q) [N(S(q) - 1)]. \tag{9.50}$$

We can understand this in another way by noting that the interaction term depends on the position of one electron with respect to others. This is the information in the pair correlation function, p, of Section 3.4. We can write:

$$\frac{e^2}{2} \int d\mathbf{r} \, d\mathbf{s} \frac{p(\mathbf{r}, \mathbf{s})}{|\mathbf{r} - \mathbf{s}|} = \frac{1}{2\Omega} \sum_q \tilde{v}(q) \int d\mathbf{r} \, d\mathbf{s} \, p(\mathbf{r}, \mathbf{s}) e^{i\mathbf{q} \cdot (\mathbf{r} - \mathbf{s})}. \tag{9.51}$$

But Eq. (3.51) says that $N(S(q) - 1)$ is the Fourier transform of $p(\mathbf{r}, \mathbf{s})$. This gives the result above.

From Eq. (8.13) and Eq. (8.23) we can transform Eq. (9.50) to:

$$E_{\text{int}} = \langle G | V_{\text{int}} | G \rangle = -\sum_q \frac{2\pi N e^2}{\Omega q^2} - \sum_q \int \frac{d\omega}{2\pi} \text{Im} \frac{1}{\epsilon(\mathbf{q}, \omega)}. \tag{9.52}$$

We also need the kinetic energy. This is not easy. However we use a trick which allows us to get both the interaction energy and the kinetic energy together. Recall the Hellmann–Feynman theorem: it says that if $\hat{\mathcal{H}}$ depends on a parameter, α (e^2 in our case) then the derivative of the energy is:

$$\frac{\partial E}{\partial \alpha} = \frac{\partial}{\partial \alpha} \langle \Psi | \hat{\mathcal{H}} | \Psi \rangle$$

$$= \left\langle \Psi \left| \frac{\partial \hat{\mathcal{H}}}{\partial \alpha} \right| \Psi \right\rangle + \left\langle \Psi | \hat{\mathcal{H}} | \frac{\partial \Psi}{\partial \alpha} \right\rangle + \left\langle \frac{\partial \Psi}{\partial \alpha} | \hat{\mathcal{H}} | \Psi \right\rangle$$

$$= \left\langle \Psi \left| \frac{\partial \hat{\mathcal{H}}}{\partial \alpha} \right| \Psi \right\rangle + E \frac{\partial \langle \Psi | \Psi \rangle}{\partial \alpha} = \left\langle \Psi \left| \frac{\partial \hat{\mathcal{H}}}{\partial \alpha} \right| \Psi \right\rangle. \tag{9.53}$$

The last equality follows from the constancy of the normalization of the wavefunction.

In our case the only term in $\hat{\mathcal{H}}$ that involves e^2 is the interaction term. Now let us suppose that the coupling constant $\alpha = e^2$ is a variable which turns on from 0 to its physical value. Then by the Hellmann–Feynman theorem:

$$\frac{\partial E_\circ}{\partial \alpha} = E_{\text{int}} / \alpha. \tag{9.54}$$

Thus:

$$E_\circ = E_\circ(\alpha = 0) + \int_0^{e^2} \frac{d\alpha}{\alpha} E_{\text{int}}(\alpha). \tag{9.55}$$

We know the ground-state energy for $\alpha = 0$ – it is just the energy of the Sommerfeld gas, $3NE_F/5$. The remaining term can be worked out from Eq. (9.52).

The integrals can be done with some difficulty. The result is:

$$E_\circ/N = \frac{2.21}{r_s^2} - \frac{0.916}{r_s} + 0.062\ln(r_s) - 0.142... \tag{9.56}$$

This differs from the diagrammatic expansion, Eq. (9.49), only in the constant term.

Quasi-particles

It is possible to find the quasi-particle energy and the mass in the RPA directly from the formulas above. However, the physical nature of the result can be gotten more simply. We simply appeal to screening to fix the problems in HF. We simply take the HF expression for F and screen it:

$$F(k, s; K, s') = \frac{\delta^2 E_\circ}{\delta f_{k,s}\delta f_{K,s'}} = -\delta_{,s,s'}\frac{4\pi e^2}{\Omega|(\mathbf{k} - \mathbf{K})|^2 + \lambda^2}. \tag{9.57}$$

This expression has been shown to give an accurate account of the quasi-particle properties in the RPA. In particular, the integral that gives the mass:

$$m/m^* = 1 + C\int_{-1}^{1} dx\frac{x}{1 - x + L}, \tag{9.58}$$

now converges. Here L is proportional to λ^2.

9.4.2 Electrons at metallic densities

The RPA calculation of the properties of the electron gas is accurate at high enough densities ($r_s < 1$). However, metals are not this dense, but have r_s in the range 2–6. Many groups have extended the diagrammatic methods of Gell-Mann and Brueckner to this regime; see Mahan (2000). The general features that we have seen above are preserved for metallic densities.

One of the most interesting ways to calculate the ground-state energy of this system is to use Monte Carlo simulation. This is not an easy task: boson systems are fairly easily simulated because boson wavefunctions are symmetric and the ground-state does not have nodes. This is not true for fermion systems. There are regions of positive and negative wavefunction, and it is difficult to prevent the system from minimizing its energy by eliminating the nodes and becoming a boson system. This is known as the "sign problem." Various methods have been devised to get around the problem. Ceperley & Alder (1980) used one of them to calculate the ground-state energy of jellium for r_s up to 200.

One interesting feature of the results is that at low density (far below those in any metal) the electron gas turns into a ferromagnet. We see this effect in Problem 4 in the HF approximation. Since the exchange interaction lowers the energy of electrons with parallel spins, it becomes energetically favorable to have more spin-up electrons (for example) if the cost in extra kinetic energy is not too large. However, HF vastly overestimates the effect because the exchange interaction needs to be screened. HF gives a ferromagnetic transition in the range of metallic densities, but the actual result is that $r_s \approx 75$ at the transition.

9.4.3 Wigner crystallization

For very low density the jellium model has another instability: the electrons localize and form a lattice. This was pointed out by E. Wigner and is known as Wigner crystallization. We can get an idea of how this comes about by noting that keeping electrons apart lowers their Coulomb energy. However, localizing them raises their kinetic energy. If the density is low enough, the Coulomb energy wins.

We can estimate the melting point of the crystal following Nozières & Pines (1958). First consider an electron in a uniform background at the center of its Wigner–Seitz cell. As it moves away from the origin by radius u it will feel an electric field from the charge inside that radius according to a result of elementary electrostatics. The potential energy will be:

$$v(r) = \frac{e^2}{u} n_+ \frac{4\pi u^3}{3} = \frac{4\pi e^2 n}{3} u^2 = \frac{1}{3} \omega_p^2 u^2. \tag{9.59}$$

Thus the electron will be in a spherical harmonic oscillator potential with natural frequency given by $\omega_o^2 = 2\omega_p/3$. The quantum fluctuations in the potential are given by the quantum mechanics of a simple harmonic oscillator:

$$\sqrt{\langle u^2 \rangle} = \sqrt{\frac{3\hbar}{2m\omega_o}} = 1.1 \, r_s^{3/4} a_o, \tag{9.60}$$

using Eq. (9.48). The distance between electrons is of order

$$c = (3/4\pi n)^{1/3} = r_s a_o.$$

Now we use the Lindemann criterion from Section 2.2.3 to predict the melting density:

$$\sqrt{\langle u^2 \rangle}/c = 1.1 \, r_s^{-1/4} = 0.1. \tag{9.61}$$

If r_s is smaller than this, i.e. if the density is too large, the crystal melts. Actually, using 0.1 in this equation gives too small a value – 0.3 would agree better with the actual result, $r_s \approx 100$, which is given by the Monte Carlo method described in the previous section. This is far away from the metallic range. However, it seems that the corresponding transition in the two-dimensional electron gas in inversion layers (as in Section 7.7) has been observed.

9.5 Inhomogeneous electron systems

Electrons in atoms and solids are not in a homogenous environment. We need to understand how to extend the ideas above to such cases. This is the subject of the *density functional theory* of W. Kohn and collaborators; see Kohn (1999).

9.5.1 Kohn–Hohenberg theory

Kohn, along with P. Hohenberg, (Hohenberg & Kohn 1964), considered an electron system in an external potential, v. The density depends on v, of course, but they showed something far more subtle: the density, $n(\mathbf{r})$, is a *unique functional* of v. The ground-state energy is the minimum of another functional of density, and the minimum is attained for the exact density. This allows us, in principle, to circumvent any discussion of the wavefunction, and use $n(\mathbf{r})$ as the variable.

The first statement is easy to prove. The Hamiltonian is the sum of the external potential, the kinetic and interaction terms, $\hat{\mathcal{H}} = \hat{T} + \hat{V}_{\text{int}} + v$. The operators $\hat{T}$ and $\hat{V}_{\text{int}}$ are the same for all systems. Thus a knowledge of v determines $\hat{\mathcal{H}}$ and thus, in principle, the ground-state wavefunction, Ψ_G. From this it is easy to construct n since $n(\mathbf{r}) = \langle \Psi_G | \hat{\rho} | \Psi_G \rangle$.

The converse relation follows from a proof by contradiction: suppose two potentials which differ by more than a constant give the same $n(\mathbf{r})$. The two potentials give two different wavefunctions, Ψ_1 and Ψ_2 since there are now two Hamiltonians, $\hat{\mathcal{H}}_{1,2} = \hat{T} + \hat{V}_{\text{int}} + v_{1,2}$. Suppose the ground-state is not degenerate. Then:

$$E_1 = \langle \Psi_1 | \hat{\mathcal{H}}_1 | \Psi_1 \rangle < \langle \Psi_2 | \hat{\mathcal{H}}_1 | \Psi_2 \rangle, \tag{9.62}$$

because Ψ_2 is not the ground-state wavefunction of $\hat{\mathcal{H}}_1$. However,

$$\begin{aligned}
\langle \Psi_2 | \hat{\mathcal{H}}_1 | \Psi_2 \rangle &= \langle \Psi_2 | \hat{T} + \hat{V}_{\text{int}} + v_1 | \Psi_2 \rangle \\
&= \langle \Psi_2 | \hat{T} + \hat{V}_{\text{int}} + v_2 | \Psi_2 \rangle + \langle \Psi_2 | v_2 - v_1 | \Psi_2 \rangle \\
&= E_2 + \int d\mathbf{r}\, (v_2(\mathbf{r}) - v_1(\mathbf{r})) n(\mathbf{r}).
\end{aligned} \tag{9.63}$$

Thus:

$$E_1 < E_2 + \int d\mathbf{r}\, (v_2(\mathbf{r}) - v_1(\mathbf{r})) n(\mathbf{r}). \tag{9.64}$$

Now interchange 1 with 2 and add the results. We get $E_1 + E_2 < E_1 + E_2$ which is a contradiction.

Thus the ground-state energy is a functional of $n(\mathbf{r})$. We make the following definition:

$$
E_G = E[n(\mathbf{r})] = T[n(\mathbf{r})] + E_{\text{int}}[n(\mathbf{r})] + \int d\mathbf{r} \, n(\mathbf{r}) v(\mathbf{r})
$$

$$
\equiv F[n(\mathbf{r})] + \int d\mathbf{r} \, n(\mathbf{r}) v(\mathbf{r}). \tag{9.65}
$$

The unknown functional, F, is the same for all electron systems. This observation is the key to what follows.

Now we can show that $E[n(\mathbf{r})]$ satisfies a minimum principle, that is, the minimum of E over all densities corresponding to the same number of electrons is the correct density. This follows at once by supposing there is a correct density, $n(\mathbf{r})$ and another candidate $n'(\mathbf{r})$ which corresponds, by the above, to a different wavefunction than Ψ_G, call it Ψ'. Now:

$$
E_G = E[n(\mathbf{r})] < \langle \Psi' | \hat{\mathcal{H}} | \Psi' \rangle = E[n'(\mathbf{r})]. \tag{9.66}
$$

Thus any incorrect density gives a higher energy than the correct one. That is, the correct density is the solution of:

$$
\delta E[n] = 0; \quad \int d\mathbf{r} \, n(\mathbf{r}) = N. \tag{9.67}
$$

The variation is with respect to all densities subject to the condition in the second equation, namely that there are the proper number of electrons in the system.

We can construct a simple example of a density functional by using Thomas–Fermi–Dirac theory. We write expressions for the kinetic, exchange, and Hartree energies and the interaction with v:

$$
\begin{aligned}
E_{\text{TFD}} = &\int d\mathbf{r} \, \frac{3}{5} E_F(n(\mathbf{r})) n(\mathbf{r}) - \int d\mathbf{r} \, \frac{3e^2}{4\pi} k_F(n(\mathbf{r})) n(\mathbf{r}) \\
&+ \frac{e^2}{2} \int d\mathbf{r} \, d\mathbf{s} \frac{n(\mathbf{r}) n(\mathbf{s})}{|\mathbf{r} - \mathbf{s}|} + \int d\mathbf{r} \, n(\mathbf{r}) v(\mathbf{r}) \\
= &\int d\mathbf{r} \, [An(\mathbf{r})^{5/3} + Bn(\mathbf{r})^{4/3}] \\
&+ \frac{e^2}{2} \int d\mathbf{r} \, d\mathbf{s} \frac{n(\mathbf{r}) n(\mathbf{s})}{|\mathbf{r} - \mathbf{s}|} + \int d\mathbf{r} \, n(\mathbf{r}) v(\mathbf{r}),
\end{aligned} \tag{9.68}
$$

where A and B are constants.

9.5.2 The Kohn–Sham equation

It is tempting to proceed with the theory in the form of the previous section, and, indeed, some interesting results were derived that way. However, there is a better way which came from the work of Kohn with L. J. Sham (Kohn & Sham 1965). They recast density-functional theory in the form of a self-consistent differential equation which resembles the Hartree–Fock equation, Eq. (9.32).

The motivation for doing this is that many of the striking features of electrons in condensed matter arise from the fact that the density looks a good deal like that of the free-electron model. For example, Friedel oscillations arise from a sharp Fermi surface. Density functional theory for ground-states should have these features because response functions are derivatives of the ground-state energy which is what the theory is supposed to give.

The proposal of Kohn and Sham was to constuct a set of orbitals for putative independent electrons in a self-consistent potential, then construct the density as the sum of the squares of the orbital wavefunctions. This gives the density functional for the kinetic energy via the intermediate step of introducing a set of orbitals for fictitious independent electrons.

The density of electrons and the Hartree energy in this formulation are exactly as in the Hartree–Fock approximation, Eq. (9.28) and the (non-interacting) kinetic energy is the first term of Eq. (9.24). The remainder of the kinetic energy and the exchange and correlation energy are lumped into a functional E_{xc}. Thus we have:

$$E_{KS} = T_0 + \int d\mathbf{r}\, n(\mathbf{r})v(\mathbf{r})$$

$$+ \frac{e^2}{2} \int d\mathbf{r}_1\, d\mathbf{r}_2 \frac{n(\mathbf{r}_1)n(\mathbf{r}_2)}{|\mathbf{r}_1 - \mathbf{r}_2|} + E_{xc}(n(\mathbf{r})). \tag{9.69}$$

The important issue here is that the obviously non-local terms, the Hartree and kinetic terms, are explicit in the equation. The remainder is likely to be local (because of screening, for example). We can take the variation of the functional with respect to the orbitals of the non-interacting system, and introduce a Lagrange multiplier, as in the Hartree–Fock case. Following the steps above we find the Kohn–Sham self-consistent equations:

$$-\frac{\hbar^2}{2m}\nabla^2\phi_i + V_H\phi_i + V_{xc}\phi_i + v\phi_i = \epsilon_i\phi_i$$

$$V_{xc} = \frac{\delta E_{xc}}{\delta n(\mathbf{r})}. \tag{9.70}$$

As in the case of Hartree–Fock (see Problem 2), we can write down the ground-state energy:

$$E_0 = \sum_i \epsilon_i + E_{xc}(n(\mathbf{r})) - \int d\mathbf{r}\, V_{xc}n(\mathbf{r}) - \frac{e^2}{2}\int d\mathbf{r}\, d\mathbf{s}\frac{n(\mathbf{r})n(\mathbf{r})}{|\mathbf{r}-\mathbf{s}|}. \tag{9.71}$$

9.5.3 Exchange-correlation functional

At this point the reader may wonder what has been accomplished. We seem to have replaced a mystery – the many-electron problem in an inhomogeneous system – with an enigma, E_{xc}. The unknown functional is supposed to have all the exchange, correlation, and some of the kinetic energy of the system. The key to the success and extraordinary popularity of the method is that a series of serviceable approximations have been developed for E_{xc}. The necessary approximations are somewhat problem dependent, but enough experience has been gained that a great number of practical situations can now be treated in a reliable manner.

To guide us to what must be done, we give, first, an interpretation of E_{xc} in terms of the pair correlation function, p, and the pair distribution function, g; see Section 3.4. We need the interaction energy minus the Hartree energy.

$$
\begin{aligned}
E_{\text{int}} - E_{\text{H}} &= \frac{e^2}{2} \left(\int d\mathbf{r} \, d\mathbf{s} \, \frac{p(\mathbf{r}, \mathbf{s})}{|\mathbf{r} - \mathbf{s}|} - \int d\mathbf{r} \, d\mathbf{s} \, \frac{n(\mathbf{r})n(\mathbf{s})}{|\mathbf{r} - \mathbf{s}|} \right) \\
&= \frac{e^2}{2} \int d\mathbf{r} \, n(\mathbf{r}) \int d\mathbf{s} \, \frac{(g(\mathbf{r}, \mathbf{s}) - 1)n(\mathbf{s})}{|\mathbf{r} - \mathbf{s}|} \\
&\equiv \frac{e^2}{2} \int d\mathbf{r} \, n(\mathbf{r}) \int d\mathbf{s} \, \frac{n_{xc}(\mathbf{r}, \mathbf{s})}{|\mathbf{r} - \mathbf{s}|}.
\end{aligned}
\tag{9.72}
$$

The last line defines the *exchange-correlation hole*. This is the generalization of the exchange hole of Problem 2 of Chapter 6. From the results in Section 3.4 we have:

$$
\int d\mathbf{s} \, n_{xc} = -1.
\tag{9.73}
$$

However, we need another part of E_{xc}, the difference between the kinetic energy of the interacting system and the non-interacting one. This is the same problem that we faced in Section 9.4.1 and it is handled the same way, via a coupling constant integration of α from 0 to e^2. We need to be a little more careful, and make sure that the external potential also depends on α in order to keep the density constant throughout. In principle this can be done, so we can write:

$$
E_{xc} = \frac{e^2}{2} \int d\mathbf{r} \, n(\mathbf{r}) \int d\mathbf{s} \, \frac{\bar{n}_{xc}(\mathbf{r}, \mathbf{s})}{|\mathbf{r} - \mathbf{s}|}; \quad \bar{n}_{xc}(\mathbf{r}, \mathbf{s}) = \int_0^{e^2} d\alpha \, n_{xc}(\mathbf{r}, \mathbf{s}, \alpha),
\tag{9.74}
$$

where $n_{xc}(\mathbf{r}, \mathbf{s}, \alpha)$ is defined by Eq. (9.72), except for a different coupling constant.

This expression can give us insight on how to form approximations. The simplest thing to do is to assume that the density varies slowly on the scale of the exchange-correlation hole, so that E_{xc} is a local object, as in Thomas–Fermi:

$$
E_{xc}^{\text{LDA}} = \int d\mathbf{r} \, n(\mathbf{r})\epsilon_{xc}(n(\mathbf{r})).
\tag{9.75}
$$

This is called the *local density approximation* (LDA). The function $\epsilon_{xc}(n)$ is the exchange-correlation energy per particle of the uniform electron gas. This is known, and was discussed above. For example, Eq. (9.56) minus the first term gives the result at high densities, and the Monte Carlo results discussed above can be used at lower densities. Perdew & Zunger (1981) have given an expression that fits the Monte Carlo results which is widely used. An older low density expression, due to Wigner, is not too far off:

$$
\epsilon_{xc} = -\frac{0.916}{r_s} - \frac{0.88}{r_s + 7.8} \text{ (Ryd)}.
\tag{9.76}
$$

The first term is the exchange energy.

The LDA is exact for uniform electron gases, of course, and is remarkably accurate for solids and molecules. Within the LDA we have an explicit expression for V_{xc}:

$$V_{xc} = \frac{\delta}{\delta n(\mathbf{r})} \int d\mathbf{s} \, n(\mathbf{s})\epsilon_{xc}(n(\mathbf{s})) = \epsilon_{xc}(n(\mathbf{r})) + n(\mathbf{r})\frac{d\epsilon_{xc}}{dn}. \tag{9.77}$$

The LDA gives ionization energies of atoms and cohesive energies with 10–20% accuracy. It predicts the correct ground-state crystal structure for many materials and phase transitions as a function of pressure. The values it predicts for lattice constants of molecules and solids are accurate to a few percent.

One of the earliest applications of the theory was to the surface energy of metals. This is certainly a problem where the inhomogeneous nature of the electrons plays a central role. In Chapter 4 we talked about adatoms moving on surfaces. The calculation of the energy landscape in which they move is often done by density-functional theory.

For the purposes of quantum chemistry the LDA is not sufficient. The next approximation is to assume that the exchange-correlation hole is sensitive to some variations of $n(\mathbf{r})$, though we expect that the hole will be fairly small – it contains one electron, so that in most cases we could replace n by its average plus a first derivative term. This is the motivation for the gradient expansion:

$$E_{xc} = E_{xc}^{LDA} + \int d\mathbf{r} \, g_2(n)|\nabla n|^2 + \cdots \tag{9.78}$$

We could get the functional g_2 by using the long-wavelength response function as in the original Kohn–Sham paper. However, this is not a practical advance since the gradients in real materials are often too large. There is a way to proceed called the generalized gradient approximation (Perdew, Burke & Ernzerhof 1996) which uses sum rules and other constraints to give a useful form. With these additions the Kohn–Sham approach is quite accurate for chemical binding problems.

9.5.4 Ab initio pseudopotentials

We have not interpreted the eigenvalues ϵ_i in any direct way. It is tempting to identify them with the band structure. This is often done, but it is not quite correct. In particular the band gaps of solids are not given very well by the usual approximations.

The source of this problem is the second term in Eq. (9.77) which shows that the potential changes with the density. Now the kinetic energy changes discontinuously by the band-gap energy when a single electron is added to an insulator. This discontinuity should also be present in E_{xc}. That is, in the Kohn–Sham formulation, the energy of every electron changes when one electron is added in the conduction band. This is part of the formalism, and is not due to a particular approximation. What is not clear is how big the errors are in adopting functionals that are not discontinuous.

Nevertheless, the Kohn–Sham eigenvalues are almost always used in modern band structure calculations. A common method is to use them to construct a pseudopotential. For a solid composed of some atom like Si the first step is to solve the Kohn–Sham equation for the Si atom including all the electrons.

For the outer valence electrons (four of them in Si) it is necessary to construct a pseudopotential so that the pseudowavefunction is smooth inside the core and matches the real wavefunction outside. For example, one could choose the pseudowavefunction, for each angular momentum l, to be $\phi_l(r) = r \exp(p(r))$ where $p(r)$ is a polynomial whose coefficients are adjusted to make the ϕ_l match the value and derivative of the true wavefunction outside the core radius and also to give the same integrated charge inside. (The last condition is called norm-conservation, and is important so that the results be transferrable to the solid.) Then a potential is constructed by insisting that it is the potential that yeilds ϕ_l. That is:

$$
\epsilon\phi_l = -\frac{\hbar^2}{2m}\left[\phi_l''(r) + \frac{l(l+1)}{2r^2}\phi_l(r)\right] + \tilde{v}_l(r)\phi_l(r),
$$
$$
\tilde{v}_l(r) = \epsilon - \frac{\hbar^2}{2m}\left(l(l+1)/2r^2 - \phi''/\phi_l\right). \tag{9.79}
$$

Here ϵ is the eigenvalue of the atomic problem.

This isn't the right potential yet. We want the potential of a "bare ion" so that we can redo the Kohn–Sham problem in a solid with the correct density. So we subtract:

$$
v_l(r) = \tilde{v}_l(r) - V_H(n_\phi) - V_{xc}(n_\phi). \tag{9.80}
$$

Here n_ϕ is the density associated with the pseudowavefunctions, ϕ. Note that v_l is dependent on l, so that it is non-local. It is easy to project out the partial waves from plane waves, so that the action of v_l in a plane-wave basis can be written down.

We have described here one of many possible variants of the calculation; for a comprehensive view see Martin (2004). This subject has been very carefully investigated, and calculations of this sort are now routine. Software packages are available which do all the computations.

9.5.5 Extensions

In the discussion above we have, for simplicity, neglected the role of spin. If it is required a separate density for spin-up and spin-down electrons can be defined. The resulting formalism, called *spin-density functional theory* has been successful, for example, in calculations of the Pauli susceptibility of real metals. There is also a version of the theory which applies at finite temperatures and gives a functional for the free energy. This has not been much exploited.

9.6 Electrons and phonons

We now consider a real material with electrons that move in a lattice that can vibrate. We will look at three topics: how the electrons determine the lattice frequencies, electron scattering from phonons, and an effective electron-electron interaction induced by phonons.

9.6.1 Phonon frequencies

As we remarked in Chapter 5, according to the Born–Oppenheimer approximation we need to treat the electron total energy as determining part of the potential energy of the ions when they move.

Simple metals

We start by looking at this question in simple (sp-bonded) metals. In Section 5.1.6 we introduced a model for the ion vibrations in such a metal. We think first of "bare" ions in a sea of fixed electrons. The vibrations have an unexpected dispersion relation, namely that for longitudinal sound the frequency did not go to zero but rather to the ion plasma frequency, $\Omega_p = \sqrt{4\pi n Z^2 e^2/M}$, where Z is the valence and M is the ion mass. We now see how to correct this: we should screen the Coulomb interaction in the equation of motion for the ions, Eq. (5.46). We can use the long-wavelength dielectric constant of Eq. (8.37) at zero frequency (because ions move slowly compared to electrons – the Born–Oppenheimer approximation again). This amounts to setting:

$$\omega^2(q) = \Omega_p^2/\epsilon(q,0)$$
$$\rightarrow \frac{4\pi n Z^2 e^2}{M} \frac{q^2}{\lambda^2}$$
$$= \frac{Zmv_F^2}{3M} q^2. \tag{9.81}$$

We have used the fact that the electron density is Z times the ion density. Note that the frequency now goes to zero linearly, as it should. We have a prediction for the velocity of sound in simple metals:

$$c^2 = \frac{Zm}{3M} v_F^2. \tag{9.82}$$

This relation is due to D. Bohm and T. Staver. The Bohm-Staver relation is quite a good prediction for the velocity of sound in simple metals.

For wavelengths that are shorter we will still have $\omega(q) \propto [\epsilon(q)]^{-1/2}$. However, now we should use the RPA dielectric function of Eq. (8.44). For a simple metal this function has a weak singularity at $k = 2k_F$ which gives rise to Friedel oscillations. W. Kohn pointed out that these should affect the phonon frequencies. In fact, this effect is observed in the form of "kinks" in $\omega(k \approx 2k_F)$.

If there is nesting, we might expect a giant Kohn anomaly. A large susceptibility could reduce the phonon frequency by a large amount. This is nothing more than the Peierls distortion of Section 8.4.5.

Frozen phonons

In materials other than simple metals dielectric functions are harder to come by, so this approach is less simple, though it has been used. Another way is to use the *frozen phonon* approximation together with density-functional theory. This method goes as follows. According to the Born–Oppenheimer approximation we need the energy of the electrons for a fixed ion configuration to get the inter-ion potential. This is exactly what density-functional theory gives, Eq. (9.71). Once we know how to solve the band-structure problem we have the ϵ_i and the rest of the parts of the energy are easily computed. From Eq. (5.15) we recall that the elements of the dynamical matrix, $G(s, l, j; s'j'l')$ are the components of the force of the lth atom in unit cell s due to a unit displacement of the atom at $s'l'$. Thus we can find G and the phonon spectrum from derivatives of the electronic energy.

One way to proceed is to find the dynamical matrix via *supercells*. We want to displace an atom, and find the forces on all the other atoms. However, if we displace an atom in a unit cell, and assume the crystal is still periodic, we are implicitly solving for only the modes at $\mathbf{k} = 0$. If the system is not periodic, we cannot use the methods of band structure theory unless we take a unit cell the size of a wavelength. To get around this is we can join together unit cells to make a *supercell* which is larger than twice the range of the forces. We find G by displacing atoms *within each supercell* and find the forces on the other atoms. Since the crystal is still periodic (with a larger periodicity) we can still do a band structure calculation. For example, to compute longitudinal phonons in the [100] direction, we glue together unit cells in the x-direction; see Figure 9.3. Then we displace each inequivalent plane of ions in the cell in turn and solve for the total energy.

To find the forces on each atom (including the displaced one) we could take numerical derivatives of the total energy with respect to motions of each atom. We can avoid this by

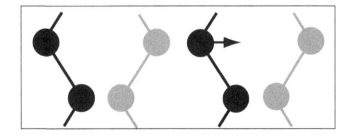

Fig. 9.3 The supercell used by Yin & Cohen (1982) to calculate the phonon dispersion relation in Si. Si atoms are pictured in the [110] plane, black, and projected on the plane, gray. One atom is displaced and the forces on all the others are found by the Hellman–Feynman theorem.

using the Hellman–Feynman theorem. The term in the energy of Eq. (9.69) that depends on the position of the ions is the external potential. Suppose we represent it by a pseudopotential, v_ϕ at each atom site:

$$E_v = \sum_{\mathbf{R}} \int d\mathbf{r} \, v_\phi(\mathbf{r} - \mathbf{R}) n(\mathbf{r}). \tag{9.83}$$

Now put:

$$v_\phi(\mathbf{r} - \mathbf{R}) = \int \frac{d\mathbf{k}}{2\pi^3} \, e^{i\mathbf{k}\cdot(\mathbf{r}-\mathbf{R})} v_\phi(\mathbf{G}); \; n(\mathbf{r}) = \sum_{\mathbf{G}} n(\mathbf{G}) e^{i\mathbf{G}\cdot\mathbf{r}}. \tag{9.84}$$

Then we have:

$$E_v = \sum_{\mathbf{G},\mathbf{R}} \hat{v}_\phi(\mathbf{G}) \, n(-\mathbf{G}) e^{i\mathbf{G}\cdot\mathbf{R}}. \tag{9.85}$$

Therefore the Hellman–Feynman force on the atom at $\mathbf{R}$ is:

$$\mathbf{F}_{\mathbf{R}} = -\sum_{\mathbf{G}} i\mathbf{G} \, \hat{v}_\phi(\mathbf{G}) \, n(-\mathbf{G}) e^{i\mathbf{G}\cdot\mathbf{R}}. \tag{9.86}$$

For a non-local pseudopotential the formula is much more complicated. If the ions carry a charge we need to add the Coulomb forces.

9.6.2 Electron-phonon interaction

In Chapter 7 we considered the scattering of electrons from lattice vibrations using the idea (appropriate for simple metals) that the effect of a lattice vibration is to carry an atom, represented by a screened pseudopotential, to a new position. Here we look at this phenomenon in a different way. Suppose we represent the interaction of the electron with the ions as:

$$\sum_{\mathbf{R}} v(\mathbf{r} - \mathbf{R} - \mathbf{u}(\mathbf{R})) \approx V_\circ - \sum_{\mathbf{R}} \mathbf{u}(\mathbf{R}) \cdot \nabla_{\mathbf{r}} v(\mathbf{r} - \mathbf{R})$$

$$= V_\circ - \sum_{\mathbf{R},\mathbf{k}} \frac{\tilde{v}(\mathbf{k})}{\Omega} i\mathbf{k} \cdot \mathbf{u}(\mathbf{R}) e^{i\mathbf{k}\cdot(\mathbf{r}-\mathbf{R})}. \tag{9.87}$$

Here $V_\circ$ is the potential of the static lattice and $\tilde{v}$ is the Fourier transform of the ion potential. To evaluate the new term we use Eq. (7.39). After a bit of algebra we find:

$$\delta\hat{\mathcal{H}} = -i\sqrt{N} \sum_{\mathbf{q},\beta,\mathbf{G}} \tilde{v}(\mathbf{q} + \mathbf{G})(\mathbf{q} + \mathbf{G}) \cdot \mathbf{U}_\beta(\mathbf{q}) q_{\mathbf{q},\beta} \frac{e^{i(\mathbf{q}+\mathbf{G})\cdot\mathbf{R}}}{\Omega}$$

$$= \sum_{\mathbf{q},\beta,\mathbf{G}} M_{\mathbf{q},\mathbf{G},\beta}\, \hat{\rho}_{\mathbf{q}+\mathbf{G}}\, (\hat{a}_{\mathbf{q},\beta} + \hat{a}^{+}_{-\mathbf{q},\beta}).$$

$$M_{\mathbf{q},\mathbf{G},\beta} = -i\tilde{v}(\mathbf{q}+\mathbf{G})\sqrt{\frac{N\hbar}{2M\omega_{\mathbf{q},\beta}}}(\mathbf{q}+\mathbf{G})\cdot\mathbf{U}_{\beta}(\mathbf{q}). \qquad (9.88)$$

In the second line we have gone to second-quantized notation for the electrons and used the operator for the electron density $\hat{\rho}(\mathbf{q}+\mathbf{G})$ from Eq. (8.26) and the operators $\hat{a}^{+}$, $\hat{a}$ for the phonons. The electron-phonon matrix element gives the coupling between the electron density and the phonon amplitude. This equation shows that the electrons scatter by creating and destroying phonons. The results of Section 7.4.2 can be rederived from this point of view.

The electron-phonon matrix element is a complicated object. It is often useful to use macroscopic reasoning to arrive at a Hamiltonian of the form of Eq. (9.88). One way to do this is through the idea of a *deformation potential*. Suppose the solid deforms so that it locally changes density. Then the electron energy will, in general, change. For example, in a metal changing the density changes the kinetic energy. Define the dilatation by $\Delta = \delta\Omega/\Omega$, or locally by $\delta n(\mathbf{r}) = -\Delta n$. Then the change in the Fermi level will be $-\frac{2}{3}E_{\mathrm{F}}\Delta$. In a semiconductor the deformation potential interaction for carriers in the conduction band will involve the change in band gap per unit dilatation. However, the dilatation is easily seen to be $\partial u_{x}/\partial x + \partial u_{y}/\partial y + \partial u_{z}/\partial z$. Therefore we have a coupling linear in the phonon operators, and a form for the electron-phonon coupling.

9.6.3 Effective electron-electron coupling

The electron-phonon coupling induces an interaction between electrons. Bardeen & Pines (1955) showed that the interaction can be *attractive*. This remarkable phenomenon is responsible for superconductivity in many materials.

We will not follow the original derivation, but rather use an intuitive, method. Suppose we return to the Coulomb lattice model of a simple metal and try to write the dielectric function including the fact that the motion of both electrons and ions can screen a potential. In order to include both effects we must add the polarizabilities of the two systems. That is we write:

$$\epsilon_{\mathbf{tot}}(\mathbf{q},\omega) = 1 + \frac{\lambda^{2}}{q^{2}} - \frac{\Omega_{\mathrm{p}}^{2}}{\omega^{2}} = \epsilon_{\mathrm{TF}}\left(1 - \frac{\omega^{2}(q)}{\omega^{2}}\right); \quad \epsilon_{\mathrm{TF}} = 1 + \lambda^{2}/q^{2}. \qquad (9.89)$$

Here $\omega(q)$ is the renormalized phonon frequency from Eq. (9.81). As usual, the zero of the dielectric function gives the longitudinal normal mode frequency.

Now if we have a pair of electrons in $\mathbf{k}$ and $\mathbf{k}'$ whose interaction interests us we can associate a frequency with this excitation of the system by putting $\hbar\omega = \epsilon_{\mathbf{k}} - \epsilon_{\mathbf{k}'}$. So we can guess that the effective interaction is:

$$v_{\mathrm{eff}}(\mathbf{k},\mathbf{k}') = \frac{\tilde{v}(\mathbf{k}-\mathbf{k}')}{\Omega\epsilon_{\mathbf{tot}}(\mathbf{k}-\mathbf{k}',\omega)}$$

$$= \frac{1}{\Omega} \frac{4\pi e^2}{|\mathbf{k} - \mathbf{k}'|^2 + \lambda^2} \left(\frac{\omega^2}{\omega^2 - \omega_{\mathbf{k}-\mathbf{k}'}^2} \right)$$

$$\omega^2 = (\epsilon_{\mathbf{k}} - \epsilon_{\mathbf{k}'})^2 / \hbar^2. \tag{9.90}$$

This interaction is *negative* if $\epsilon_{\mathbf{k}} - \epsilon_{\mathbf{k}'}$ is very small. Note that the frequency dependence of the interaction is crucial. The usual intuitive explanation for this effect is that an electron passes through a region of lattice and polarizes it. Later another electron is attracted to the place where the first one *was*.

9.7 Strong interactions and magnetism in metals

So far we have dealt with systems where the interaction effects are relatively small. These wide band materials are very well described by band theory and density functional theory. However, there are other kinds of materials with narrow bands, arising, for example, from d or f shell electrons where the picture is more complicated, and correlation effects can be quite strong. We will give a brief discussion of these effects here. In order to treat this subject properly, advanced techniques are a necessity, and lie beyond the scope of this book.

9.7.1 Hubbard model

The essential physics of strong interactions is simple, in essence. Electrons repel, and if this is the dominant interaction, they will localize, as in the Wigner crystal. Consider, for example, a system with one electron per atom. To minimize the interaction energy, we can localize one electron on each site. In this case we pay a price in kinetic energy for the localization. This physics is commonly represented in a model due to J. Hubbard, in which electrons repel only when they are on the same site, and have kinetic energy in the tight-binding form:

$$\hat{\mathcal{H}} = -t \sum_{i,\delta,\sigma} (\hat{c}_{i+\delta,\sigma}^+ \hat{c}_{i,\sigma} + \hat{c}_{i,\sigma}^+ \hat{c}_{i+\delta,\sigma}) + U \sum_i \hat{n}_{i+} \hat{n}_{i-}. \tag{9.91}$$

Here i runs over the sites, and δ over nearest neighbors. The quantity, U, is of the order of several electron volts. The operators at the sites, $\hat{c}_i^+, \hat{c}_i$ are related to the running electron waves in the usual way:

$$\hat{c}_i^+ = \frac{1}{\sqrt{N}} \sum_{\mathbf{k}} \hat{c}_{\mathbf{k}}^+ e^{i\mathbf{k} \cdot \mathbf{R}_i}. \tag{9.92}$$

Using this $\hat{\mathcal{H}}$ can be rewritten:

$$\hat{\mathcal{H}} = \sum_{\mathbf{k},\sigma} \epsilon_\circ(\mathbf{k}) \hat{c}_{\mathbf{k},\sigma}^+ \hat{c}_{\mathbf{k},\sigma} + U \sum_i \hat{n}_{i+} \hat{n}_{i-}. \tag{9.93}$$

The ground-state of the Hubbard model is not known analytically except in one dimension. However, various limiting cases can be discussed.

If the Coulomb integral, U, is very large, we can start our theory with the ground-state for infinite U, namely one electron localized at each site. The only degree of freedom left is the arrangement of spins, and all such arrangements have the same energy. However, we lift the degeneracy if we use the kinetic energy as a perturbation.

For simplicity, consider just two spins. Second-order perturbation theory involves matrix elements between the ground-state and excited states. In this case the perturbation transfers an electron from one site to the other. However, if spins are parallel the transfer is forbidden by the Pauli principle. If they are antiparallel, we do have a matrix element, and the energy is lowered by the square of the matrix element divided by the energy denominator, namely:

$$\Delta E = -t^2/U. \tag{9.94}$$

This can be represented by an effective spin Hamiltonian:

$$\hat{\mathcal{H}} = \frac{2t^2}{U}(s_1^z s_2^z - 1/4). \tag{9.95}$$

That is, the kinetic energy induces an antiferromagnetic coupling between adjacent spins with an exchange integral whose size is of order t^2/U. For a lattice that can be broken down into two sublattices; the ground-state is antiferromagnetic. The symmetry is broken, so that the unit cell is doubled, and the material is an insulator.

As we make U/t smaller the effects of correlations will be smaller. If t is large enough the interaction term is negligible, and we return to the problem of band theory. However, in this case the material must be a metal.

Comparing the two limits we see that, as a function of U/t, there will be a *metal-insulator transition*. This is called the Mott transition (Mott 1990), or, in this context, the Mott–Hubbard transition; compare Problem 2 of Chapter 7. There are indications that many transition-metal oxides and sulfides are insulators for this reason. The most famous example is NiO which should be a metal by the rules of band theory, but is, in fact, an insulator whose ground-state is antiferromagnetic.

9.7.2 Stoner criterion and itinerant magnetism

In our discussion of magnetism so far we have implicitly been talking about magnetic insulators with well-localized spins on lattice sites. However, the most common ferromagnets, Fe and Ni, are metals. In fact, for these materials the magnetic moment per lattice site (as measured by the magnetization, for example) does not correspond to an integral number of electrons so that electron transfer between sites must be important. Materials of this sort are called itinerant magnets.

We can understand this effect, at a qualitative level, by continuing to use the Hubbard Hamiltonian. In this case both terms are important. To get an understanding of the situation, we use a version of the Hartree–Fock approximation in which electrons of each spin see the average effects of the other spin . Suppose that there is an average value of $\hat{n}_{i\sigma}$, and that fluctuations about the average are small. For example, for a ferromagnetic state we can put $\langle \hat{n}_{i\sigma} \rangle = n_\sigma$. Then we can set:

$$\hat{n}_{i\sigma} = n_\sigma + \hat{\delta}_{i\sigma}, \tag{9.96}$$

where $\hat{\delta}$ is small. Now substitute this expression into the Hubbard Hamiltonian and drop terms of order δ^2. Then we have, using Eq. (9.92) and Eq. (9.93):

$$\hat{\mathcal{H}} = \sum_{\mathbf{k}}(\epsilon_\circ(\mathbf{k}) + Un_-)\hat{n}_{\mathbf{k}+} + \sum_{\mathbf{k}}(\epsilon_\circ(\mathbf{k}) + Un_+)\hat{n}_{\mathbf{k}-} - UNn_+n_-. \tag{9.97}$$

The last term is a constant.

The rest of $\hat{\mathcal{H}}$ has a simple interpretation: spin up electrons have their band energies shifted by an amount that depends on the occupancy of the spin down band, and *vice versa*. The occupancies depend on the band positions as shown in Figure 9.4. The occupancies of the bands are determined self-consistently, and need not be an integer. This self-consistency criterion can only be satisfied for sufficiently large U.

The picture that each spin band is rigidly shifted by the exchange interaction was introduced by E. Stoner in 1939. We can see when a solution can occur by looking at the paramagnetic state. As usual, ferromagnetism is signaled by a divergence of the magnetic susceptibility. We have band energies that are given by:

$$\epsilon_+ = \epsilon_\circ + Un_- - \gamma\hbar H; \quad \epsilon_- = \epsilon_\circ + Un_+ + \gamma\hbar H; \tag{9.98}$$

The total number of up electrons is:

$$N_+ = \frac{1}{2}\int d\epsilon \mathcal{D}(\epsilon)f(\epsilon_+), \tag{9.99}$$

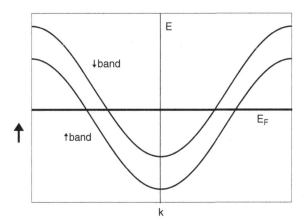

Fig. 9.4   **Hartree–Fock treatment of ferromagnetism. The spin-down band is higher than the spin-up band by $U(n_\uparrow - n_\downarrow)$. Since the Fermi level of the two bands must be the same there are more up states occupied than down states, as shown.**

and similarly for the down spins. If we work at $T = 0$, and assume that the density of states is slowly varying we can write:

$$N_+ - N_- = \frac{1}{2}\mathcal{D}(E_F) \int d\epsilon \, [f(\epsilon_+) - f(\epsilon_-)]$$

$$= \frac{1}{2}\mathcal{D}(E_F)[U(n_+ - n_-) + 2\gamma\hbar H]. \tag{9.100}$$

Now $n_\sigma = N_\sigma/N$ so that we can solve:

$$M = \gamma\hbar(N_+ - N_-) = \frac{(\gamma\hbar)^2 \mathcal{D}(E_F)}{1 - U\mathcal{D}(E_F)/2N} H. \tag{9.101}$$

From this we can read the susceptibility. There is a divergence if the denominator passes through zero as we increase U. This gives the *Stoner criterion* for ferromagnetism:

$$U\mathcal{D}(E_F)/2N > 1. \tag{9.102}$$

The Stoner criterion gives a good qualitative account of which metals are ferromagnetic.

To calculate in a more realistic way we need to resort to the spin-density functional theory mentioned above. It is quite similar to our Hartree–Fock approach: each spin sees the exchange-correlation potential of the other. Considerable success has been achieved in getting the correct moments of transition metals this way. A calculation for the separate spin-up and spin-down Fermi surfaces in Fe is shown in Figure 9.5. Similar calculations have been done for antiferromagnetism in band theory.

However, we need to be wary of the Hartree–Fock approach. For example, a purely band theory approach to NiO seems to work well only for the ground-state. At finite temperatures we can only understand experimental results by assuming that effects of the Mott–Hubbard gap persist even when the material is not magnetic ; see Mott (1990). Sorting all this out, and

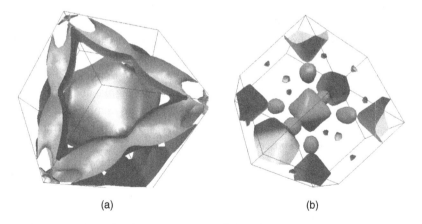

(a)   (b)

Fig. 9.5   **Computed Fermi surfaces for the majority spins, (a), and the minority spins, (b) for magnetic Fe. From www.phys.ufl.edu/fermisurface, the Fermi surface database at the University of Florida.**

going beyond Hartree–Fock theory has proved to be very difficult. Solutions of the Hubbard model, for example, are not easy to come by, and different plausible approximations give different results.

Suggested reading

There are many books on electron-electron interactions.

Nozieres & Pines (1999)
Fetter & Walecka (1971)
Mahan (2000).
For the Luttinger liquid see:
Giamarchi (2004)
The Hubbard model, itinerant magnetism and strong interactions is a very active field. A promising approach is:
Georges, Kotliar, Krauth & Rozenberg (1996).

Problems

1. The electron-hole liquid: Semiconductors illuminated by intense optical radiation are populated by large densities of electrons and holes. Sometimes the carriers condense into a "liquid": i.e. there are regions of the crystal populated by lots of carriers. (a) Write down the ground-state energy, E_{tot}, of a system of n electrons and n holes cm^{-3}. Note that you can use the jellium model because the electrons are a background for the holes and vice-versa. Use Hartree–Fock suitably modified to include effective masses m_e and m_H for the two species, and with a Coulomb interaction modified by putting $e^2 \rightarrow e^2/\epsilon$ where ϵ is the dielectric constant of the host crystal.

   Hint: the ground-state energy in HF was derived above. It is of the form $E/N = An^{2/3} - Bn^{1/3}$.

   (b) Droplets form when the carriers readjust to minimize their energy. Prove that this means that $\partial E_{tot}/\partial n = 0$. (c) Calculate, using the above, the equilibrium ground-state energy and density for the liquid. (d) Put in numbers for Ge where $m_e, m_H = 0.1m$, where m is the mass of the electron, $\epsilon = 16$. What temperatures, would you guess, are necessary to observe the phenomenon?

2. (a) Show that the energy of an atom in the Hartree–Fock approximation is given by:

$$E_0 = \sum_k \epsilon_k - \langle \hat{\mathcal{H}}_2 \rangle,$$

   where the second term is given by Eq. (9.27) and k includes both a spin and orbital index. We need the second term to avoid double-counting the Hartree and exchange energies.

Hint: multiply Eq. (9.32) by ϕ_k, integrate and sum on k.

(b) Suppose the atom is ionized by removing an electron from state k. Show the ionization energy is ϵ_k. This is known as Koopman's theorem.

Hint: the ionization energy is the difference of ground-state energies with and without the occupancy of orbital k.

3. Derive Eq. (9.42).

4. (a) Another way to get the Pauli susceptibility (but there's a reason to do it this way, as we will see): suppose that the number of spin up electrons is not the same as spin down. Write:

$$N_\pm = (N/2)(1 \pm \alpha); \quad |\alpha| < 1.$$

Write the ground-state energy of the non-interacting electron gas as :

$$E_\circ = (3/5)N_+ E_F^+ - \mu_B H + (3/5)N_- E_F^- + \mu_B H,$$

where $E_F^\pm = (\hbar^2/2m)(6\pi^2 n_\pm)^{2/3}$ and $n_\pm = N_\pm/\Omega$. Explain where this comes from. Now minimize w.r.t. α (for $|\alpha| << 1$) and show that you get the classic answer for the Pauli susceptibility:

$$\chi_P = 3N\mu_B^2/2E_F; \quad N = N_+ + N_-; \quad E_F = (\hbar^2/2m)(3\pi^2 n)^{2/3}.$$

(b) Now add the Hartree–Fock exchange term for each spin:

$$E_\circ \rightarrow E_\circ - (3e^2/4\pi)N_+ k_F^+ - (3e^2/4\pi)N_- k_F^-; \quad k_F^\pm = (6\pi^2 n_\pm)^{1/3}.$$

Minimize again, and show that:

$$\chi = \chi_P/(1 - Q n^{-1/3}).$$

Find Q. For small density there is a zero of the denominator. Interpret this. At what r_s does this occur?

5. Consider the functional in Eq. (9.68) without exchange. It is to be minimized with the constraint that the number of particles is fixed. Enforce this condition with a Lagrange multiplier. (a) Take its variational derivative to find an equation obeyed by the density. Your result should be of the form:

$$an(\mathbf{r})^{2/3} + b \int d\mathbf{r}' \, \frac{n(\mathbf{r}')}{|\mathbf{r} - \mathbf{r}'|} + v(\mathbf{r}) = \mu.$$

(b) Use the result of (a) to derive Eq. (9.33). You may need to recall from electrostatics that $\nabla_\mathbf{r}^2 |\mathbf{r} - \mathbf{r}'|^{-1} = -4\pi\delta(\mathbf{r} - \mathbf{r}')$.

6. Derive Eq. (9.71). Start with the hint in Problem 2.

In 1911 H. Kamerlingh Onnes was investigating the electrical properties of metals such as Pt and Au in the range of a few K (Kammerlingh Onnes 1911). He observed that the resistance is determined by the number of impurities in this regime. (This is now called Mattheissen's rule, Eq. (7.24).) In order to eliminate impurities he turned to mercury which was available in a very pure form. Instead of a lower resistance, which he expected, he found a *sudden jump* to a vanishingly small resistance at about 4.2 K. This temperature is now called the transition temperature, T_c. His statement was: "Mercury has passed into a new state which on account of its extraordinary electrical properties may be called the superconductive state." He also found transitions to zero resistance for lead and tin with different T_c's, also in the range of a few K. (We should note that later work showed that superconductivity is, in fact, not affected very much by impurities.)

The superconductive state is now known to occur for around forty elements and hundreds of compounds. The transition temperature is less than about 100 K for all known cases. The superconductivity of copper oxides discovered by Bednorz & Muller (1986) is of particular current interest, and gives the highest known T_c's.

Supercurrents flow without friction and thus are persistent: they have been observed to flow around a ring without decay for the better part of a year. Superconductors have some technological applications such as high-field magnets (an application already foreseen by Kamerlingh Onnes). However, our focus here will be on the implications of the existence of the "new state." This state is an example of a *macroscopic quantum phenomenon*.

Another kind of of a macroscopic quantum phenomenon is the superfluidity of liquid helium-4. It was discovered by Allen & Misener (1938), and simultaneously by Kapitza (1938). They found that He-4 liquid abruptly loses its viscosity for temperatures less than 2.17 K. This temperature is called the lambda transition temperature, T_λ, because the specific heat as a function of T looks like a backwards λ close to 2.17 K. Superfluidity means that the liquid can flow through very tiny capillary tubes or through porous media, which would ordinarily stop flow. These are referred to as superleaks. Note that the atoms in He-4 are bosons, in contrast to electrons in metals which are fermions. Superfluidity in this case is related to the Bose–Einstein condensation (BEC). However, to complicate the picture, at much lower temperatures, in the millikelvin regime, the fermion system He-3 becomes superfluid too.

Still another example is found in extremely dilute alkali gases in magnetic traps in the nanokelvin temperature range. Superfluid flow has been observed both for bosonic and fermionic atoms. This set of experiments started with the observation of the BEC of trapped atoms by C. Weiman, E. Cornell, (Anderson, Ensher, Matthews, Wieman & Cornell 1995) and W. Kettele (Davis, Mewes, Andrews, Vandruten, Durfee, Kurn & Ketterle 1995) and collaborators. These systems, which are not, of course condensed matter (they are very dilute vapors), give us valuable insights about the systems we are interested in here.

We will start with boson systems which are conceptually a bit simpler.

10.1 Bose–Einstein condensation and superfluidity

10.1.1 Non-interacting bosons

The Bose–Einstein condensation was worked out by A. Einstein in 1924 following a suggestion by S. Bose. We review it briefly: it is very well treated in all texts on statistical physics. Bose showed that the average thermal occupancy of orbitals at temperature T for non-interacting bosons is:

$$\langle \hat{n}_j \rangle_T = f_B(\epsilon_j) = \frac{1}{e^{\beta(\epsilon_j - \mu)} - 1}. \tag{10.1}$$

Note the resemblance to Eq. (6.31) with the crucial difference of a minus sign in the denominator. It is clear that f_B cannot diverge for any energy; thus we require that $\mu < 0$ for non-interacting bosons. In contrast, for fermions $\mu > 0$ for low temperatures and approaches E_F as $T \to 0$. Now consider Eq. (6.32) for free bosons:

$$N = \sum_j f(\epsilon_j) = \frac{\Omega}{2\pi^3} \int 4\pi k^2 dk \, f_B(\hbar^2 k^2 / 2M). \tag{10.2}$$

Note that the number of particles in the state $\mathbf{k} = 0$ is zero.

It is easy to see that the integral converges for three dimensions even if we set $\mu = 0$. Thus there is a maximum value of N that can be described by this equation at a given temperature. If the actual number in the system is larger than this number, or if the temperature is too low, there is an apparent paradox. It is an easy calculation (see problems) to show that the temperature, T_c, for which the paradox sets in is given by:

$$N/\Omega = 2.612 (M k_B T_c / 2\pi \hbar^2)^{3/2}. \tag{10.3}$$

Einstein realized that the way out of the difficulty is to understand that the remainder of the particles are condensed in the lowest energy state, $\mathbf{k} = 0$. For the particles in the condensate the passage from the sum to an integral is not correct. The density in the condensate is:

$$n_0 = N_0 / \Omega = \frac{N}{\Omega} \left(1 - \left[\frac{T}{T_c} \right]^{3/2} \right). \tag{10.4}$$

We have two strange phenomena, BEC and the onset of superfluidity. Further, if we put in numbers for He-4 we get $T_c = 3.14$K, near the observed T_λ. Furthermore, the fermion system, He-3, which is chemically identical to He-4 does not show superfluidity unless it is cooled to the millikelvin range.

This led F. London and L. Tisza to propose that the BEC transition caused superfluidity. (For a remarkable review of the early history of the subject see Balibar, 2003.) However superfluid helium is not a weakly interacting gas, but a strongly interacting liquid so that others (notably L. Landau) found the identification unconvincing. Since the observation of BEC in dilute atomic vapors all lingering doubts have been laid to rest. In fact, the atom trap experiments showed the existence of the condensate quite directly: after the BEC took place, the experimenters turned off the trap, and looked at the momentum distribution of the free atoms. In exact correspondence to the theory they found a bump in the distribution for $\mathbf{k} = 0$.

The bosons in the condensate all have the same wavefunction. Thus a quantum mechanical wavefunction, ψ_0, is associated with a finite fraction of the macroscopic number of atoms in the system. This is a the prototype of a macroscopic quantum phenomenon.

10.1.2 Two-fluid model

Once we have a condensate we can understand superfluidity as being a mass current carried by particles in the condensate. This current can flow without friction, as we will see below. If there is a finite current ψ_0 is no longer a constant, as it is in the ground-state. We will normalize ψ_0 so that its squared integral is N_0. We can write:

$$\psi_0 = \sqrt{n_0}e^{i\theta(\mathbf{r})}. \tag{10.5}$$

The quantum mechanical expression for number current density is:

$$\hat{\mathbf{j}}_s = \frac{\hbar}{2Mi}(\psi_0^* \nabla \psi_0 - \psi_0 \nabla \psi_0^*) = n_0 \frac{\hbar}{M} \nabla \theta(\mathbf{r}). \tag{10.6}$$

The last expression is correct if n_0 is constant in space, but the phase varies. The supercurrent is *proportional to the gradient of the phase of* ψ_0. The velocity for the condensate particles is

$$\mathbf{v}_s = (\hbar/M)\nabla\theta. \tag{10.7}$$

In addition, at finite temperature there is current carried by the thermally excited states. This is called the normal current. In the two-fluid model it is customary to define a mass density of superfluid, by writing $M\mathbf{j}_s = \rho_s \mathbf{v}_s$. Note that ρ_s is not the same as the condensate density. At $T = 0$ all of the current is supercurrent, but, as we will see, the condensate density is not the entire density in an interacting superfluid.

This is a remarkable conclusion. There are *two independent velocity fields* in a superfluid, that of the condensate, and that associated with the average momentum of the excited states (the normal current). Tisza expanded on this idea in his *two-fluid* model of He. There are several conclusions that are immediate. One is that there can be a new kind of sound (second sound) in superfluid He where the two fluids oscillate with respect to one another. (Compare the optical modes of Chapter 5.) We can think of this as a propagating wave of temperature since the condensate, a single wavefunction, carries no entropy, but the normal fluid does. In

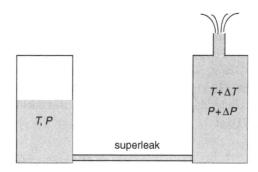

Fig. 10.1 **The fountain effect. The increased temperature on the right causes an increase in pressure.**

ordinary fluids temperature obeys the diffusion equation, not the wave equation (cf. Chapter 4). Second sound is easily observed.

Suppose two containers of He below the λ transition are connected by a thin capillary; then only the superfluid can pass between them. Since the superfluid fraction carries no entropy, the entropy per unit volume in the receiving container decreases. Therefore the temperature goes down. The superleak acts as an entropy filter. Note that since the superfluid carries no entropy, the usual condition for equilibrium between two containers, $T_1 = T_2$, no longer applies.

If, on the other hand, the temperature of one container is held above that of the other, the pressure will increase on the hot side and fluid will flow in. If the difference in temperatures is large enough, the fluid on the hotter side will spray out of the container: this is called the *fountain effect*; see Figure 10.1.

The fountain effect is easy to explain from simple thermodynamics. When mass flows between the two containers we need the chemical potentials to be equal. However, thermodynamics says that $d\mu = -sdT + vdP$ where s is the entropy per particle and v the volume per particle. Thus the change in pressure is given by:

$$\Delta P = \frac{s}{v}\Delta T. \tag{10.8}$$

The pressure difference can be quite large for small temperature gradients: this is how fountains are produced.

There are important direct consequences of Eq. (10.7). Since the superfluid velocity is the gradient of the phase, it follows that $\nabla \times \mathbf{v}_s = \kappa = 0$. In fluid dynamics κ is called the vorticity. The integral of κ over a surface is called the circulation:

$$\int d\mathbf{S} \cdot \nabla \times \mathbf{v} = \oint d\mathbf{r} \cdot \mathbf{v}. \tag{10.9}$$

This is Stokes' theorem. The second integral is around the edge of the surface, and clearly measures the circulating motion.

Taken at face value $\kappa = 0$ means that the superfluid cannot rotate: cold helium in a rotating bucket should remain at rest. This does happen for very low rotation rates, but the fluid can be set in rotation by a surprising route. It develops *quantized vortex lines*. This is a state with κ very large along a singularity line and zero everywhere else, see Figure 10.2. The fluid swirls around the vortex line. The region near the vortex line, the vortex core, is thought to be empty of superfluid, and is of atomic dimensions.

Since $\nabla \times \mathbf{v}_s = 0$ outside the vortex core, the circulation (expressed by the right-hand side of Eq. (10.9)) is independent of path, except for depending on the number of times the path circles the singularity. For a path that encloses the singularity once we see from Eq. (10.7) that :

$$\oint d\mathbf{r} \cdot \mathbf{v}_s = \frac{\hbar}{M} \Delta\theta. \tag{10.10}$$

Here $\Delta\theta$ is the change in the phase of the wavefunction around the loop. However, this must be $2\pi n$ for some integer n in order to have a single-valued wavefunction. This result is called *quantization of circulation*:

$$\oint d\mathbf{r} \cdot \mathbf{v}_s = \frac{nh}{M}. \tag{10.11}$$

Near a long straight vortex line singly excited vortex line we can write, for the velocity itself:

$$\mathbf{v}_s = \frac{A}{2\pi r} \hat{\theta}, \tag{10.12}$$

where $A = h/M$ and $\hat{\theta}$ is the unit vector in the azimuthal direction. Vortex lines were predicted by L. Onsager and R. Feynman. They were observed shortly thereafter.

Vortex lines cannot end in the fluid so they must be attached to the walls or the surface, as in Figure 10.2 or in closed loops like smoke rings. One of the most interesting ways to

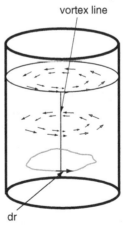

vortex line

dr

Fig. 10.2   **A single vortex line in a rotating bucket of liquid He-4. The flow lines for the superfluid are shown. If the rotation is large there will be many such lines and the surface profile will approach the parabolic profile of a classical rotating liquid. Also shown (bottom) is an integration contour for finding the circulation.**

observe these excitations is due to Rayfield & Reif (1964). They attached ions to closed vortex loops and accelerated them with an electric field.

10.1.3 Landau criterion for superfluidity

Landau formulated an alternate version of the two-fluid theory which clarified an important point: it is not enough to have a condensate that can carry current to have stable superflow. It is also necessary to have a suitable spectrum of excitations.

The non-interacting Bose gas has excitations above the condensate that have the dispersion relation of free-particles, $\epsilon(k) = \hbar^2 k^2 / 2M$. However, in an interacting system we must have phonons (i.e. sound waves) at long wavelengths. In fact, for liquid He, Landau was able to deduce that the spectrum must be of this form by examining specific heat measurements at very low temperatures, i.e. $C_v \propto T^3$, cf. Eq. (5.105). At higher temperatures there was a contribution which looked like $\exp(-\Delta/k_B T)$ with $\Delta/k_B = 8.7$K. Landau deduced that the spectrum looked like Figure 10.3. Neutron scattering experiments analogous to those described in Chapter 5 subsequently verified the dispersion relation.

Suppose we have established a superflow associated with the condensate. If we were to have friction the fluid must dissipate kinetic energy. For small T this can occur only by transferring energy to an excitation. We will work out the kinematics of this transfer.

Consider a state with superflow and no excitations and go to a frame of reference moving with the flow. The walls are moving backwards with $\mathbf{v}$. The total momentum of the liquid in the moving frame is zero. Now return to the laboratory reference frame. The kinetic energy, $\mathcal{T}$, (of the entire liquid) transforms as:

$$\mathcal{T} = \frac{1}{2} \sum_i M(\mathbf{v}_i + \mathbf{v})^2$$

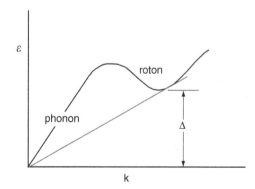

Fig. 10.3 The excitation spectrum of liquid He. The small k modes are phonons. Near the minimum the modes are known as rotons. The minimum is at $k = 1.9 \times 10^8$ cm^{-1}. The slope of the gray line is the critical velocity multiplied by $\hbar$ (see text).

$$= \mathcal{T}_0 + \sum_i \mathbf{p}_i \cdot \mathbf{v} + \frac{1}{2} \sum_i M v^2$$

$$= \mathcal{T}_0 + \frac{1}{2} N M v^2, \tag{10.13}$$

since $\sum \mathbf{p}_i = 0$. Here NM is the total mass of the liquid and $\mathcal{T}_0$ is the kinetic energy in the moving frame. The potential energy of the liquid is unchanged by the reference change so that the total energy in the laboratory frame is:

$$E = E_0 + \frac{1}{2} N M v^2. \tag{10.14}$$

If there is an excitation with energy $\epsilon(\mathbf{p})$ the liquid has net momentum $\mathbf{p}$ in the moving frame and energy $E_0 + \epsilon$. Following the same steps as above, and noting that $\sum \mathbf{p}_i = \mathbf{p}$ we find:

$$E = E_0 + \epsilon + \mathbf{p} \cdot \mathbf{v} + \frac{1}{2} N M v^2. \tag{10.15}$$

Thus the change in the energy associated with making an excitation is $\epsilon + \mathbf{p} \cdot \mathbf{v}$. The criterion for the superflow to slow down is that this change be negative. This first happens if $\mathbf{p}$ and $\mathbf{v}$ are antiparallel and:

$$v = \epsilon/p. \tag{10.16}$$

Thus we have a limit on $\mathbf{v}_s$. The largest velocity that can occur without producing excitations is proportional to the slope of the line in Figure 10.3 which is tangent to the dispersion relation. Note that for a free particle spectrum the tangent slope is zero: a non-interacting Bose gas is not a superfluid even if there is a BEC. In practice the Landau criterion gives too large a critical velocity. In most experiments the excitation that slows the superfluid is a vortex loop attached to a wall.

10.1.4 Weakly interacting Bose gas

We now show that weak interactions can be taken into account for a system of bosons. The result is, as we have anticipated, a phonon branch for the spectrum at long wavelengths. We do not get rotons – they occur in strongly interacting systems. The method, which is due to N. Bogoliubov (1947) is of considerable interest; a similar calculation can be done for superconductors; for an accessible treatment see Landau *et al.* (1980).

The method uses second-quantized notation. Second quantization for bosons is exactly the same as for fermions except that boson wavefunctions are symmetric. The phase factors in Section 6.1.1 are replaced by unity, and the boson operators, which we call $\hat{a}^+, \hat{a}$ have commutation relations rather than anticommutation relations. The formalism looks like that for phonons in Chapter 5. However, the content is different: we are dealing here with boson particles, not collective excitations.

The Hamiltonian which we will use describes a set of spinless bosons with orbitals labeled by $\mathbf{k}$ which interact via a weak repulsive interaction of strength g. We assume that the interaction potential is independent of momentum (a point interaction). In terms of the scattering length, b (cf. Chapter 2) $g = 4\pi\hbar^2 b/M$. With these assumptions we write:

$$\hat{\mathcal{H}} = \sum_{\mathbf{k}} \frac{\hbar^2 k^2}{2M} \hat{a}_{\mathbf{k}}^+ \hat{a}_{\mathbf{k}} + \frac{g}{2\Omega} \sum_{k,K,q} \hat{a}_{k+q}^+ \hat{a}_{K-q}^+ \hat{a}_K \hat{a}_k. \tag{10.17}$$

In fact, the assumption of a point interaction is too crude to allow us to compute the ground-state of this Hamiltonian: it involves divergent sums over $\mathbf{k}$. A suitable cutoff fixes this problem. However, we are only interested in the spectrum of excitations, so we ignore this complication.

Now we build in the idea of a condensate. If there is macroscopic occupation in $\mathbf{k} = 0$ as Einstein proposed, the operators $\hat{a}_0^+, \hat{a}_0$ are in some sense classical quantities. That is $\langle \hat{a}_0^+ \hat{a}_0 \rangle = N_0$ is much bigger than the commutator, $[\hat{a}_0, \hat{a}_0^+] = 1$. Thus, we can replace any operator with $\mathbf{k} = 0$ by the *number* $\sqrt{N_0}$. Further, an operator with $\mathbf{k} \neq 0$ is much "smaller" than $\hat{a}_0^+$ or $\hat{a}_0$. This leads us to classify the terms in Eq. (10.17) according to the number of condensate operators that are present.

The largest term is for $\mathbf{k} = \mathbf{K} = \mathbf{q} = 0$ and gives N_0^2. There is no term with three condensate operators because then momentum conservation means the fourth one is zero too, and already counted. The next terms have two of the momenta equal to zero and are proportional to N_0. We neglect the rest of the terms. The Hamiltonian now is:

$$\hat{\mathcal{H}} = \frac{N_0^2 g}{2\Omega} + \sum_{\mathbf{k}} \epsilon_0(k) \hat{a}_{\mathbf{k}}^+ \hat{a}_{\mathbf{k}} + \frac{N_0 g}{2\Omega} \sum_{\mathbf{k} \neq 0} (\hat{a}_{\mathbf{k}} \hat{a}_{-\mathbf{k}} + \hat{a}_{\mathbf{k}}^+ \hat{a}_{-\mathbf{k}}^+ + 4\hat{a}_{\mathbf{k}}^+ \hat{a}_{\mathbf{k}}). \tag{10.18}$$

There are several things to notice about this form of $\hat{\mathcal{H}}$. First it does not conserve number. We will work in the grand canonical ensemble to fix the average number to be N. Also $\hat{\mathcal{H}}$ is quadratic in the operators, so it can be exactly diagonalized, as we will see.

In the non-interacting gas $N_0 = N$ at $T = 0$. This is no longer true for the interacting gas; the difference (the depletion of the condensate) depends on g. We will prove below that the depletion is proportional to $g^{3/2}$. The ground-state energy, to lowest order in g, is the first term in Eq. (10.18). Thus we can find the leading term in the chemical potential of the interacting system at $T = 0$:

$$\mu = \frac{\partial}{\partial N} \frac{N_0^2 g}{2\Omega} \approx \frac{N g}{\Omega} = gn. \tag{10.19}$$

Here n is the density of the gas. (For an interacting Bose system the chemical potential can be positive.)

In the grand canonical ensemble we replace $\hat{\mathcal{H}}$ by $\hat{\mathcal{K}} = \hat{\mathcal{H}} - \mu \hat{N}$, where the number operator, $\hat{N}$ is given by:

$$\hat{N} = N_0 + \sum_{\mathbf{k} \neq 0} \hat{a}_{\mathbf{k}}^+ \hat{a}_{\mathbf{k}}. \tag{10.20}$$

The expectation value of $\hat{\mathcal{K}}$ is the grand potential of thermodynamics from which all ground-state properties can be derived. We have:

$$
\hat{\mathcal{K}} = \frac{N_0^2\, g}{2\Omega} - \mu N_0 + \sum_{\mathbf{k} \neq 0}(\epsilon_0(k) - \mu)\hat{a}_{\mathbf{k}}^+\hat{a}_{\mathbf{k}}
$$

$$
+ \frac{gn}{2}\sum_{\mathbf{k} \neq 0}(\hat{a}_{\mathbf{k}}\hat{a}_{-\mathbf{k}} + \hat{a}_{\mathbf{k}}^+\hat{a}_{-\mathbf{k}}^+ + 4\hat{a}_{\mathbf{k}}^+\hat{a}_{\mathbf{k}})
$$

$$
= \frac{N_0^2\, g}{2\Omega} - \mu N_0 + \sum_{\mathbf{k} \neq 0}(\epsilon_0(k) + gn)\hat{a}_{\mathbf{k}}^+\hat{a}_{\mathbf{k}}
$$

$$
+ \frac{gn}{2}\sum_{\mathbf{k} \neq 0}(\hat{a}_{\mathbf{k}}\hat{a}_{-\mathbf{k}} + \hat{a}_{\mathbf{k}}^+\hat{a}_{-\mathbf{k}}^+). \tag{10.21}
$$

We will abbreviate $(\epsilon_0(k) + gn)$ by $\zeta_{\mathbf{k}}$ in what follows. Our goal is to eliminate the "dangerous" (i.e. off-diagonal) terms, $\hat{a}_{\mathbf{k}}\hat{a}_{-\mathbf{k}} + \hat{a}_{\mathbf{k}}^+\hat{a}_{-\mathbf{k}}^+$. We do this by a canonical transformation:

$$
\hat{a}_{\mathbf{k}} = u_{\mathbf{k}}\hat{b}_{\mathbf{k}} - v_{\mathbf{k}}\hat{b}_{-\mathbf{k}}^+ \tag{10.22}
$$

$$
\hat{a}_{-\mathbf{k}} = u_{\mathbf{k}}\hat{b}_{-\mathbf{k}} - v_{\mathbf{k}}\hat{b}_{\mathbf{k}}^+. \tag{10.23}
$$

The unknown functions u, v are even in $\mathbf{k}$. It is easy to see that the new operators $\hat{b}^+, \hat{b}$ have the correct commutation relations if we take $u_{\mathbf{k}}^2 - v_{\mathbf{k}}^2 = 1$, or, equivalently:

$$
u_{\mathbf{k}} = \cosh(\theta_{\mathbf{k}}); \quad v_{\mathbf{k}} = \sinh(\theta_{\mathbf{k}}). \tag{10.24}
$$

It is straightforward, though a bit tedious, to express $\hat{\mathcal{K}}$ in terms of the new operators. The result is:

$$
\hat{\mathcal{K}} = C + \frac{1}{2}\sum_{\mathbf{k} \neq 0}(\zeta_{\mathbf{k}}\cosh(2\theta_{\mathbf{k}}) - gn\sinh(2\theta_{\mathbf{k}}))\hat{b}_{\mathbf{k}}^+\hat{b}_{\mathbf{k}}
$$

$$
+ \frac{1}{2}\sum_{\mathbf{k} \neq 0}(gn\cosh(2\theta_{\mathbf{k}}) - \zeta_{\mathbf{k}}\sinh(2\theta_{\mathbf{k}}))
$$

$$
\times (\hat{b}_{\mathbf{k}}\hat{b}_{-\mathbf{k}} + \hat{b}_{\mathbf{k}}^+\hat{b}_{-\mathbf{k}}^+). \tag{10.25}
$$

The first term is the grand potential in the ground-state, $E_0 - \mu N$.

The dangerous terms in the last line are eliminated if we put:

$$
\tanh(2\theta_{\mathbf{k}}) = \frac{gn}{\zeta_{\mathbf{k}}} = \frac{gn}{\epsilon_0(k) + gn}. \tag{10.26}
$$

Now substituting into the first term, and using some identities for hyperbolic functions gives:

$$\hat{\mathcal{K}} = C + \sum_{\mathbf{k}} E_{\mathbf{k}} \hat{b}_{\mathbf{k}}^{+} \hat{b}_{\mathbf{k}}$$

$$E_{\mathbf{k}} = \sqrt{\zeta_{\mathbf{k}}^2 - (gn)^2} = \sqrt{\epsilon_{\circ}(k)^2 + 2gn\,\epsilon_{\circ}(k)}. \tag{10.27}$$

We have achieved our goal. The operator $\hat{\mathcal{K}}$ is in the form of the sum of independent boson excitations whose energy is $E_{\mathbf{k}}$. Further, at long wavelengths only the second term in the square root is important:

$$E_{\mathbf{k}} \to k \sqrt{gn\,\hbar^2/M}. \tag{10.28}$$

This is a phonon dispersion relation, and the Landau criterion says that this system can be superfluid.

We now calculate the depletion of the condensate by using Eq. (10.20) and Eq. (10.23). We take the expectation value in the ground-state of the $\hat{b}$ operators. A bit of algebra shows:

$$N = \langle \hat{N} \rangle = N_{\circ} + \sum_{\mathbf{k} \neq 0} v_{\mathbf{k}}^2 = N_{\circ} + \frac{1}{2} \sum_{\mathbf{k} \neq 0} \left(\frac{\epsilon_{\circ}(k) + gn}{E_{\mathbf{k}}} - 1 \right). \tag{10.29}$$

Converting the sum to a somewhat nasty integral gives the result we need, after considerable algebra; see (Fetter & Walecka 1971). It is usually expressed in terms of the scattering length mentioned above:

$$\frac{N - N_{\circ}}{N} = \frac{8}{3} \left(\frac{nb^3}{\pi} \right)^{1/2}; \quad b = Mg/4\pi\hbar^2. \tag{10.30}$$

10.2 Helium-3

London based his confidence that the superfluidity of He-4 was a result of the BEC on the fact that liquid He-3, in which the atoms obey Fermi statistics, did not show superfluidity. In fact, He-3 is a superfluid at much lower temperatures. This is a phenomenon very like superconductivity in metals which is the subject of the next section, though more complicated. It works because at sufficiently low temperatures, pairs of He-3 atoms bind together to form the Cooper pairs that we will see below.

The experiments that discovered the effect were Osheroff, Richardson & Lee (1972) and Osheroff, Gully, Richardson & Lee (1972). An informal and accessible account of the phenomena that lead to the very complicated phases of the superfluid, by one of the major theoretical contributors is Leggett (2004). There are many references to the literature in this article.

10.3 Superconductivity

We now turn to superconductivity. The experiment of Kammerlingh Onnes was recognized as important and puzzling, but, at first, very little progress was made in understanding it. The next important advance was 22 years after the original discovery in the work of W. Meissner and R. Ochsenfeld.

For superconductors the supercurrent is charged, and gives rise to magnetic fields. We should recall the classical effect of eddy currents: if we try to apply a magnetic field to a conductor, the changing flux induces an emf and currents that cancel the applied field. Very good conductors freeze out (exclude) magnetic fields. Superconductors do this. However, very good conductors also freeze *in* fields that are already there. For example, magnetic field lines are frozen into the Sun's atmosphere. Superconductors do not do this, but something else which is quite startling.

10.3.1 Meissner effect

Meissner & Ochsenfeld (1933) showed that a superconductor is more than merely a perfect electrical conductor. Instead of trapping flux it *expels* it. This effect shows that the supercurrent is intrinsically quantum mechanical in nature.

Meissner and Ochsenfeld put two parallel cylinders of superconductor (they did the experiment with both lead and tin) in a weak external field perpendicular to their axes. Above T_c the field lines pass through the material because the metals are very weakly magnetic. Then the metals were cooled below T_c. The field between the cylinders increased; see Figure 10.4. Calculations for this geometry showed that the amount of increase corresponded to the field lines being *totally excluded* from the metals. The metals acted as a *perfect diamagnet* so that $\mathbf{B} = 0$ inside them. We now know (see below) that the field penetrates a very short distance into the conductors; this is the region where the eddy currents flow.

A separate experiment showed that flux can be trapped inside a lead tube. The field was applied for $T > T_c$, the temperature lowered, and then the external field was turned off. The field inside remained because current continued to flow in the metal. This is an example of a persistent current.

There are several things to notice about flux exclusion. One is that the diamagnetic moment of a superconductor must be very large.

$$\mathbf{B} = \mathbf{H} + 4\pi\mathbf{M} = 0; \quad \chi_D = M/H = -1/4\pi. \tag{10.31}$$

This is orders of magnitude larger than the moment of any normal diamagnetic metal.

Since the magnetic field inside a superconductor is independent of history (i.e. flux cannot be frozen in) we can apply thermodynamics to the system with $\mathbf{B}$ as a variable. If we do work on a long cylindrical superconductor, for example by moving it in an inhomogeneous,

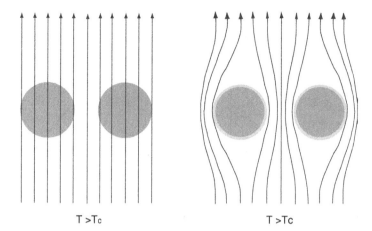

T >Tc   T >Tc

Fig. 10.4 The Meissner–Ochsenfeld experiment. Two cylinders of lead or tin are shown end-on. When the temperature is reduced below T_c the flux is excluded from the metal, and thus increased between the cylinders. This increase is what Meissner and Ochsenfeld measured.

external **B** field parallel to its axis, the work done is

$$dW = -\Omega \int \mathbf{M} \cdot d\mathbf{B}.$$

However, in this case **H** inside the metal is the same as **B** outside because tangential components of **H** are continuous. Also $dF = -SdT + dW$. Using Eq. (10.31) we find:

$$F(H) - F(0) = -\Omega \int \mathbf{M} \cdot d\mathbf{H} = \frac{\Omega}{4\pi} \int H dH = \Omega \frac{H^2}{8\pi}. \qquad (10.32)$$

Excluding the field costs free energy $H^2/8\pi$ per unit volume.

From this we can expect, and it is observed, that an external magnetic field *destroys superconductivity* because it raises the free energy of the superconducting state. Indeed, there is a characteristic magnetic field $H_c(T)$ which will drive a superconductor normal. From Eq. (10.32) we see that:

$$\frac{H_c^2(T)}{8\pi} = \frac{F_n - F_s}{\Omega}, \qquad (10.33)$$

so that the critical field directly measures the condensation free energy of the superconductor. Empirically the critical field is well represented by:

$$\frac{H_c(T)}{H_c(0)} = 1 - \left(\frac{T}{T_c}\right)^2. \qquad (10.34)$$

This gives the phase diagram of a superconductor, as shown in Figure 10.5.

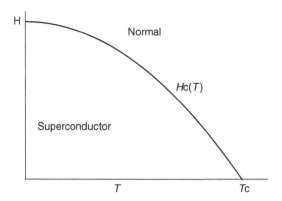

Fig. 10.5 **The phase diagram of a superconductor.**

10.3.2 London Equations

F. London and H. London (London & London 1935) gave a description of the interaction of a superconductor with electric and magnetic fields which included the Meissner effect. First, a superconductor is a perfect conductor. We need something other than Ohm's law in this case. We expect that electric fields will accelerate electrons since there is no scattering. Suppose, in the spirit of the two-fluid model, that there is a fraction of the electrons with density n_s which carry the supercurrent. We can assume that at $T = 0$ all of the electrons are superfluid so that $n_s(0) = n$. Then, as in Drude theory:

$$m\frac{d\mathbf{v}}{dt} = e\mathbf{E}; \quad \frac{d\mathbf{j}}{dt} = \frac{n_s e^2}{m}\mathbf{E}. \tag{10.35}$$

Thus:

$$\mathbf{E} = \frac{\partial}{\partial t}\left(\frac{4\pi\lambda_L^2}{c^2}\mathbf{j}\right); \quad \lambda_L^2 = \frac{c^2 m}{4\pi n_s e^2}. \tag{10.36}$$

This equation defines the *London penetration depth*, λ_L. At the moment we only know its value at $T = 0$:

$$\lambda_L^2(T = 0) = c^2 m/4\pi n e^2, \tag{10.37}$$

which turns out to of the order of 1000 Å. The significance of this characteristic length will be explained shortly.

Now we use Maxwell's equations to deal with the magnetic field. We want to describe the supercurrents that run along the surface of the metal and produce a magnetic field that cancels the external field. So far we have dealt with the macroscopic (Maxwell) fields $\mathbf{B}$ and $\mathbf{H}$ which are defined to be averages over many atomic spacings. However, here we want to see the way in which the field is screened very near the surface. We have an unusual situation because the field associated with the supercurrents varies on scales of $\lambda_L(0)$. This is long on an atomic scale, but not really macroscopic either; we may call it mesoscopic. We will follow Tinkham (1996) by introducing a special symbol, **h**, for these mesoscopic fields.

The relevant Maxwell equations are:

$$\nabla \times \mathbf{E} = -\frac{1}{c}\frac{\partial \mathbf{h}}{\partial t}; \quad \nabla \times \mathbf{h} = \frac{4\pi \mathbf{j}}{c}. \tag{10.38}$$

Combining these with Eq. (10.35) gives:

$$\frac{\partial}{\partial t}\left(\nabla \times \left[\frac{4\pi\lambda_L^2}{c^2}\mathbf{j}\right] + \frac{\mathbf{h}}{c}\right) = 0. \tag{10.39}$$

There are two types of solution to this equation. First, we can have **j** and **h** independent of time, and $\nabla \times \mathbf{h} = 4\pi\mathbf{j}/c$. This is a static solution and represents frozen-in flux as in a merely perfect conductor. However, we can also have the sum of the two terms give zero, so that:

$$\nabla \times \left[\frac{4\pi\lambda_L^2}{c^2}\mathbf{j}\right] + \frac{\mathbf{h}}{c} = 0. \tag{10.40}$$

The Londons proposed that this is the proper description of the Meissner effect.

We can see how this works by eliminating **j** from Eq. (10.40) using the second equation of 10.38. The result is:

$$\nabla \times (\nabla \times \mathbf{h}) + \mathbf{h}/\lambda_L^2 = 0. \tag{10.41}$$

Using the vector identity $\nabla \times \nabla \times \mathbf{h} = -\nabla^2\mathbf{h} + \nabla(\nabla \cdot \mathbf{h})$ and $\nabla \cdot \mathbf{h} = 0$ we find:

$$\nabla^2\mathbf{h} = \mathbf{h}/\lambda_L^2. \tag{10.42}$$

Now consider the situation shown in Figure 10.6. The field **h** is everywhere in the same direction, and its magnitude satisfies Eq. (10.42). Its value at the surface is the same as B outside. Therefore we find:

$$h(x) = Be^{-x/\lambda_L}. \tag{10.43}$$

That is, λ_L gives the depth of penetration of the field into the superconductor. For depths large compared to λ_L the field and the supercurrent are essentially zero.

F. London made a very significant step towards understanding the London equations and the basis of superconductivity in 1950 (London 1961). He assumed that the supercurrent is a quantum effect, and looked at the quantum mechanical expression for the current. First we note that for no external field the current vanishes. Thus

$$\langle \hat{\mathbf{j}} \rangle = \frac{e\hbar}{2mi}\langle \Psi^*\nabla\Psi - \Psi\nabla\Psi^* \rangle = 0. \tag{10.44}$$

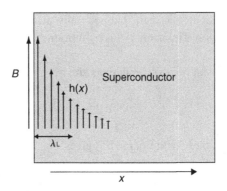

Fig. 10.6 **A superconductor with a tangential magnetic field. Outside $\mathbf{B} = \mathbf{H}$. The tangential component of $\mathbf{H}$ is continuous so $\mathbf{B} = \mathbf{h}(0)$.**

Now turn on the field. There is another term, the diamagnetic current, proportional to the vector potential, $\mathbf{A}$.

$$\langle \hat{\mathbf{j}} \rangle = \frac{e\hbar}{2mi}\langle \Psi^* \nabla \Psi - \Psi \nabla \Psi^* \rangle - \frac{e^2}{mc}\mathbf{A}\langle |\Psi|^2 \rangle. \tag{10.45}$$

London said that Ψ is in some sense "rigid" so that the first term (the paramagnetic current) is still zero. Thus we can take:

$$\mathbf{j} = -\frac{n_{\mathrm{s}}e^2}{mc}\mathbf{A} = -\frac{c}{4\pi\lambda_{\mathrm{L}}^2}\mathbf{A}. \tag{10.46}$$

Now n_{s} has the interpretation of the normalization of the wavefunction. The curl of this equation is Eq. (10.40). The notion of wavefunction rigidity is now viewed as an inspired guess. We will see its interpretation in microscopic theory below.

We need to be careful about the vector potential here because Eq. (10.46) is not gauge-invariant. The equation is only valid in the *London gauge* for which $\mathbf{A}$ and $\mathbf{j}$ have the same boundary conditions and satisfy $\nabla \cdot \mathbf{A} = 0$ (incompressible flow). In any other gauge, the wavefunction is not rigid, but acquires a phase in response to the magnetic field.

The London equations make a definite prediction for the low-temperature penetration depth, Eq. (10.36), if we assume that all the electrons are superconducting at $T = 0$. The penetration depth can be measured; it turns out that the London prediction is qualitatively correct, but very often too small, particularly in pure superconductors.

A. B. Pippard showed how to describe penetration depths correctly by introducing another characteristic length (Pippard 1950). He reasoned by analogy to normal metals with long mean free paths for which the effects of external fields can be non-local. In this case fields at one point give rise to currents at another point less than a mean free path away; cf. Problem 6 of Chapter 7. Pippard took the quantum nature of the effect into account by suggesting that the new length, the coherence length, ξ_o, was related to the binding energy of the superconductor. He supposed that electrons near the Fermi surface were affected by the binding. The range of the effect is estimated by putting:

$$\frac{1}{2m}(p_F + \delta p)^2 = kT_c + E_F. \tag{10.47}$$

This gives $\delta p = kT_c/v_F$. Then using the uncertainty principle we have:

$$\xi_o = \frac{\hbar}{\delta p} = a\frac{\hbar v_F}{kT_c}. \tag{10.48}$$

The factor a is a number of order unity. We will soon see that ξ_0 is the size of the fundamental constituent of the superconducting state, the Cooper pair. It is of order 1000–10,000 Å.

Now if this is the range of non-locality we can make a generalization of Eq. (10.46):

$$\mathbf{j}(\mathbf{r}) = -\frac{c}{4\pi\lambda_L^2}\int d\mathbf{s}\, K(\mathbf{r} - \mathbf{s})\mathbf{A}(\mathbf{s}). \tag{10.49}$$

The kernel, K, is a function whose range is ξ_0 and whose integral is unity. Its exact form is predicted by the microscopic theory below, but is not important for our purposes here.

If $\mathbf{A}$ varies on a scale large compared to ξ_0, it can be taken out of the integral and we recover Eq. (10.46). A material of this type, with $\lambda_L \gg \xi_0$ is called a London superconductor. In the opposite limit, $\lambda_L \ll \xi_0$, the penetration depth will be different from λ_L; call it λ_e. In this case we can estimate the integral for the geometry of Figure 10.6 by noting that $\mathbf{A}$ is non-zero only in a layer of thickness λ_e, and thus for $\mathbf{r}$ near the surface the integral is of order $\lambda_e\mathbf{A}(\mathbf{r})/\xi_0$. We are now back to a local equation and the coefficient of $\mathbf{A}$ gives the actual penetration depth. From Eq. (10.49) we find:

$$\mathbf{j}(\mathbf{r}) = -\frac{c}{4\pi\lambda_e^2}\mathbf{A}(\mathbf{r})$$

$$\lambda_e^2 = \lambda_L^2\frac{\xi_0}{\lambda_e}. \tag{10.50}$$

Solving for λ_e gives $(\lambda_L^2\xi_0)^{1/3}$. This works quite well for tin and aluminum if the parameter a in Eq. (10.48) is taken to be 0.15. The penetration depth is quite a lot larger than λ_L.

Another way to increase penetration depths is to make the superconductor "dirty," that is, introduce non-magnetic impurities. Pippard showed that the data could be understood by replacing ξ_0 by a new length:

$$\frac{1}{\xi_e} = \frac{1}{\xi_0} + \frac{1}{\mathcal{L}},$$

where $\mathcal{L}$ is the mean free path in the normal state. We will not go into detail here, but only remark that dirty superconductors have large penetration depths.

10.4 Microscopic theory

The microscopic explanation of superconductivity was given in 1957 by J. Bardeen, L. Cooper, and R. Schrieffer (BCS) (Bardeen, Cooper & Schrieffer 1957). There were clues to the mechanism in two significant experiments. First, Fröhlich (1950) pointed out that the mechanism probably involved the electron-phonon coupling because of the *isotope effect*. This is the observation that if we make a superconductor from different isotopes of the same element the transition temperature shifts. In fact, $T_c \propto M^{-1/2} \propto \Theta$, where M is the mass of the isotope, and Θ is the Debye temperature. However, attempts to use this insight by Fröhlich, Bardeen, and others failed because $T_c \ll \Theta$, so that there is a subtle mechanism at work. Still, this indicates that the Bardeen–Pines interaction, Eq. (9.90), is somehow involved.

Second, experiments showed that there was a gap in the spectrum of superconductors, almost like the band gap of an insulator. For example, the heat capacity is not linear in T, as in a normal metal, but is activated, $C_V \propto e^{-1.5T_c/T}$ indicating an energy gap of order $k_B T_c$. And, infrared absorption experiments showed no absorption below a gap of order $3.5kT_c$. (The factor of around 2 arises from the fact that when we excite optically we make a pair of excitations – in the case of an insulator, an electron and a hole).

10.4.1 Cooper pairs

An important step was the work of Cooper (1956) who considered a simple problem: suppose we have a weak attractive interaction between electrons in a Fermi sea. The strength of the interaction is $v_{eff} \ll E_F$. If v_{eff} is large, we could expect to bind electrons into pairs. Then we would have a gas of bosons, and presumably a BEC. This explanation will not work for conventional superconductors however because the attractive part of the potential of Eq. (9.90) is very weak, too weak to expect a bound state for an isolated pair of electrons. What Cooper showed is remarkable: in the presence of the Fermi sea, any attractive potential will bind.

Now consider two electrons whose wavefunctions have momentum components $k > k_F$ which will make up our pair. The states for $k < k_F$ are occupied and will not participate. In order to be as similar as possible to the condensate in BEC we take the center-of-mass momentum of the pair to be zero so that we combine states with $\mathbf{k}$ and $-\mathbf{k}$. The most general wavefunction will be:

$$\psi(\mathbf{r}_1, \mathbf{r}_2) = \sum_{k > k_F} g(\mathbf{k}) e^{i\mathbf{k} \cdot \mathbf{s}}; \quad \mathbf{s} = \mathbf{r}_1 - \mathbf{r}_2. \tag{10.51}$$

We will assume that we have a spin singlet state so that the spatial wavefunction, g, is symmetric. The Schrödinger equation now reads:

$$\sum_{k > k_F} [2\epsilon_o(\mathbf{k}) + v_{eff}(\mathbf{r}_1, \mathbf{r}_2) - E]g(\mathbf{k}) e^{i\mathbf{k} \cdot \mathbf{s}} = 0. \tag{10.52}$$

Now apply the operator $\int d\mathbf{r}_1 \, d\mathbf{r}_2 \exp(-i\mathbf{K} \cdot \mathbf{s})$ to this equation. We find:

$$[2\epsilon_\circ(\mathbf{K}) - E]g(\mathbf{K}) + \sum_{k < k_F} V(\mathbf{K}, \mathbf{k})g(\mathbf{k}) = 0,$$

$$V(\mathbf{K}, \mathbf{k}) = \frac{1}{\Omega^2} \int d\mathbf{r}_1 \, d\mathbf{r}_2 e^{i(\mathbf{k}-\mathbf{K})\cdot\mathbf{s}} v_{\text{eff}}(\mathbf{r}_1, \mathbf{r}_2). \qquad (10.53)$$

The actual form of the Bardeen–Pines interaction is complicated, and this equation is hard to solve in general. The most important effect is that the potential is attractive if both $\mathbf{k}$ and $\mathbf{K}$ are close to the Fermi surface. Cooper took a simple approximation which has the correct overall behavior:

$$V(\mathbf{K}, \mathbf{k}) = -V/N, \quad \text{if} \quad |\epsilon_\circ(\mathbf{K}) - E_F| < \hbar\omega_D$$

$$\text{and } |\epsilon_\circ(\mathbf{k}) - E_F| < \hbar\omega_D$$

$$= 0 \qquad \text{otherwise.} \qquad (10.54)$$

Here V measures the overall strength of the potential. Now suppose that $|\epsilon_\circ(\mathbf{K}) - E_F| < \hbar\omega_D$. Then we have:

$$g(\mathbf{K}) = \frac{V}{2\epsilon_\circ(\mathbf{K}) - E} \frac{1}{N} \hat{\sum} g(\mathbf{k}). \qquad (10.55)$$

The notation $\hat{\sum}$ means a sum over the region $E_F < \epsilon_\circ(\mathbf{k}) < E_F + \hbar\omega_D$. This equation is solved by summing both sides over the region near the Fermi surface:

$$\frac{1}{V} = \frac{1}{N} \hat{\sum} \frac{1}{2\epsilon_\circ(\mathbf{k}) - E} = \frac{1}{N} \int_{E_F}^{E_F + \hbar\omega_D} d\epsilon \, \mathcal{D}(\epsilon) \frac{1}{2\epsilon - E}. \qquad (10.56)$$

We can replace $\mathcal{D}(\epsilon)/N$ with $\mathcal{D}(E_F)/2N \equiv D_\circ$ and do the integral. (Note: we must take the density of states for one spin since there is no spin sum. The symbol $D_\circ$ means the density of states per particle for one spin.) Solving for E gives:

$$E = 2E_F - \frac{2\hbar\omega_D}{e^{2/D_\circ V} - 1} \approx 2E_F - 2\hbar\omega_D \, e^{-2/D_\circ V}. \qquad (10.57)$$

The last approximation is for weak coupling. There is always a bound state with respect to the minimum energy of the pair of electrons. The binding energy is $E_B = 2\hbar\omega_D \, e^{-2/D_\circ V}$. This energy is not analytic in the strength of the potential and cannot be calculated in perturbation theory.

Now we can construct the wavefunction from Eq. (10.55):

$$g(\mathbf{k}) = C \frac{V}{2\epsilon_\circ(\mathbf{K}) - E}; \quad \psi = C \hat{\sum} \frac{\cos(\mathbf{k} \cdot \mathbf{s})}{2\epsilon_\circ(\mathbf{k}) - 2E_F + E_B}. \qquad (10.58)$$

Here C is a constant. From this we can estimate the size of the wavefunction by finding the $\mathbf{k}$ necessary to double the denominator. This gives Pippard's result:

$$2\left(\frac{\hbar^2(k_F + \delta k)^2}{2m} - E_F\right) \approx E_B; \quad \frac{1}{\delta k} \approx \xi_\circ = \frac{\hbar v_F}{E_B}. \qquad (10.59)$$

The Cooper calculation signals an instability of the ground-state. It is not a theory of superconductivity because if one pair forms, then many will.

10.4.2 BCS theory

In their celebrated paper Bardeen *et al.* (1957) formulated a theory with many pairs. Since the pairs are large and overlap, we cannot neglect the exchange interaction between the electrons in the pairs. This is taken into account in a kind of Hartree–Fock theory for pairs.

We define an operator which creates a pair from electrons with $\mathbf{k}, -\mathbf{k}$ and opposite spins.

$$\hat{P}_{\mathbf{k}}^+ = \hat{c}_{\mathbf{k}\uparrow}^+ \hat{c}_{-\mathbf{k}\downarrow}^+. \tag{10.60}$$

This is not a boson operator except in the case of widely separated pairs, as we will see in the problems. We are focussing on states with pairing. To make this more straightforward we pick out of the Hamiltonian the terms that involve only the pair operators. Thus we make the truncation of the general form for electron-electron interactions:

$$\sum_{\mathbf{k},\mathbf{K},\mathbf{q},s,s'} v(\mathbf{k},\mathbf{K},\mathbf{q},s,s')\hat{c}_{\mathbf{k}+\mathbf{q},s}^+ \hat{c}_{\mathbf{K}-\mathbf{q},s'}^+ \hat{c}_{\mathbf{K},s'}\hat{c}_{\mathbf{k},s} \Rightarrow \sum_{\mathbf{k},\mathbf{q}} V(\mathbf{k},\mathbf{q})\hat{P}_{\mathbf{k}}^+\hat{P}_{\mathbf{q}}. \tag{10.61}$$

With this approximation we form the operator $\hat{\mathcal{K}}$, which is referred to as the pairing Hamiltonian. It is convenient to measure energies with respect to the Fermi energy, so we will call $\epsilon_\circ(\mathbf{k}) - \mu = \xi_{\mathbf{k}}$ as in Eq. (10.21).

$$\hat{\mathcal{K}} = \sum_{\mathbf{k},s} \xi_{\mathbf{k}}\hat{c}_{\mathbf{k}s}^+\hat{c}_{\mathbf{k}s} + \sum_{\mathbf{k},\mathbf{q}} V(\mathbf{k},\mathbf{q})\hat{P}_{\mathbf{k}}^+\hat{P}_{\mathbf{q}}. \tag{10.62}$$

Ground-state

The strategy of BCS was to propose a variational form for the ground-state wavefunction. Minimizing the expectation value of $\hat{\mathcal{K}}$ gives the best approximation to the actual ground-state. Their form is:

$$|\psi_\circ\rangle = \prod_{\mathbf{k}} (u_{\mathbf{k}} + v_{\mathbf{k}}\hat{P}_{\mathbf{k}}^+)|0\rangle. \tag{10.63}$$

Here $u_{\mathbf{k}}, v_{\mathbf{k}}$ are variational functions to be determined. The wavefunction is normalized if $|u_{\mathbf{k}}|^2 + |v_{\mathbf{k}}|^2 = 1$. Note that if

$$v_{\mathbf{k}} = 1, \ k < k_{\mathrm{F}}; \quad v_{\mathbf{k}} = 0, \ k > k_{\mathrm{F}}$$

we have the ordinary Hartree–Fock ground-state used in Eq. (9.40). The BCS state allows there to be a quantum mechanical amplitude for partial occupancy of a given states with a pair. As in the case of Hartree–Fock for atoms, each state sees an average exchange interaction produced by the other pairs.

Once more, as in the BEC case, we are in a situation where number is not conserved. The BCS ground-state is a mixture of states with different numbers of electrons, and the average number is fixed by the chemical potential. However, the fluctuations are small, as we will see in the problems.

It is a simple exercise to write down the expectation value of $\hat{\mathcal{K}}$:

$$\langle \psi_0 | \hat{\mathcal{K}} | \psi_0 \rangle = \sum_{\mathbf{k}} 2\xi_k |v_{\mathbf{k}}|^2 + \sum_{\mathbf{k},\mathbf{q}} V(\mathbf{k},\mathbf{q}) u_{\mathbf{k}} v_{\mathbf{k}}^* u_{\mathbf{q}}^* v_{\mathbf{q}}. \qquad (10.64)$$

For the ground-state we can take u, v to be real. (In current carrying states they acquire a relative phase.). Further, we can build in the normalization condition by setting:

$$u_{\mathbf{k}} = \sin(\theta_{\mathbf{k}}); \quad v_{\mathbf{k}} = \cos(\theta_{\mathbf{k}}).$$

Therefore the quantity to be minimized is:

$$\langle \psi_0 | \hat{\mathcal{K}} | \psi_0 \rangle = \sum_{\mathbf{k}} \xi_k (1 + \cos(2\theta_{\mathbf{k}})) + \frac{1}{4} \sum_{\mathbf{k},\mathbf{q}} V(\mathbf{k},\mathbf{q}) \sin(2\theta_{\mathbf{k}}) \sin(2\theta_{\mathbf{q}}). \qquad (10.65)$$

Now we minimize with respect to $\theta_{\mathbf{k}}$.

$$\frac{\partial \langle \hat{\mathcal{K}} \rangle}{\partial \theta_{\mathbf{k}}} = -2\xi_{\mathbf{k}} + \sum_{\mathbf{q}} V(\mathbf{k},\mathbf{q}) \cos(2\theta_{\mathbf{k}}) \sin(2\theta_{\mathbf{q}}) = 0,$$

$$\tan(2\theta_{\mathbf{k}}) = \frac{\sum_{\mathbf{q}} V(\mathbf{k},\mathbf{q}) \sin(2\theta_{\mathbf{q}})}{2\xi_{\mathbf{k}}}. \qquad (10.66)$$

We define a quantity $\Delta_{\mathbf{k}}$ which is known as the *gap*; it turns out to be the gap in the excitation spectrum, as we will see shortly.

$$\Delta_{\mathbf{k}} \equiv -\frac{1}{2} \sum_{\mathbf{q}} V(\mathbf{k},\mathbf{q}) \sin(2\theta_{\mathbf{q}}) = -\sum_{\mathbf{q}} V(\mathbf{k},\mathbf{q}) u_{\mathbf{q}} v_{\mathbf{q}}$$

$$\tan(2\theta_{\mathbf{k}}) = -\frac{\Delta_{\mathbf{k}}}{\xi_{\mathbf{k}}}. \qquad (10.67)$$

The gap measures the overlap of states that have pairs and those which don't. We can manipulate the last equation to find:

$$\sin(2\theta_{\mathbf{k}}) = \Delta_{\mathbf{k}} / \sqrt{\xi_{\mathbf{k}}^2 + \Delta_{\mathbf{k}}^2}$$

and put this into the first line of Eq. (10.67). The result called the BCS gap equation:

$$\Delta_{\mathbf{k}} = -\frac{1}{2} \sum_{\mathbf{q}} \frac{\Delta_{\mathbf{q}}}{E_{\mathbf{q}}} V(\mathbf{k},\mathbf{q}); \quad E_{\mathbf{q}} = \sqrt{\xi_{\mathbf{q}}^2 + \Delta_{\mathbf{q}}^2}. \qquad (10.68)$$

The new function, $E_{\mathbf{k}}$, turns out to be the excitation spectrum of the interacting system as we will see below. A little more manipulation gives:

$$u_{\mathbf{k}}^2 = \frac{1}{2}(1 + \xi_{\mathbf{k}}/E_{\mathbf{k}}); \quad v_{\mathbf{k}}^2 = \frac{1}{2}(1 - \xi_{\mathbf{k}}/E_{\mathbf{k}}). \tag{10.69}$$

We can solve the gap equation explicitly if we use the potential of Eq. (10.54). In that case Δ is a constant near the Fermi surface and zero otherwise:

$$\Delta_{\mathbf{k}} = \Delta \quad |\xi_{\mathbf{k}}| \leq \hbar\omega_{\mathrm{D}};$$
$$= 0 \quad |\xi_{\mathbf{k}}| > \hbar\omega_{\mathrm{D}}. \tag{10.70}$$

The solution is:

$$1 = \frac{V}{N} \hat{\sum} \frac{1}{2E_{\mathbf{q}}} = D_0 V \int_0^{\hbar\omega_{\mathrm{D}}} \frac{dx}{\sqrt{\Delta^2 + x^2}} = D_0 V \sinh^{-1}\left(\frac{\hbar\omega_{\mathrm{D}}}{\Delta}\right)$$
$$\Delta = \frac{\hbar\omega_{\mathrm{D}}}{\sinh(1/D_0 V)} \approx 2\hbar\omega_{\mathrm{D}} e^{-1/D_0 V}. \tag{10.71}$$

We can now use this solution to find the value of $\langle \hat{\mathcal{K}} \rangle$:

$$\langle \psi_0 | \hat{\mathcal{K}} | \psi_0 \rangle = \sum_{\mathbf{k}} (\xi_{\mathbf{k}} - \xi_{\mathbf{k}}^2/E_{\mathbf{k}}) - \Delta^2/V. \tag{10.72}$$

Now we can find a physically meaningful result. The difference between this expression and its value for $V = 0$ is the difference between the ground-state energy of the superconductor and that of the normal metal, i.e., what we have called the condensation energy, Eq. (10.32). For the normal metal we sum the kinetic energy as usual. Thus:

$$E_{\mathrm{s}} - E_{\mathrm{n}} = \sum_{\mathbf{k}} (\xi_{\mathbf{k}} - \xi_{\mathbf{k}}^2/E_{\mathbf{k}}) - \frac{\Delta^2}{V} - 2 \sum_{k < k_{\mathrm{F}}} \xi_{\mathbf{k}}$$
$$\approx 2 \sum_{k > k_{\mathrm{F}}} (\xi_{\mathbf{k}} - \xi_{\mathbf{k}}^2/E_{\mathbf{k}}) - \frac{\Delta^2}{V}$$
$$= -\frac{N}{2} D_0 \Delta^2. \tag{10.73}$$

The second line results from the fact that near $\xi = \epsilon - E_{\mathrm{F}} = 0$ we can take ξ to be linear in $\mathbf{k}$. The last step results from converting the sum to an integral with the density of states, as above. As we have seen above this is a prediction for $H_{\mathrm{c}}(T = 0)$. For example, H_{c} should have the same isotopic dependence as ω_{D}, namely $H_{\mathrm{c}} \sim M^{-1/2}$, which is observed.

Excitations

In the BCS paper a somewhat complicated method was used to construct the excitation spectrum above $|\psi_0\rangle$, and many ground-state and finite-temperature properties were calculated; all of them agreed remarkably well with experiment. Here we will construct the excited states in a different way following the work of N. Bogoliubov (Bogoliubov 1958) (which was in the same spirit as his calculation for the BEC case). A similar method was proposed by J. Valatin (Valatin 1958).

We return to the pairing Hamiltonian and proceed as we did in Section 10.1.4. The idea, again, is that the pair occupancy is in some sense large, and nearly classical. We put:

$$\hat{P}_{\mathbf{k}}^{+} = b_{\mathbf{k}} + (\hat{c}_{\mathbf{k}\uparrow}^{+}\hat{c}_{-\mathbf{k}\downarrow}^{+} - b_{\mathbf{k}}) = b_{\mathbf{k}} + \hat{\delta}, \tag{10.74}$$

and similarly for $\hat{P}_{\mathbf{k}}$. We assume that $\hat{\delta}$ is small. We will eventually find the number $b_{\mathbf{k}}$ by demanding that $\langle \hat{P}_{\mathbf{k}}^{+} \rangle = b_{\mathbf{k}}$. Now we can put this into Eq. (10.62) and keep terms that are first order in $\hat{\delta}$. A brief calculation gives:

$$\hat{\mathcal{K}} = \sum_{\mathbf{k},s} \xi_{\mathbf{k}} \hat{c}_{\mathbf{k}s}^{+}\hat{c}_{\mathbf{k}s} + \sum_{\mathbf{k},\mathbf{q}} V(\mathbf{k},\mathbf{q})(b_{\mathbf{q}}\hat{c}_{\mathbf{k}\uparrow}^{+}\hat{c}_{-\mathbf{k}\downarrow}^{+} + b_{\mathbf{k}}\hat{c}_{-\mathbf{q}\downarrow}\hat{c}_{\mathbf{q}\uparrow} - b_{\mathbf{k}}b_{\mathbf{q}}). \tag{10.75}$$

We can write this in terms of the gap. We will see that the following definition is consistent with the one we already used:

$$\Delta_{\mathbf{k}} = -\sum_{\mathbf{q}} V(\mathbf{k},\mathbf{q})b_{\mathbf{q}}. \tag{10.76}$$

Then we can write:

$$\hat{\mathcal{K}} = \sum_{\mathbf{k},s} \xi_{\mathbf{k}} \hat{c}_{\mathbf{k}s}^{+}\hat{c}_{\mathbf{k}s} - \sum_{\mathbf{k}} (\Delta_{\mathbf{k}}\hat{c}_{\mathbf{k}\uparrow}^{+}\hat{c}_{-\mathbf{k}\downarrow}^{+} + \Delta_{\mathbf{k}}\hat{c}_{-\mathbf{k}\downarrow}\hat{c}_{\mathbf{k}\uparrow} - \Delta_{\mathbf{k}}b_{\mathbf{k}}). \tag{10.77}$$

We are again in the situation we saw in Section 10.1.4: $\hat{\mathcal{K}}$ is quadratic in the fermion operators so it can be exactly diagonalized by a canonical transformation. We need to introduce two new operators now because we have two sets of fermion operators, $\hat{c}_{\mathbf{k}\uparrow}, \hat{c}_{\mathbf{k}\downarrow}$. We write:

$$\hat{\gamma}_{\mathbf{k}1} = u_{\mathbf{k}}\hat{c}_{\mathbf{k}\uparrow} - v_{\mathbf{k}}\hat{c}_{\mathbf{k}\downarrow}^{+}$$
$$\hat{\gamma}_{\mathbf{k}2} = u_{\mathbf{k}}\hat{c}_{-\mathbf{k}\downarrow} + v_{\mathbf{k}}\hat{c}_{\mathbf{k}\uparrow}^{+}. \tag{10.78}$$

We get the correct commutation relations if we take $|u_{\mathbf{k}}|^2 + |v_{\mathbf{k}}|^2 = 1$. For the ground-state we can take u, v to be real so that we put, exactly as above, $u_{\mathbf{k}} = \sin(\theta_{\mathbf{k}})$, $v_{\mathbf{k}} = \cos(\theta_{\mathbf{k}})$.

We now need to substitute these relations into Eq. (10.77). The algebra is tedious, and the result is:

$$\hat{\mathcal{K}} = C + \sum_{\mathbf{k}} [\xi_{\mathbf{k}}(u_{\mathbf{k}}^2 - v_{\mathbf{k}}^2) + 2\Delta_{\mathbf{k}}u_{\mathbf{k}}v_{\mathbf{k}}](\hat{\gamma}_{\mathbf{k}1}^{+}\hat{\gamma}_{\mathbf{k}1} + \hat{\gamma}_{\mathbf{k}2}^{+}\hat{\gamma}_{\mathbf{k}2})$$
$$+ \sum_{\mathbf{k}} [2u_{\mathbf{k}}v_{\mathbf{k}}\xi_{\mathbf{k}} + \Delta_{\mathbf{k}}(v_{\mathbf{k}}^2 - u_{\mathbf{k}}^2)](\hat{\gamma}_{\mathbf{k}1}\hat{\gamma}_{\mathbf{k}2} + \hat{\gamma}_{\mathbf{k}1}^{+}\hat{\gamma}_{\mathbf{k}2}^{+}). \tag{10.79}$$

The first term is the constant ground-state value of the grand potential that we have already calculated. The second line has the dangerous operators that we want to eliminate by the proper choice of u, v. We require:

$$2u_{\mathbf{k}}v_{\mathbf{k}}\xi_{\mathbf{k}} + \Delta_{\mathbf{k}}(v_{\mathbf{k}}^2 - u_{\mathbf{k}}^2) = 0. \tag{10.80}$$

Expressing things in terms of $\theta_{\mathbf{k}}$ and doing some manipulation we find:

$$\tan(2\theta_{\mathbf{k}}) = -\Delta_{\mathbf{k}}/\xi_{\mathbf{k}}, \tag{10.81}$$

which is exactly Eq. (10.67).

Now we look at the first term in Eq. (10.79) and use Eq. (10.80):

$$\xi_{\mathbf{k}}(u_{\mathbf{k}}^2 - v_{\mathbf{k}}^2) + 2\Delta_{\mathbf{k}} u_{\mathbf{k}} v_{\mathbf{k}} = -\xi_{\mathbf{k}}\cos(2\theta_{\mathbf{k}}) + \Delta_{\mathbf{k}}\sin(2\theta_{\mathbf{k}})$$

$$= \sqrt{\xi_{\mathbf{k}}^2 + \Delta_{\mathbf{k}}^2} \equiv E_{\mathbf{k}}. \tag{10.82}$$

Thus:

$$\hat{\mathcal{K}} = \sum_{\mathbf{k}} E_{\mathbf{k}}(\hat{\gamma}_{\mathbf{k}1}^+ \hat{\gamma}_{\mathbf{k}1} + \hat{\gamma}_{\mathbf{k}2}^+ \hat{\gamma}_{\mathbf{k}2}) + C. \tag{10.83}$$

The excitations are independent fermions and their spectrum is given by the function $E_{\mathbf{k}}$. There is a gap, Δ, between the ground-state and the lowest-lying excitations, see Figure 10.7. The gap in the spectrum can be measured by infrared spectroscopy and compared to Eq. (10.71).

In the following we will need to sum functions of E. A useful way is to convert the sum to an integral over energy. However, the spectrum has changed so we need to recalculate the density of states. Since we move states around, but don't lose them when we turn on the gap we can write, for the density of states per particle per spin for each branch of the dispersion relation:

$$\mathcal{D}_n(\xi)d\xi = \mathcal{D}_s(E)dE.$$

Thus:

$$\mathcal{D}_s(E) = D_0 \frac{d\xi}{dE} = D_0 \frac{E}{\sqrt{E^2 - \Delta^2}}; \quad E > \Delta. \tag{10.84}$$

There are no states for $E < \Delta$ and the density of states is zero. Thus our integrals will start at Δ.

We have not yet imposed consistency on the theory in order to determine $b_{\mathbf{k}}$. We do this by returning to Eq. (10.76):

$$\Delta_{\mathbf{k}} = -\sum_{\mathbf{q}} V(\mathbf{k}, \mathbf{q}) b_{\mathbf{q}} = -\sum_{\mathbf{q}} V(\mathbf{k}, \mathbf{q}) \langle \hat{c}_{\mathbf{q}\uparrow}^+ \hat{c}_{-\mathbf{q}\downarrow}^+ \rangle. \tag{10.85}$$

The operators $\hat{c}^+$ can be expressed in terms of the $\hat{\gamma}$'s by using Eq. (10.78). This gives:

$$\Delta_{\mathbf{k}} = -\sum_{\mathbf{q}} V(\mathbf{k}, \mathbf{q}) u_{\mathbf{q}} v_{\mathbf{q}} \langle 1 - \hat{\gamma}_{\mathbf{k}1}^+ \hat{\gamma}_{\mathbf{k}1} - \hat{\gamma}_{\mathbf{k}2}^+ \hat{\gamma}_{\mathbf{k}2} \rangle$$

$$= -\sum_{\mathbf{q}} V(\mathbf{k}, \mathbf{q}) \frac{\sin(2\theta_{\mathbf{q}})}{2}(1 - 2f(E_{\mathbf{q}}))$$

$$= -\sum_{\mathbf{q}} V(\mathbf{k}, \mathbf{q}) \frac{\Delta_{\mathbf{q}}}{2E_{\mathbf{q}}}(1 - 2f(E_{\mathbf{q}})). \tag{10.86}$$

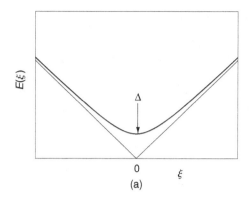

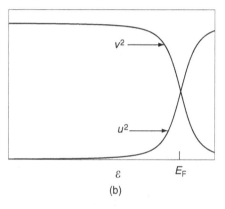

Fig. 10.7   **(a) The function $E(\epsilon - \mu)$, the dispersion relation for excitations above the BCS ground-state. The straight lines are the energies of an electron or a hole measured from E_F. (b) The functions $u^2(\epsilon), v^2(\epsilon)$ which characterize the BCS wavefunction.**

Here f is the Fermi distribution function. At $T = 0$ this is the BCS gap equation, Eq. (10.71). For $T > 0$ the gap decreases with temperature because of the last factor on the right-hand side. In particular, at T_c the gap goes to zero. Using the simplified potential we find:

$$1 = \frac{V}{N}\hat{\sum_{\mathbf{q}}}\frac{\tanh(\xi_{\mathbf{q}}/2k_B T_c)}{\xi_{\mathbf{q}}}. \tag{10.87}$$

Now we have:

$$\frac{1}{V} = D_0 \int_0^{\hbar\omega_D} dx \frac{\tanh(x/2k_B T_c)}{x}$$

$$= D_0 \int_0^{\hbar\omega_D/2k_B T_c} \frac{dy}{y}\tanh(y). \tag{10.88}$$

The integral can be evaluated by an integration by parts. Note that the upper limit of the integral is very large since T_c is much less than the Debye temperature for weak coupling superconductors. Thus:

$$\int_0^u dy \frac{\tanh(y)}{y} \approx \ln(u)\tanh(u) - \int_0^\infty dy \ln(y)\mathrm{sech}^2(y) \approx \ln(u) + 0.819,$$

where $u = \hbar\omega_D/2k_B T_c$. (The easiest way to get the definite integral is by numerical integration, though it can be expressed in terms of Euler's constant.) Solving for T_c gives:

$$k_B T_c = 1.13 \ \hbar\omega_D \ e^{-1/D_0 V}. \tag{10.89}$$

There is a testable prediction of these equations. In Eq. (10.71) we saw that at $T = 0$ the gap is given by an expression of the same form as Eq. (10.88) except with a factor of 2 in front. Thus:

$$\Delta(T = 0) = 1.76 \ k_B T_c. \tag{10.90}$$

This result is remarkably accurate for the weak-coupling superconductors that we have been discussing. The temperature dependence of Δ is given by Eq. (10.86):

$$\frac{1}{D_0 V} = \int_0^{\hbar\omega_D} dx \ \frac{1}{\sqrt{x^2 + \Delta^2}} \tanh\left(\frac{\sqrt{x^2 + \Delta^2}}{2k_B T}\right). \tag{10.91}$$

This implicit equation has to be worked out numerically. As T approaches T_c there is a square root singularity: $\Delta \propto \sqrt{T_c - T}$. (See Figure 10.8.)

Zero resistance and Meissner effect

With the wavefunction and excitation spectrum in hand, we should be able to show that superconductors actually have zero resistance and exclude flux. BCS attacked the first problem by using linear response theory. The details are complicated. We can see, qualitatively, that we have zero resistance by using the Landau argument of Eq. (10.16). If we measure energies from the chemical potential, as we should, we see in Figure 10.7 that we have a critical velocity.

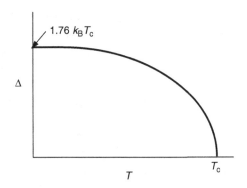

Fig. 10.8   **A sketch of the temperature dependence of the gap.**

The Meissner effect can be derived as well. We will restrict ourselves to the London limit, i.e. nearly uniform fields, and try to derive the London equation. We need to apply a vector potential, $\mathbf{A}$ and form a perturbation Hamiltonian. This is standard quantum mechanics:

$$\delta\hat{\mathcal{H}} = -\frac{ie\hbar}{2mc}\sum_i(\nabla_i \cdot \mathbf{A}(\mathbf{r}_i) + \mathbf{A}(\mathbf{r}_i) \cdot \nabla_i). \tag{10.92}$$

We pass to second quantization in the usual way:

$$\delta\hat{\mathcal{H}} = -\frac{ie\hbar}{2mc}\sum_{\mathbf{k},\mathbf{q},s}M_{\mathbf{k},\mathbf{q}}\,\hat{c}^+_{\mathbf{k}+\mathbf{q}s}\hat{c}_{\mathbf{q}s}$$

$$M_{\mathbf{k},\mathbf{q}} = \langle\mathbf{k}+\mathbf{q}|\nabla\cdot\mathbf{A}+\mathbf{A}\cdot\nabla|\mathbf{k}\rangle = i(2\mathbf{k}+\mathbf{q})\cdot\mathbf{A}(\mathbf{q}). \tag{10.93}$$

In the London gauge $\mathbf{q}\cdot\mathbf{A}(\mathbf{q}) = 0$. We will restrict ourselves to uniform vector potentials so that only the term $\mathbf{q} = 0$ survives. Also we can express $\sum_s \hat{c}^+_{\mathbf{k}s}\hat{c}_{\mathbf{q}s}$ in terms of the $\hat{\gamma}$'s. The result is:

$$\delta\hat{\mathcal{H}} = -\frac{e\hbar}{mc}\mathbf{A}(0)\cdot\sum_{\mathbf{k}}\mathbf{k}(\hat{\gamma}^+_{\mathbf{k}1}\hat{\gamma}_{\mathbf{k}1} - \hat{\gamma}^+_{\mathbf{k}2}\hat{\gamma}_{\mathbf{k}2}). \tag{10.94}$$

Now we are in a very simple situation: the perturbation is diagonal in the $\hat{\gamma}$'s so that $\delta\hat{\mathcal{H}}$ just changes the energies of the two kinds of fermions:

$$E_{\mathbf{k},1} = E_{\mathbf{k}} - \frac{e\hbar}{mc}\mathbf{k}\cdot\mathbf{A}; \quad E_{\mathbf{k},2} = E_{\mathbf{k}} + \frac{e\hbar}{mc}\mathbf{k}\cdot\mathbf{A}. \tag{10.95}$$

What we want to calculate is the paramagnetic current density whose classical expression is $ne\mathbf{v}$. In quantum mechanics the corresponding operator is (for $\mathbf{q} = 0$):

$$\hat{\mathbf{j}}_p = \frac{e}{\Omega}\sum\frac{\hbar\mathbf{k}}{m}\hat{c}^+_k\hat{c}_{\mathbf{k}} = \frac{e\hbar}{m\Omega}\sum_{\mathbf{k}}\mathbf{k}(\hat{\gamma}^+_{\mathbf{k}1}\hat{\gamma}_{\mathbf{k}1} - \hat{\gamma}^+_{\mathbf{k}2}\hat{\gamma}_{\mathbf{k}2}). \tag{10.96}$$

The $\hat{\gamma}$'s were introduced in the same way as above. Therefore:

$$\langle\mathbf{j}_p\rangle = \frac{e\hbar}{m\Omega}\sum\mathbf{k}[f(E_{\mathbf{k},1}) - f(E_{\mathbf{k},2})] = \frac{2e^2\hbar^2}{m^2c\,\Omega}\sum_{\mathbf{k}}[\mathbf{A}\cdot\mathbf{k}]\mathbf{k}\left(-\frac{\partial f}{\partial\epsilon}\right). \tag{10.97}$$

We now convert the sum to an integral, and look at the current in a particular direction and use isotropy (so we replace the angular average of k_x^2 by $k_F^2/3$). In order to evaluate the integral we use Eq. (10.84). Assembling all of this we find:

$$\begin{aligned}
\mathbf{j}_p &= \frac{2e^2\hbar^2k_F^2}{3m^2c\,\Omega}\sum_{\mathbf{k}}\left(-\frac{\partial f}{\partial\epsilon}\right)\mathbf{A} \\
&= \frac{4e^2}{3mc}\frac{E_F}{\Omega}\frac{\mathcal{D}(E_F)}{2}2\int_\Delta^\infty dE\left(-\frac{\partial f}{\partial\epsilon}\right)\frac{E}{\sqrt{E^2-\Delta^2}}\mathbf{A} \\
&= \frac{c}{4\pi\lambda_L^2(0)}2\int_\Delta^\infty dE\left(-\frac{\partial f}{\partial\epsilon}\right)\frac{E}{\sqrt{E^2-\Delta^2}}\mathbf{A}.
\end{aligned} \tag{10.98}$$

The factor of 2 in the second line comes from the two branches of the dispersion relation.
  The total current is:

$$\mathbf{j} = \mathbf{j}_p + \mathbf{j}_d = -\frac{c}{4\pi\lambda_L^2(0)} \left[-2\int_\Delta^\infty dE \left(-\frac{\partial f}{\partial \epsilon}\right) \frac{E}{\sqrt{E^2 - \Delta^2}} + 1 \right] \mathbf{A}$$

$$\equiv -\frac{c}{4\pi\lambda(T)^2} \mathbf{A}. \tag{10.99}$$

We can identify the object in brackets as $\lambda_L^2/\lambda^2(T)$. For example, at $T = 0$ the negative derivative of the Fermi function is $\delta(E)$, as usual. However, this is outside of the range of the integral, so the first term vanishes, and we recover the London equation. The integral vanishes because of the gap; this is the precise meaning of London rigidity. If the gap is zero, then the integral is $2f(0) = 1$ and the paramagnetic and diamagnetic currents cancel, as they should, in the normal state. For larger temperatures the penetration depth increases and goes to infinity as the gap goes to zero at $T = T_c$. There is an empirical fit to data which works pretty well:

$$\lambda_L(0)/\lambda(T) = \sqrt{1 - (T/T_c)^4}. \tag{10.100}$$

Note that $\lambda(T) \sim (T_c - T)^{-1/2}$ as $T \to T_c$. We now have a precise definition of $n_s(T)$ from Eq. (10.36), namely $n_s \propto 1/\lambda^2(T)$.

Other quantities

The BCS theory can predict a large number of physical quantities with startling success. For example, I. Giaver did a tunneling experiment between a normal metal and a superconductor. Elastic tunneling can take place only when there are states for electrons to tunnel in to. Said another way, if V is the voltage across the junction, there will be no tunneling unless $|eV| > \Delta$ because we need to create an excitation in the superconductor. The experiment measures the density of states of Eq. (10.84) directly. This is a remarkable success of the theory and allows the gap to be measured by measuring the I-V characteristic of a tunneling junction.

  There are superconductors for which the theory is not directly applicable. For example, there is a very large class of materials which behave differently from what we have described in large magnetic fields. These are called Type II superconductors, see the next section. If the coupling is strong we cannot get away with the simple Cooper approximation. Detailed work based partly on numerics and using the formalism of Green's functions often works very well.

  Superconductors based on copper oxides are a special case. These have very high transition temperatures, of order 100 Kelvins, and, still pose a puzzle for theorists 20 years or more after their discovery. A few things seem clear: these materials do seem to have Cooper pairs, but they do not form in the simple way that we have described. Experiments have shown that the pairs have internal angular momentum – they are d states. This is probably related to the fact that the coupling that binds the pairs is not the electron-phonon interaction, but some exotic version of the electron-electron interaction. The coherence lengths in these materials are very small, and their magnetic properties are quite complex. There has been a suggestion that in this strong coupling case pairs could be bound above T_c. In fact A. Leggett and others have looked at the crossover from BCS theory to the BEC of strongly bound pairs (Nozieres & Schmittrink 1985). The BCS theory can be simply extended to this regime. The qualitative result is that there is a smooth crossover. Some workers have tried to understand peculiar properties of copper oxide superconductors in these terms.

The crossover from BCS to BEC has taken on new interest in the context of atom traps. It is possible, using an ingenious technique (the Feshbach resonance) to tune the atom-atom interaction strength. There are intricate and fascinating effects which are under investigation as this is being written.

10.5 Ginsburg–Landau theory

Thus far we have treated uniform superconductors. There are interesting effects that occur when the gap and other quantities vary in space. In general, this is a difficult regime to deal with, though the Green's function theory of L. Gorkov did make considerable progress.

There is one situation where the theory is simple and tractable, namely for $T \approx T_c$. This has a special significance because Δ is small and λ is large in this regime. The latter fact means that superconductors are local near T_c because the Pippard coherence length ξ_0 is independent of temperature.

V. Ginsburg and L. Landau gave a phenomenological theory of superconductors near T_c based on the Landau theory of second-order phase transitions which was outlined in Section 2.1.3. For superconductors we need to identify an order parameter which goes to zero at T_c. An obvious choice is the number of pairs $n_p = n_s/2$. However, to discuss currents we need phases: Ginsburg and Landau introduced a "wavefunction", Ψ whose phase gives the supercurrent. We will normalize the order parameter so that for a uniform situation $|\Psi| = \sqrt{n_p}$. Gorkov's Green's function theory derived the equations we are about to discuss and found that Ψ should be proportional to the gap. The gap, the number of pairs, and Ψ all approach zero as $T \rightarrow T_c$.

We now proceed as in Section 2.1.3 except that we need to look at the free energy density, $f(\mathbf{r})$. The total free energy, F is given by $F = \int d\mathbf{r}\, f(\mathbf{r})$. Since the free energy depends on the order parameter we must allow Ψ to vary in space. There will be three parts to the expression: the first, as in Section 2.1.3, is a Taylor expansion for small Ψ. The second is the kinetic energy density of a supercurrent that flows because of the spatial variation of $\Psi(\mathbf{r})$, and the third, the energy density of the magnetic field:

$$f = f_n + a|\Psi|^2 + b|\Psi|^4 + \frac{1}{2m^*}\left|\left(\frac{\hbar}{i}\nabla - \frac{e^*}{c}\mathbf{A}\right)\Psi\right|^2 + \frac{h^2}{8\pi}$$

$$a = a_\circ(T - T_c). \tag{10.101}$$

Here $e^* = 2e$ is the charge of a pair, and $m^* = 2m$ is the mass of a pair. (We are being ahistorical here: the Ginsburg–Landau paper predated Cooper and BCS.)

We note that the order parameter can be written in terms of n_s and therefore in terms of $\lambda(T)$. However, for $T < T_c$ we have, from Section 2.1.3, $|\Psi|^2 = -a/2b$. Also, the condensation energy, $H_c^2(T)/8\pi$ is $-a^2/4b$, (Problem 5 of Chapter 2). That means that a and b can be written in terms of measurable quantities. A bit of algebra gives:

$$a = -\frac{2e^2}{mc^2}H_c^2(T)\lambda^2(T)$$

$$b = \frac{8\pi e^4}{m^2 c^4}H_c^2(T)\lambda^4(T). \tag{10.102}$$

Now we form the free energy and vary with respect to Ψ^* and set it to zero. The result is:

$$a\Psi + 2b|\Psi|^2\Psi + \frac{1}{2m^*}\left(\frac{\hbar}{i}\nabla - \frac{e^*}{c}\mathbf{A}\right)^2 \Psi = 0. \tag{10.103}$$

As usual we have integrated by parts to get the last term.

Similarly we set the variation with respect to $\mathbf{A}$ to zero. After a few integrations by parts we regain the familiar expression:

$$\mathbf{j} = \frac{c}{4\pi}\nabla \times \mathbf{h} = \frac{e^*\hbar}{2m^*i}\left(\Psi^*\nabla\Psi - \Psi\nabla\Psi^*\right) - \frac{e^2}{mc}\mathbf{A}|\Psi|^2, \tag{10.104}$$

as in Eq. (10.45). For example, if $|\Psi|$ is constant:

$$\mathbf{j} = \frac{e^*}{m^*}|\Psi|^2(\hbar\nabla\theta - \frac{e^*}{c}\mathbf{A}). \tag{10.105}$$

The gradient of the phase gives the current as in Eq. (10.6). These are the Ginsburg–Landau (GL) equations. The boundary condition at a vacuum interface is that the normal component of $\mathbf{j}$ vanishes.

10.5.1 Flux quantization

Since the supercurrent is the gradient of a phase, we are back in the situation that we had in a superfluid, Section 10.1.2, where we found quantization of circulation. In this case we will find that the magnetic flux threading a hole in a superconductor is quantized.

Suppose we have a ring of superconductor with magnetic flux, Φ through the hole. Then we can integrate Eq. (10.105) around a contour well within the material where the current is zero. Using the fact that the wavefunction must be single-valued we have:

$$\oint d\mathbf{r} \cdot \mathbf{j} = 0 = \frac{e^* n_s}{m^*}\oint d\mathbf{r} \cdot \left(\hbar\nabla\theta - \frac{e^*}{c}\mathbf{A}\right)$$

$$\oint d\mathbf{r} \cdot \mathbf{A} = \frac{\hbar c}{2e}\oint d\mathbf{r} \cdot \nabla\theta$$

$$\Phi = n\frac{hc}{2e} \equiv n\Phi_s, \tag{10.106}$$

where n is an integer. That is, the flux is quantized with a flux quantum, $hc/2|e| = 2.07 \times 10^{-7}$ gauss/cm^2 corresponding to the charge on a Cooper pair.

When a persistent current runs in a ring there is an integer number of flux quanta trapped inside. The decay rate of the current is given by the rate for a flux quantum to escape. This could happen if there is a fluctuation of the superconductor to be normal across the ring. These fluctuations are exceedingly rare, and persistent currents basically do not decay.

10.5.2 Coherence length

There is another length scale in the theory now. It is called the coherence length (or the the theory of phase transitions, the correlation length). It is not the same as Pippard's ξ_0, but it reduces to it at low temperatures. It is gotten by noting that for $\mathbf{A} = 0$ there is a natural scale on which Ψ varies. Suppose we are near a vacuum interface. Write $\Psi_\infty = (-a/2b)^{1/2}$ and $g(x) = \Psi/\Psi_\infty$. Then the first GL equation is:

$$-\frac{\hbar^2}{2m^*}\Psi + a\Psi + 2b|\Psi|^2\Psi = 0$$

$$-\xi^2\frac{d^2g}{dx^2} + g^3 = g; \ \xi^2 = \frac{\hbar^2}{2m^*|a|}. \tag{10.107}$$

The boundary conditions are $g = 0, x = 0; \ g \to 1, x \to \infty$. The new length, ξ, has the significance that it is the length over which Ψ recovers from a disturbance. We can find an exact solution to the non-linear equation to demonstrate this:

$$g(x) = \tanh(x/\sqrt{2}\xi), \tag{10.108}$$

as can be easily checked. See Figure 10.9 for the behavior of Ψ near a vacuum interface. The wavefunction Ψ is suppressed over a region of width ξ near the surface. This costs energy, a surface energy for the superconductor. We could calculate it by integrating Eq. (10.101) to get the total F. The surface energy is the term $|\nabla\Psi|^2$.

We can estimate the surface energy by pointing out that we lose the condensation energy $H_c^2/8\pi$ over a depth of order ξ so that the surface energy is of order $\xi H_c^2/8\pi$.

Note that:

$$\xi = \sqrt{\hbar^2/2m^*|a|} \propto (T_c - T)^{-1/2}. \tag{10.109}$$

The GL coherence length diverges at T_c.

10.5.3 Type II superconductors

We now have two lengths which diverge in the same way, λ and ξ. Their ratio is an interesting quantity:

$$\kappa = \frac{\lambda(T)}{\xi(T)}. \tag{10.110}$$

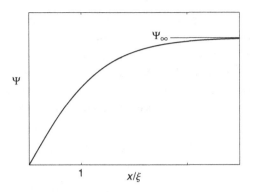

Fig. 10.9   **The behavior of the superconducting wavefunction near a vacuum interface according to the Ginsburg–Landau theory.**

In a pure superconductor κ is usually quite small. However, as we have remarked, in dirty superconductors penetration lengths are large and κ can be larger than unity.

The significance of this ratio was pointed out by A. A. Abrikosov. When κ is large (specifically when $\kappa > 1/\sqrt{2}$) superconductors behave entirely differently from what we have described so far. They are called Type II superconductors – the ones we have been discussing are Type I. A Type II superconductor in a large magnetic field does not exclude flux entirely, rather the flux enters in bundles to make the *mixed state*. The bulk of the material remains superconducting.

To see the reason for this we return to the surface energy of the previous section. Suppose there is a magnetic field of order H_c parallel to the surface. For a Type I superconductor this means that the field penetrates a short distance, λ into the material, and we recover an energy of order $\lambda H_c^2/8\pi$ per unit area. However, near the surface Ψ is depressed from its bulk value over length ξ, see Figure 10.9. Thus the surface energy is of order $(\xi - \lambda)H_c^2/8\pi$ per unit area. For a Type I superconductor this is a positive number and a surface costs energy. See Figure 10.10.

However, for a strongly Type II material, $\lambda \gg \xi$ the situation is reversed: we gain energy through a depth λ and lose through ξ. The system lowers its energy every time it makes a surface. The system is unstable against processes where the magnetic field penetrates the bulk and makes "tubes" of normal material. These are called vortex cores. The process has a limit: the field divides up until each normal core is associated with one quantum of magnetic flux, see Eq. (10.106). The process produces an array of magnetic vortices which allow all the external flux to pass through the material. (The use of the word vortex in this context is in close analogy to the case in superfluids because there is a circulating supercurrent that surrounds each core; see the problems.) Between the vortices the material is still superconducting.

Lower critical field

A Type II material does not have a complete Meissner effect for large fields. For small enough fields, however, the flux is still excluded. The field necessary to start the creation

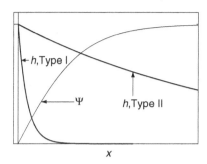

Fig. 10.10   Type I and Type II superconductors. For Type I the magnetic penetration length is less than the coherence length. For Type II the situation is reversed.

of vortices is called the lower critical field, H_{c1}. We can estimate it as follows. We need the smallest field that will lead to the creation of a single vortex. It does cost energy to make a vortex because there is a tube of radius ξ in which superconductivity is destroyed. Thus for a vortex of length, L, we pay an energy $\epsilon_1 L$, where the line energy ϵ_1 is of order $\xi^2 H_c^2$. On the other hand, we gain energy by inserting the vortex, as we see from Eq. (10.32). The magnetization is decreased by $L\Phi_s/\Omega$ so that the energy gain is $\Phi_s H/4\pi$. These two contributions balance when:

$$H_{c1} = \frac{4\pi\epsilon_1}{\Phi_s}. \tag{10.111}$$

It is possible to show that for large κ we have $H_{c1} < H_c$ where H_c is the usual thermodynamic critical field defined by the condensation energy.

Abrikosov lattice and upper critical field

As we increase the external field above H_{c1} more and more vortices penetrate the superconductor. At a certain point their interactions become important; it turns out that vortices repel. It is not hard to believe that as they become crowded they will form a close-packed lattice in two dimensions. The triangular lattice of vortices is called the Abrikosov lattice.

There is a limit to the crowding, however. Even in Type II superconductors sufficiently high fields destroy superconductivity. The largest field that can be supported is called the upper critical field, H_{c2}. It is greater than H_c so that we have the situation shown in Figure 10.11.

We can estimate H_{c2} as follows. Suppose we are just above H_{c2} so that superconductivity has been destroyed. The order parameter is zero everywhere. Now lower the field a bit so that superconductivity can nucleate. Since Ψ is small everywhere we can drop the non-linear terms in the GL equation and write:

$$\frac{1}{2m^*}\left(\frac{\hbar}{i}\nabla - \frac{e^*}{c}\mathbf{A}\right)^2 \Psi = -a\Psi. \tag{10.112}$$

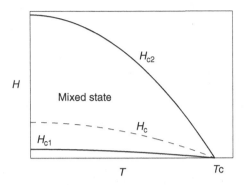

Fig. 10.11 **Phase diagram of a Type II superconductor. Between H_{c1} and H_{c2} the material is in the mixed state and flux penetrates. Below H_{c1} the flux is excluded as in the usual Meissner state. The thermodynamic critical field is shown for reference, but it does not separate phases.**

This is an equation we have seen before: it is the Schrödinger equation for a free particle in a magnetic field, Eq. (7.53), with the GL parameter, $-a$, playing the role of the eigenvalue. However, we know the lowest eigenvalue for this problem from Eq. (7.57):

$$\epsilon_0 = \frac{\hbar |e^*| H}{2m^* c} = |a|.$$
(10.113)

We can solve for H and use Eq. (10.102) to give:

$$H_{c2} = \frac{2m^* c}{|e^*| \hbar} |a| = \sqrt{2} \kappa H_c(T).$$
(10.114)

Thus for $\kappa > 1/\sqrt{2}$ we have an upper critical field greater than the thermodynamic critical field and the mixed state is possible.

Superconducting magnets

Type II superconductors are used in one of the most common applications of superconductivity, that of making a high-field magnet. The mixed state can support quite high fields as Figure 10.11 illustrates. However, there is an important point: we must be sure that the vortices in the mixed state do not move. There is certainly a force on them since there are currents surrounding each one, and a Lorentz force on the currents. If the vortices do move they will dissipate energy and drive the material normal.

The trick is to *pin* the vortices. If there are impurities where superconductivity is suppressed already, vortices will have lower energy there, and will be pinned. A good deal of engineering effort has gone into making suitable high-field magnets this way.

10.6 Josephson effect

Josephson (1962) predicted a remarkable effect that occurs when two superconductors are coupled by a weak link or a tunneling junction. His purely theoretical ideas were quickly verified by experiments, and are the basis of what is now known as the Josephson effect.

We have already discussed tunneling of electrons between a normal metal and a superconductor. Josephson considered, in effect, the tunneling of Cooper pairs. This is not as unlikely as it sounds. In effect, what we need is that there be some coupling matrix element between the ground-state wavefunctions of the two superconductors. Then we can do a tunneling calculation as in the quantum mechanics of a double-well system. This was Josephson's approach.

Our approach will be somewhat different: we will base it on GL theory. Suppose we have the situation shown in Figure 10.12 where there is a short weak link between two bulk superconductors. By short we mean that $L \ll \xi$. In the junction we can use Eq. (10.107), $\xi^2 g'' + g - g^3 = 0$. Further, we can estimate the order of magnitude of the terms. The first is of order $g\xi^2/L^2$, and is much bigger than the second. The third term is small because g is small in the weak link. Thus we have $g'' \approx 0$ so that $g = ux + v$ where u, v are constants.

Now we match boundary conditions. The normalized wavefunction, g must be unity at $x = 0$, and acquire a phase, $\Delta\theta$ for $x = L$. The correct solution is:

$$g = (1 - x/L) + e^{i\Delta\theta}x/L. \tag{10.115}$$

Now we calculate the total current through a junction of area A using Eq. (10.104). We get the *first Josephson equation*:

$$I = I_c \sin\Delta\theta; \quad I_c = \frac{A}{L}\frac{|e|\hbar\Psi_\infty^2}{m}. \tag{10.116}$$

Now if we carry a Cooper pair across the junction it will change its energy by $\Delta E = |e^*|V$ where V is the voltage across the junction. The extra phase of the wavefunction is $e^{i\Delta Et/\hbar}$. Thus the phase obeys:

$$\frac{d\Delta\theta}{dt} = \omega_J; \quad \omega_J = \frac{2|e|V}{\hbar}. \tag{10.117}$$

This is the second Josephson equation.

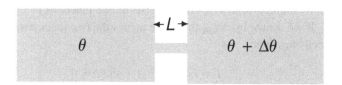

A weak link between two superconductors can make a Josephson junction. The phases of the wavefunctions on the two sides of the link differ by $\Delta\theta$.

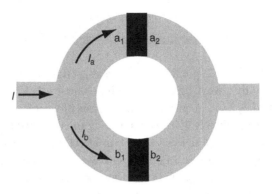

Fig. 10.13 **A simple form of SQUID. The two black regions are Josephson junctions. The current that passes through the device depends sensitively on the magnetic field through the hole in the center.**

Now we have a remarkable situation: if $V = 0$ we can have a supercurrent in the interval $[-I_c, I_c]$. If we impose a dc voltage, however, we have an alternating current, because the solutions to the two equations are:

$$\Delta\theta(t) = \omega_J t + \theta_o; \quad I = I_c \sin(\omega_J t + \theta_o). \tag{10.118}$$

The Josephson frequency is given by Eq. (10.117). It has been used to establish voltage standards.

In a magnetic field we need to generalize the Josephson equations because the phase, as we have already seen, is not a gauge-invariant quantity. The proper generalization is suggested by looking at Eq. (10.105), namely:

$$\Delta\theta \to \gamma = \Delta\theta - \frac{e^*\hbar}{c}\int \mathbf{A}\cdot d\mathbf{s} = \Delta\theta - \frac{2\pi}{\Phi_s}\int \mathbf{A}\cdot d\mathbf{s}. \tag{10.119}$$

The integration is across the weak link. If there is significant flux in the weak link the extra term gives rise to a kind of diffraction pattern in the current; see the problems. The Josephson equation now reads:

$$I = I_c \sin(\gamma). \tag{10.120}$$

10.6.1 The SQUID

Josephson junctions are used in a gadget called the Superconducting Quantum Interference Device or SQUID. SQUIDs give extremely accurate measurements of small magnetic fields. In its simplest form a SQUID consists of two junctions in a ring as in Figure 10.13.

If we denote by $\Delta\theta_{a,b}$ the phase across the two junctions, Eq. (10.116) implies that the total current is:

$$I = I_c[\sin(\Delta\theta_a) + \sin(\Delta\theta_b)]$$
$$= 2I_c \sin\left(\frac{\Delta\theta_a + \Delta\theta_b}{2}\right)\cos\left(\frac{\Delta\theta_a - \Delta\theta_b}{2}\right). \tag{10.121}$$

We have assumed that the flux through the junctions is negligible.

Inside the superconductor the supercurrent vanishes so that Eq. (10.105) implies $\nabla\theta = 2e\mathbf{A}/\hbar c$. Integrating this we find:

$$\Delta\theta_a - \Delta\theta_b = -\frac{2e}{\hbar c}\left(\int_{a_1}^{b_1} \mathbf{A}\cdot d\mathbf{s} + \int_{b_2}^{a_2} \mathbf{A}\cdot d\mathbf{s}\right)$$

$$= -\frac{2e}{\hbar c}\oint \mathbf{A}\cdot d\mathbf{s} = -\frac{2e}{\hbar c}\,\Phi = -2\pi\,\frac{\Phi}{\Phi_s}. \qquad (10.122)$$

The sum of the phase differences can be taken as fixed by the external circuit. The maximum current through the SQUID is:

$$2I_c|\cos(\pi\Phi/\Phi_s)|. \qquad (10.123)$$

Thus the current changes from a maximum to a minimum for a flux change of $\Phi_s/2$. SQUID magnetometers based on this principle have extraordinary precision.

Suggested reading

The classic work of London was very influential:
  London (1961)
General references:
  Wilks & Betts (1986)
  Tinkham (1996)
  de Gennes (1989)
  Landau, Lifshitz & Pitaevskii (1980), volume 2.
  Schrieffer (1964)

Problems

1. Use Eq. (10.2) with $\mu = 0$ to prove Eq. (10.3).
2. Work out the commutation relations for $\hat{P}_\mathbf{k}^+$ and $\hat{P}_\mathbf{k}$. Show, in particular, that

$$[\hat{P}_\mathbf{k},\hat{P}_\mathbf{k}^+] = 1 - \hat{n}_{\mathbf{k}\uparrow} - \hat{n}_{\mathbf{k}\downarrow}.$$

3. Use the BCS ground-state. (a) Show that

$$\bar{N} \equiv \langle\hat{N}\rangle = 2\sum_\mathbf{k} v_\mathbf{k}^2.$$

Here $\hat{N}$ is the operator for the total number of electrons. (b) Show that

$$\delta N^2 \equiv \langle (\hat{N} - \bar{N})^2 \rangle = 4 \sum_{\mathbf{k}} u_{\mathbf{k}}^2 v_{\mathbf{k}}^2.$$

(c) Estimate the sums to prove that: $\delta N / \bar{N} = \mathcal{O}(\bar{N}^{-1/2})$, so that the number fluctuations are small for a macroscopic sample.

4. Structure of a vortex in a type-II superconductor: Write the London equation in the form $\nabla^2 \mathbf{h} = \mathbf{h}/\lambda^2$. Solve this outside a normal core of small radius, ξ. (Remember for type-II, $\xi \ll \lambda$.) Use cylindrical coordinates, and fix constants in the solution so that the flux in the normal core is a flux quantum. Your solution should involve the modified Bessel function $K_\circ$. Work out the supercurrent and the magnetic field.

   Hint: (partial answer): Let ρ be the distance from the normal core. Then

$$j \propto \frac{1}{\rho}; \quad \rho \ll \lambda.$$

   The $1/\rho$ behavior is the reason this structure is called a vortex. It has the same pattern as a vortex in a fluid, Eq. (10.12).

5. A Josephson junction is in a magnetic field parallel to the plane of the junction. The junction is a square of side L and thickness T and lies in the $x-y$ plane. The magnetic field is of magnitude B, and is in the x direction. Show that the critical current is given by:

$$I_c(B)/I_c(0) = |\sin(u)/u|; \quad u = \pi\Phi/\Phi_s,$$

   where $\Phi = BTL$ is the flux through the junction, and Φ_s is the flux quantum.

   Hints: First note that the current density depends on position in the junction. Write, for the total current, an integral over the junction:

$$I = (I_c/L^2) \int dx \int dy \, \sin[\Delta\theta + (2e/hc) \int \mathbf{A} \cdot d\mathbf{s}].$$

   The integral inside the sine (the magnetic field effect on the phase) is from one side of the junction to the other. Note that A_z can be taken to depend on y. You will get an expression that involves $\Delta\theta$. Maximize with respect to $\Delta\theta$ to find the critical current. I found the following identity useful: $\cos(b)\cos(a) = 2\sin[(a+b)/2]\sin[(a-b)/2]$.

References

Alder, B. J. & Wainwright, T. E. (1957), "Phase transition for a hard sphere system," *Journal of Chemical Physics* **27**(5), 1208–9.

Allen, J. F. & Misener, A. D. (1938), "Flow of liquid helium II," *Nature* **141**, 75.

Anderson, M. H., Ensher, J. R., Matthews, M. R., Wieman, C. E. & Cornell, E. A. (1995), "Observation of Bose-Einstein condensation in a dilute atomic vapor," *Science* **269**(5221), 198–201.

Anderson, P. W. (1972), "More is different – broken symmetry and nature of hierarchical structure of science," *Science* **177**(4047), 393–6.

Anderson, P. W. (1997), *Concepts in Solids : Lectures on the Theory of Solids*, Singapore, River Edge, NJ, World Scientific.

Ashcroft, N. W. (1968), "Electron-ion pseudopotentials in alkali metals," *Journal of Physics Part C Solid State Physics* **1**(1), 232.

Ashcroft, N. W. & Mermin, N. D. (1976), *Solid State Physics*, New York, Holt, Rinehart and Winston.

Balibar, S. (2003), "Looking back at superfluid helium," *arXiv.org* **cond-mat/0303561**.

Barabasi, A. & Stanley, H. E. (1995), *Fractal Concepts in Surface Growth*, Cambridge, Cambridge University Press.

Bardeen, J., Cooper, L. N. & Schrieffer, J. R. (1957), "Theory of superconductivity," *Physical Review* **108**(5), 1175–204.

Bardeen, J. & Pines, D. (1955), "Electron-phonon interaction in metals," *Physical Review* **99**(4), 1140–50.

Baym, G. (1990), *Lectures on Quantum Mechanics*, Redwood City, Calif., Addison-Wesley.

Bednorz, J. G. & Muller, K. A. (1986), "Possible high-Tc superconductivity in the Ba-La-Cu-O system," *Zeitschrift Fur Physik B-Condensed Matter* **64**(2), 189–93.

Binnig, G., Quate, C. F. & Gerber, C. (1986), "Atomic force microscope," *Physical Review Letters* **56**(9), 930–3.

Binnig, G., Rohrer, H., Gerber, C. & Weibel, E. (1982), "Tunneling through a controllable vacuum gap," *Applied Physics Letters* **40**(2), 178–80.

Bogoliubov, N. N. (1958), "A new method in the theory of superconductivity .1," *Soviet Physics JETP-USSR* **7**(1), 41–6.

Bolthausen, E. & Bovier, A. (2007), *Spin Glasses*, Berlin, Springer.

Born, M. & Huang, K. (1985), *Dynamical Theory of Crystal Lattices*, Oxford, New York, Oxford University Press.

Bragg, W. H. & Bragg, W. L. (1913), "Structure of the diamond," *Proceedings of the Royal Society of London* **A89**, 277.

Bragg, W. L. (1913), "Structure of some crystals as indicated by their diffraction of x-rays," *Proceedings of the Royal Society of London* **A89**, 248.

Burton, W. K., Cabrera, N. & Frank, F. C. (1951), "The growth of crystals and the equilibrium structure of their surfaces," *Philosophical Transactions of the Royal Society of London Series A-Mathematical and Physical Sciences* **243**(866), 299–358.

Ceperley, D. M. & Alder, B. J. (1980), "Ground-state of the electron-gas by a stochastic method," *Physical Review Letters* **45**(7), 566–9.

Chaikin, P. M. & Lubensky, T. C. (1995), *Principles of Condensed Matter Physics*, Cambridge, New York, Cambridge University Press.

Chui, S. T. & Weeks, J. D. (1976), "Phase transition in the two-dimensional coulomb gas, and the interfacial roughening transition," *Physical Review B* **14**(11), 4978–82.

Cohen, M. L. & Bergstresser, T. K. (1966), "Band structures and pseudopotential form factors for 14 semiconductors of diamond and zinc-blende structures," *Physical Review* **141**(2), 789.

Cooper, L. N. (1956), "Bound electron pairs in a degenerate fermi gas," *Physical Review* **104**(4), 1189–90.

Davis, K. B., Mewes, M. O., Andrews, M. R., Vandruten, N. J., Durfee, D. S., Kurn, D. M. & Ketterle, W. (1995), "Bose-Einstein condensation in a gas of sodium atoms," *Physical Review Letters* **75**(22), 3969–73.

de Gennes, P.-G. (1979), *Scaling Concepts in Polymer Physics*, Ithaca, NY, Cornell University Press.

de Gennes, P.-G. (1989), *Superconductivity of Metals and Alloys*, Redwood City, Calif., Addison-Wesley.

DiVincenzo, D. P. & Steinhardt, P. J. (1999), *Quasicrystals: the State of the Art*, Singapore, River Edge, NJ, World Scientific.

Drenth, J. & Mesters, J. (2007), *Principles of Protein X-Ray Crystallography*, New York, Springer.

Fetter, A. L. & Walecka, J. D. (1971), *Quantum Theory of Many-particle Systems*, San Francisco, McGraw-Hill.

Frauenfelder, H. (1962), *The Mössbauer Effect: A Review, with a Collection of Reprints*, New York, W. A. Benjamin.

Friedrich, W., Knipping, P. & von Laue, M. (1912), "Interferenzerscheinungen bei roentgen-strahlen," *S. B. Bayer. Akad. Wiss.* **42**, 303.

Fröhlich, H. (1950), "Theory of the superconducting state. 1. The ground state at the absolute zero of temperature," *Physical Review* **79**(5), 845–56.

Georges, A., Kotliar, G., Krauth, W. & Rozenberg, M. J. (1996), "Dynamical mean-field theory of strongly correlated fermion systems and the limit of infinite dimensions," *Reviews of Modern Physics* **68**(1), 13–125.

Giamarchi, T. (2004), *Quantum Physics in One Dimension*, Oxford, New York, Oxford University Press.

Gilvarry, J. J. (1956), "The Lindemann and Grüneisen laws," *Physical Review* **102**(2), 308–16.

Godreche, G. (1991), *Solids Far From Equilibrium*, Cambridge, New York, Cambridge University Press.

Grosso, G. & Pastori Parravicini, G. (2000), *Solid State Physics*, San Diego, Academic Press.

Grüner, G. (1994), *Density Waves in Solids*, Reading, Mass, Addison-Wesley.

Herring, C. (1940), "A new method for calculating wave functions in crystals," *Physical Review* **57**(12), 1169–77.

Hirschfelder, J. O., Curtiss, C. F. & Bird, R. B. (1965), *Molecular Theory of Gases and Liquids*, New York, Wiley.

Hohenberg, P. & Kohn, W. (1964), "Inhomogeneous electron gas," *Physical Review B* **136**(3B), B864.

Huang, K. (1987), *Statistical Mechanics*, New York, Wiley.

Jackson, J. D. (1999), *Classical Electrodynamics*, New York, Wiley.

Jain, J. K. (2000), "The composite fermion: A quantum particle and its quantum fluids," *Physics Today* **53**(4), 39–45.

Josephson, B. D. (1962), "Possible new effects in superconductive tunnelling," *Physics Letters* **1**(7), 251–3.

Kammerlingh Onnes, H. (1911), "Further experiments with liquid helium d – on the change of the electrical resistance of pure metals at very low temperatures, etc v the disappearance of the resistance of mercury," *Proceedings of the Koninklijke Akademie Van Wetenschappen Te Amsterdam* **14**, 113–5.

Kapitza, P. (1938), "Viscosity of liquid helium below the gimel-point," *Nature* **141**, 74.

Kittel, C. (1963), *Quantum Theory of Solids*, New York, Wiley.

Kittel, C. (2005), *Introduction to Solid State Physics*, Hoboken, NJ, Wiley.

Kohn, W. (1957), "Shallow impurity states in silicon and germanium," *Solid State Physics–Advances in Research and Applications* **5**, 257–320.

Kohn, W. (1999), "Nobel lecture: Electronic structure of matter-wave functions and density functionals," *Reviews of Modern Physics* **71**(5), 1253–66.

Kohn, W. & Sham, L. J. (1965), "Self-consistent equations including exchange and correlation effects," *Physical Review* **140**(4A), 1133.

Kronig, R. d. L. & Penney, W. G. (1931), "Quantum mechanics of electrons in crystal lattices," *Proceedings of the Royal Society of London* **A130**, 449.

Landau, L. D. & Lifshitz, E. M. (1977), *Quantum Mechanics: Non-relativistic Theory*, Oxford, New York, Pergamon Press.

Landau, L. D., Lifshitz, E. M., Kosevich, A. M. & Pitaevskii, L. P. (1986), *Theory of Elasticity*, Oxford, New York, Pergamon Press.

Landau, L. D., Lifshitz, E. M. & Pitaevskii, L. P. (1980), *Statistical Physics*, Oxford, New York, Pergamon Press.

Landau, L. D. & ter Haar, D. (1965), *Collected Papers of L. D. Landau*, Oxford, New York, Pergamon Press.

Langer, J. S. (1980), "Instabilities and pattern-formation in crystal-growth," *Reviews of Modern Physics* **52**(1), 1–28.

Laughlin, R. B. (1981), "Quantized Hall conductivity in two dimensions," *Physical Review B* **23**(10), 5632–33.

Laughlin, R. B. (1983), "Anomalous quantum Hall effect – an incompressible quantum fluid with fractionally charged excitations," *Physical Review Letters* **50**(18), 1395–8.

Leggett, A. J. (2004), "Nobel lecture: Superfluid He-3: the early days as seen by a theorist," *Reviews of Modern Physics* **76**(3), 999–1011.

Lieb, E. H. (1981), "Thomas-Fermi and related theories of atoms and molecules," *Reviews of Modern Physics* **53**(4), 603–41.

London, F. (1961), *Superfluids*, New York, Dover Publications.

London, F. & London, H. (1935), "The electromagnetic equations of the supraconductor," *Proceedings of the Royal Society of London Series A-Mathematical and Physical Sciences* **149**(A866), 0071–0088.

Luttinger, J. M. (1960), "Fermi surface and some simple equilibrium properties of a system of interacting fermions," *Physical Review* **119**(4), 1153–63.

Lyddane, R. H., Sachs, R. G. & Teller, E. (1941), "On the polar vibrations of alkali halides," *Physical Review* **59**(8), 673–6.

Mahan, G. D. (2000), *Many-particle physics*, New York, Kluwer Academic/Plenum Publishers.

Mandelbrot, B. B. (1982), *The Fractal Geometry of Nature*, W.H. Freeman, San Francisco.

Marder, M. P. (2000), *Condensed matter physics*, John Wiley, New York.

Martin, R. M. (2004), *Electronic Structure : Basic Theory and Practical Methods*, Cambridge University Press, Cambridge, UK, New York.

Mattis, D. C. (1988), *The Theory of Magnetism*, Berlin, New York, Springer-Verlag.

Meakin, P. (1998), *Fractals, Scaling, and Growth Far From Equilibrium*, Cambridge, Cambridge University Press.

Meissner, W. & Ochsenfeld, R. (1933), "Short initial announcements," *Naturwissenschaften* **21**, 787–8.

Mermin, N. D. (1966), "A short simple evaluation of expressions of Debye-Waller form," *Journal of Mathematical Physics* **7**(6), 1038.

Mermin, N. D. (1968), "Crystalline order in two dimensions," *Physical Review* **176**(1), 250.

Messiah, A. (1968), *Quantum Mechanics*, Amsterdam, New York, North-Holland.

Mott, N. F. (1990), *Metal-insulator Transitions*, London, New York, Taylor and Francis.

Mullins, W. W. & Sekerka, R. F. (1963), "Stability of a planar interface during solidification of a dilute alloy," *Journal of Applied Physics* **34**, 323.

Murray, C. A. & Grier, D. G. (1996), "Video microscopy of monodisperse colloidal systems," *Annual Review of Physical Chemistry* **47**, 421–62.

Nelson, D. R. & Halperin, B. I. (1979), "Dislocation-mediated melting in two dimensions," *Physical Review B* **19**(5), 2457–84.

Nozières, P. & Pines, D. (1958), "Correlation energy of a free electron gas," *Physical Review* **111**(2), 442–54.

Nozieres, P. & Pines, D. (1999), *The Theory of Quantum Liquids*, Cambridge, Mass, Perseus Books.

Nozieres, P. & Schmittrink, S. (1985), "Bose condensation in an attractive fermion gas – from weak to strong coupling superconductivity," *Journal of Low Temperature Physics* **59**(3-4), 195–211.

Osheroff, D. D., Gully, W. J., Richardson, R. C. & Lee, D. M. (1972), "New magnetic phenomena in liquid He-3 below 3 mK," *Physical Review Letters* **29**(4), 920.

Osheroff, D. D., Richardson, R. C. & Lee, D. M. (1972), "Evidence for a new phase of solid He-3," **28**(14), 885.

Peierls, R. E. (1991), *More Surprises in Theoretical Physics*, Princeton, NJ, Princeton University Press.

Pelcé, P. (2004), *New Visions on Form and Growth : Fingered Growth, Dendrites, and Flames*, Oxford, New York, Oxford University Press.

Perdew, J. P., Burke, K. & Ernzerhof, M. (1996), "Generalized gradient approximation made simple," *Physical Review Letters* **77**(18), 3865–8.

Perdew, J. P. & Zunger, A. (1981), "Self-interaction correction to density-functional approximations for many-electron systems," *Physical Review B* **23**(10), 5048–79.

Phillips, J. C. & Kleinman, L. (1959), "New method for calculating wave functions in crystals and molecules," *Physical Review* **116**(2), 287–94.

Phillips, P. (2003), *Advanced Solid State Physics*, Boulder, Colo., Westview Press.

Pines, D. (1999), *Elementary Excitations in Solids : Lectures on Protons, Electrons, and Plasmons*, Reading, Mass, Persus Books.

Pippard, A. B. (1950), "Field variation of the superconducting penetration depth," *Proceedings of the Royal Society of London Series A-Mathematical and Physical Sciences* **203**(1073), 210–23.

Plischke, M. & Bergersen, B. (1994), *Equilibrium Statistical Physics*, Singapore, River Edge, NJ, World Scientific.

Prange, R. E. & Girvin, S. M. (1990), *The Quantum Hall Effect*, New York, Springer-Verlag.

Radnoczi, G., Vicsek, T., Sander, L. M. & Grier, D. (1987), "Growth of fractal crystals in amorphous $GeSe_2$ films," *Physical Review A* **35**(9), 4012–5.

Rayfield, G. W. & Reif, F. (1964), "Quantized vortex rings in superfluid helium," *Physical Review A-General Physics* **136**(5A), 1194.

Saminadayar, L., Glattli, D. C., Jin, Y. & Etienne, B. (1997), "Observation of the e/3 fractionally charged Laughlin quasiparticle," *Physical Review Letters* **79**(13), 2526–9.

Sander, L. M. (2000), "Diffusion limited aggregation, a kinetic critical phenomenon?," *Contemporary Physics* **41**, 203–18. (42) DOE.

Schiff, L. I. (1968), *Quantum Mechanics*, New York, McGraw-Hill.

Schrieffer, J. R. (1964), *Theory of Superconductivity*, New York, W.A. Benjamin.

Squires, G. L. (1978), *Introduction to the Theory of Thermal Neutron Scattering*, Cambridge, New York, Cambridge University Press.

Sze, S. M. (1981), *Physics of Semiconductor Devices*, Wiley, New York.

Taylor, P. L. & Heinonen, O. (2002), *A Quantum Approach to Condensed Matter Physics*, Cambridge, New York, Cambridge University Press.

Tinkham, M. (1996), *Introduction to Superconductivity*, New York, McGraw Hill.

Tsui, D. C., Stormer, H. L. & Gossard, A. C. (1982), "Two-dimensional magnetotransport in the extreme quantum limit," *Physical Review Letters* **48**(22), 1559–62.

Valatin, J. G. (1958), "Comments on the theory of superconductivity," *Nuovo Cimento* **7**(6), 843–57.

Von Klitzing, K., Dorda, G. & Pepper, M. (1980), "New method for high-accuracy determination of the fine-structure constant based on quantized hall resistance," *Physical Review Letters* **45**(6), 494–7.

White, R. M. (1970), *Quantum Theory of Magnetism*, New York, McGraw-Hill.

Wilks, J. & Betts, D. S. (1986), *An Introduction to Liquid Helium*, Oxford, New York, Oxford University Press.

Witten, T. A. & Sander, L. M. (1981), "Diffusion-limited aggregation, a kinetic critical phenomenon," *Physical Review Letters* **47**, 1400.

Wyckoff, R. W. G. (1963), *Crystal Structures*, New York, Interscience Publishers.

Yin, M. T. & Cohen, M. L. (1982), "Theory of static structural-properties, crystal stability, and phase-transformations – application to Si and Ge," *Physical Review B* **26**(10), 5668–87.

Young, A. P. (1979), "Melting and the vector Coulomb gas in two dimensions," *Physical Review B* **19**(4), 1855–66.

Yu, P. Y. & Cardona, M. (2001), *Fundamentals of semiconductors : Physics and Materials Properties*, Berlin, New York, Springer.

Zangwill, A. (1988), *Physics at Surfaces*, Cambridge, New York, Cambridge.

Ziman, J. M. (1962), *Electrons and Phonons: The Theory of Transport Phenomena in Solids*, Oxford, Clarendon Press.

Ziman, J. M. (1972), *Principles of the Theory of Solids*, University Press, Cambridge.

Ziman, J. M. (1979), *Models of Disorder: The Theoretical Physics of Homogeneously Disordered Systems*, Cambridge, New York, Cambridge University Press.

Index